ROBOTIC TELESCOPES IN THE 1990s

A SERIES OF BOOKS ON RECENT DEVELOPMENTS IN ASTRONOMY AND ASTROPHYSICS

© Copyright 1992 Astronomical Society of the Pacific
390 Ashton Avenue, San Francisco, California 94112

All rights reserved

Printed by BookCrafters, Inc.

First published 1992

Library of Congress Catalog Card Number: 92-75085
ISBN 0-937707-53-8

D. Harold McNamara, Managing Editor of Conference Series
408 ESC Brigham Young University
Provo, UT 84602
801-378-2298

A SERIES OF BOOKS ON RECENT DEVELOPMENTS IN ASTRONOMY AND ASTROPHYSICS

Vol. 1-Progress and Opportunities in Southern Hemisphere Optical Astronomy: The CTIO 25th Anniversary Symposium
ed. V. M. Blanco and M. M. Phillips　　　　　　　　　　　　ISBN 0-937707-18-X

Vol. 2-Proceedings of a Workshop on Optical Surveys for Quasars
ed. P. S. Osmer, A. C. Porter, R. F. Green, and C. B. Foltz　　　　ISBN 0-937707-19-8

Vol. 3-Fiber Optics in Astronomy
ed. S. C. Barden　　　　　　　　　　　　　　　　　　　　ISBN 0-937707-20-1

Vol. 4-The Extragalactic Distance Scale: Proceedings of the ASP 100th Anniversary Symposium
ed. S. van den Bergh and C. J. Pritchet　　　　　　　　　　　ISBN 0-937707-21-X

Vol. 5-The Minnesota Lectures on Clusters of Galaxies and Large-Scale Structure
ed. J. M. Dickey　　　　　　　　　　　　　　　　　　　　ISBN 0-937707-22-8

Vol. 6-Synthesis Imaging in Radio Astronomy: A Collection of Lectures from the Third NRAO Synthesis Imaging Summer School
ed. R. A. Perley, F. R. Schwab, and A. H. Bridle　　　　　　　ISBN 0-937707-23-6

Vol. 7-Properties of Hot Luminous Stars: Boulder-Munich Workshop
ed. C. D. Garmany　　　　　　　　　　　　　　　　　　　ISBN 0-937707-24-4

Vol. 8-CCDs in Astronomy
ed. G. H. Jacoby　　　　　　　　　　　　　　　　　　　　ISBN 0-937707-25-2

Vol. 9-Cool Stars, Stellar Systems, and the Sun. Sixth Cambridge Workshop
ed. G. Wallerstein　　　　　　　　　　　　　　　　　　　ISBN 0-937707-27-9

Vol. 10-The Evolution of the Universe of Galaxies. The Edwin Hubble Centennial Symposium
ed. R. G. Kron　　　　　　　　　　　　　　　　　　　　　ISBN 0-937707-28-7

Vol. 11-Confrontation Between Stellar Pulsation and Evolution
ed. C. Cacciari and G. Clementini　　　　　　　　　　　　　ISBN 0-937707-30-9

Vol. 12-The Evolution of the Interstellar Medium
ed. L. Blitz　　　　　　　　　　　　　　　　　　　　　　ISBN 0-937707-31-7

Vol. 13-The Formation and Evolution of Star Clusters
ed. K. Janes　　　　　　　　　　　　　　　　　　　　　　ISBN 0-937707-32-5

Vol. 14-Astrophysics with Infrared Arrays
ed. R. Elston　　　　　　　　　　　　　　　　　　　　　　ISBN 0-937707-33-3

Vol. 15-Large-Scale Structures and Peculiar Motions in the Universe
ed. D. W. Latham and L. A. N. da Costa　　　　　　　　　　ISBN 0-937707-34-1

Vol. 16-Atoms, Ions and Molecules: New Results in Spectral Line Astrophysics
ed. A. D. Haschick and P. T. P. Ho　　　　　　　　　　　　ISBN 0-937707-35-X

Vol. 17-Light Pollution, Radio Interference, and Space Debris
ed. D. L. Crawford　　　　　　　　　　　　　　　　　　　ISBN 0-937707-36-8

Vol. 18-The Interpretation of Modern Synthesis Observations of Spiral Galaxies
ed. M. Duric and P. C. Crane　　　　　　　　　　　　　　　ISBN 0-937707-37-6

Vol. 19-Radio Interferometry: Theory, Techniques, and Applications, IAU Colloquium 131
ed. T. J. Cornwell and R. A. Perley　　　　　　　　　　　　ISBN 0-937707-38-4

Vol. 20-Frontiers of Stellar Evolution, celebrating the 50th Anniversary of McDonald Observatory
ed. D. L. Lambert　　　　　　　　　　　　　　　　　　　　ISBN 0-937707-39-2

Vol. 21-The Space Distribution of Quasars
ed. D. Crampton　　　　　　　　　　　　　　　　　　　　ISBN 0-937707-40-6

Vol. 22-Nonisotropic and Variable Outflows from Stars
ed. L. Drissen, C. Leitherer, and A. Nota   ISBN 0-937707-41-4

Vol. 23-Astronomical CCD Observing and Reduction Techniques
ed. S. B. Howell   ISBN 0-937707-42-4

Vol. 24-Cosmology and Large-Scale Structure in the Universe
ed. R. R. de Carvalho   ISBN 0-937707-43-0

Vol. 25-Astronomical Data Analysis Software and Systems I
ed. D. M. Worrall, C. Biemesderfer, and J. Barnes   ISBN 0-937707-44-9

Vol. 26-Cool Stars, Stellar Systems, and the Sun, Seventh Cambridge Workshop
ed. M. S. Giampapa and J. A. Bookbinder   ISBN 0-937707-45-7

Vol. 27-The Solar Cycle
ed. K. L. Harvey   ISBN 0-937707-46-5

Vol. 28-Automated Telescopes for Photometry and Imaging
ed. S. J. Adelman, R. J. Dukes, Jr., and C. J. Adelman ISBN 0-937707-47-3

Vol. 29-Workshop on Cataclysmic Variable Stars
ed. N. Vogt   ISBN 0-937707-48-1

Vol. 30-Variable Stars and Galaxies, in honor of M. S. Feast on his retirement
ed. B. Warner   ISBN 0-937707-49-X

Vol. 31-Relationships Between Active Galactic Nuclei and Starburst Galaxies
ed. A. V. Filippenko   ISBN 0-937707-50-3

Vol. 32-Complementary Approaches to Double and Multiple Star Research, IAU Colloquim 135
ed. H. A. McAlister and W. I. Hartkopf   ISBN 0-937707-51-1

Vol. 33-Research Amateur Astronomy
ed. S. J. Edberg   ISBN 0-937707-52-X

Inquiries concerning these volumes should be directed to the:
Astronomical Society of the Pacific
CONFERENCE SERIES
390 Ashton Avenue
San Francisco, CA 94112-1722
415-337-1100

ASTRONOMICAL SOCIETY OF THE PACIFIC
CONFERENCE SERIES

Volume 34

ROBOTIC TELESCOPES IN THE 1990s

Proceedings of a Symposium held as part of the 103rd
Annual Meeting of the Astronomical Society of the Pacific,
at the University of Wyoming, Laramie, Wyoming
22-24 June 1991

Edited by
Alexei V. Filippenko
Department of Astronomy, University of California,
Berkeley California 94720   U.S.A.

TABLE OF CONTENTS

Preface	x
Dedication	xii
Registered Participants	xiv

I. RESULTS FROM AUTOMATIC PHOTOELECTRIC TELESCOPES — 1

The Wisconsin APT: The First Robotic Telescope — 3
Arthur D. Code

Results from the Four College APT — 9
Robert J. Dukes, Jr.

Interrogation of Duplicitous Stars with an APT — 19
Bernard W. Bopp

Multiperiodicities in the Photometric Variability of RS CVn Binaries — What We Know After 15 Years — 27
Douglas S. Hall

Starspot Photometry: Observational Review and Interplay With Spectroscopy — 39
Klaus G. Strassmeier

II. CCD IMAGING WITH ROBOTIC TELESCOPES — 53

The Scientific Potential of Automatic CCD Imaging Telescopes — 55
Alexei V. Filippenko

A Doubly Robotic Telescope: The Berkeley Automated Supernova Search — 67
Saul Perlmutter, Richard A. Muller, Heidi J. M. Newberg, Carlton R. Pennypacker, Timothy P. Sasseen, and Craig K. Smith

Automated CCD Photometry at Behlen Observatory — 73
Edward G. Schmidt

Architecture of the Software for the Indiana CCD Automated Telescope — 77
R. Kent Honeycutt and George W. Turner

Software for CCD Photometry with the PC — 89
Ronald H. Kaitchuck

An Automated Telescope System for Undergraduate Teaching   97
    D. V. Deleo and R. L. Mutel

Progress Report on the Berkeley Automatic Imaging Telescope   105
    Michael W. Richmond, Richard R. Treffers, and Alexei V. Filippenko

Technical Description of the Berkeley Automatic Imaging Telescope   115
    Richard R. Treffers, Michael W. Richmond, and Alexei V. Filippenko

The Explosive Transient Camera — An Automatic Wide-Field Sky Monitor for Short-Timescale Optical Transients   123
    Roland K. Vanderspek, George C. Ricker, and John P. Doty

The Rapidly Moving Telescope: A Progress Report   137
    S. D. Barthelmy, T. L. Cline, B. J. Teegarden, and T. T. von Rosenvinge

III. FUTURE USES OF ROBOTIC TELESCOPES   151

The Use of Robotic Telescopes for Detecting Planetary Systems   153
    William J. Borucki and Russell M. Genet

A Direct Census of the Oort Cloud with a Robotic Telescope   171
    T. S. Axelrod, C. Alcock, K. H. Cook, and H.-S. Park

Application of Robotic Telescopes to the Physical Investigation of Comets   183
    David Jewitt

The Search for Massive Compact Halo Objects with a (Semi) Robotic Telescope   193
    C. Alcock, T. S. Axelrod, D. P. Bennett, K. H. Cook, H.-S. Park, K. Griest, S. Perlmutter, C. W. Stubbs, K. C. Freeman, B. A. Peterson, P. J. Quinn, and A. W. Rogers

2MASS: The 2 μm All Sky Survey   203
    S. G. Kleinmann

Automation of Interferometric Observations   213
    M. Bester, C. G. Degiacomi, W. C. Danchi, L. J. Greenhill, and C. H. Townes

Prospects for Automated Radial-Velocity Observations   223
    R. F. Griffin

Robotic Telescopes with Fiber-Coupled Spectrographs 227
  Lawrence W. Ramsey

Global Networking: The Australian Connection 235
  John J. Van-Vegchel

Robotic Telescope Networks 241
  Russell M. Genet

APTs for Indian Universities 249
  Ranjan Gupta

The UCSB Remote Access Astronomy Project 253
  Philip Lubin and Janet van der Veen

Photometry from Space 261
  M. J. Nelson, R. C. Bless, J. W. Percival, and R. L. White

IV. ROBOTIC TELESCOPES ON THE MOON 271

Telescopes for a Lunar Outpost 273
  Jack O. Burns

Multi-Use Lunar Telescopes 289
  Russell M. Genet, David R. Genet, David L. Talent,
  Mark Drummond, Butler Hine, Louis J. Boyd, and
  Mark Trueblood

The Lunar Transit Telescope: A Moon-Based Strip Search
  of the Universe 305
  John T. McGraw

A Lunar Polar Infrared Observatory 325
  Dan Lester

Telerobotically Deployed Lunar Farside VLF Observatory 335
  P. B. Landecker, M. A. Caylor, D. U. Choi, R. J. Drean,
  C. R. Edelsohn, J. G. Gurley, F. A. Hagen, G. W. Su
  M. L. Tillman, and C. R. Wassgren

Engineering Design of an Unmanned Lunar Radio Observatory 347
  R. J. Drean, M. A. Caylor, D. U. Choi, C. R. Edelsohn,
  J. G. Gurley, F. A. Hagen, P. B. Landecker, G. W. Su,
  M. L. Tillman, and C. R. Wassgren

Afterthoughts 359
  Virginia Trimble

Author Index 367

PREFACE

These are the *Proceedings* of a symposium on "Robotic Telescopes in the 1990s," which was held as part of the 103rd Annual Meeting of the Astronomical Society of the Pacific, 22–24 June 1991, at the University of Wyoming in Laramie. There were about 70 participants, and nearly 40 oral presentations or poster papers. This volume contains most of the contributions in written form.

Robotic telescopes have been in frequent use for roughly a decade. The advantages they offer, and the scientific results they have produced, are becoming more and more widely known in the scientific community. Although single-aperture photoelectric photometry of bright variable stars has thus far dominated the field, there are now many new techniques and a wide variety of scientific projects. Moreover, several exciting programs are planned for the near future. Thus, it seemed appropriate to review our progress in the design, construction, and use of fully automated telescopes, to summarize research results, and to consider improvements that can be made over the next decade.

The quality of the work presented at the symposium was impressive. High-precision photoelectric photometry has been used to study multiperiodicities in RS CVn binaries, starspots, multimode Cepheids, magnetic Ap stars, non-radially pulsating B stars, and interacting binary stars — to name just a few examples. During the past few years, there has also been excellent progress in automating CCD imaging systems. The group at Berkeley has 20 supernova discoveries to its credit, over 100 variable stars are being monitored at the Behlen Observatory, a truly remarkable system is available at Indiana University, and searches are being conducted for very rapid optical transients. In addition, robotic imaging telescopes are being incorporated into several University astronomy courses. Future uses of automatic or semi-automatic telescopes include detection of planetary systems, the Oort cloud, and massive compact halo objects, as well as the physical investigation of comets. Automated infrared all-sky surveys, interferometric observations, and spectroscopy are also on the horizon. Finally, networks of robotic telescopes offer many advantages; they should certainly be pursued with vigor. In the much more distant future, we can look forward to placing automatic telescopes on the Moon; there are many complementary possibilities, as discussed in this volume.

I would like to take this opportunity to thank Russ Genet, Doug Hall, and the late Harlan Smith for their assistance as members of the Scientific Organizing Committee. Harlan was particularly interested in the possibility of robotic telescopes on the Moon, and he had his heart set on attending the symposium. Unfortunately, he was forced to undergo intensive cancer therapy in Houston at that time. He lost his courageous battle against cancer in October 1991, and with him the astronomical community lost a wonderful colleague. He was the close friend and mentor of many individuals. Harlan's spirited approach to life, his contributions as a scientist and an educator, his vision, and his unflagging optimism will long be remembered by those who knew him. I dedicate these *Proceedings* to his memory.

PREFACE

My sincere thanks go to the officers of the ASP for their effective organization of the entire meeting: Andrew Fraknoi (Executive Director), Barbara Keville (Associate Director), Dan De Vries (Director of Development), Robyn Talman (Associate Director of Development and Meeting Coordinator), and Julie Lutz (President). The Local Organizing Committee (Ronald Canterna, Mel Dyck, Evelyn Haskell, and Robert Stencel) and their staff did a superb job in planning and running the day-to-day activities. The Department of Physics and Astronomy at the University of Wyoming was the official host of the meeting, which was generously co-sponsored by *Astronomy* Magazine/Kalmbach Publishing. I am also grateful to numerous organizations for providing financial or other support.

The excitement provided by my group's attempts to build a robotic imaging telescope, and the dedication required for completion and successful operation of the system, are partly to blame for my procrastination while editing this volume. I very much appreciate the patience of D. Harold McNamara, Managing Editor of the Astronomical Society of the Pacific Conference Series. Despite the delay, I trust that the book will still be useful, especially given the relative youth of the field.

Russ Genet persuaded me to organize the symposium on Robotic Telescopes in the 1990s. After all the help, encouragement, and inspiration he has given me over the past few years, it was the least I could do to repay him. Thanks, Russ.

> Alex Filippenko
> University of California
> Berkeley, CA, USA

DEDICATION

HARLAN J. SMITH

25 August 1924 — 17 October 1991

This volume is dedicated to the memory of Harlan J. Smith, formerly of the University of Texas and McDonald Observatory. Harlan originally planned to organize and chair the session on lunar astronomy at this symposium. He was unable to do so due to a worsening case of cancer which he fought with his usual vigor for the last year of his life.

Harlan visualized that astronomy in the next century would be centered on the Moon, where astronomers would finally be free of the ill effects of the Earth's atmosphere and the many difficulties of Earth orbit. Harlan also visualized that 21st-century astronomy would be more international in approach. He was actively working with Soviet astronomers in developing ideas for astronomical activities from the Moon, and he had an interest in using Soviet booster capabilities to place astronomical payloads on the Moon.

In his final few years, Harlan became convinced that small telescopes should precede large ones on the Moon, and that such telescopes should be tested extensively at very remote locations here on Earth prior to their placement on the Moon. Harlan had, in fact, decided exactly where these "lunar precursor" telescopes should be placed — a 20,000 foot high mountain in the middle of the northern Chilean desert.

Harlan had been to this site; he drove up to the top via the highest road in the world, and he spent many hours there without oxygen tanks. His numerous photographs convincingly showed that this was probably the best astronomical site in the entire world. The site is protected from clouds by high mountains to both the east and the west. Precipitation is less than once every 10 years, and there appears to be no vegetation of any sort, not even an occasional blade of grass. The area looks much like the Moon.

A couple of months before his death, Harlan sent me a photograph of the Earth taken by Galileo as it swung by the Earth to pick up speed. While enveloped by clouds, North America and South America could be seen in outline. Almost all of northern South America was covered by clouds, including Cerro Tololo and ESO (which were in what Harlan called the "rain belt"), but there was a clear spot about 100 miles wide. Harlan had triumphantly pasted an arrow on the photograph pointing to his mountain, which was near the center of the clear hole.

Harlan was a visionary who looked off into a distant future more clearly than most of us look to next year. He greatly inspired many astronomers during his productive career. I expect that most of his calmly stated predictions will indeed come to pass. It is sad that he will not be here to see this happen.

Russell M. Genet
Fairborn Observatory

HARLAN J. SMITH

REGISTERED PARTICIPANTS

Harry Albert	Ball Aerospace
Charles Alcock	Lawrence Livermore National Laboratory
Ed Anderson	Gogebic Community College
Timothy S. Axelrod	Lawrence Livermore National Laboratory
Scott D. Barthelmy	NASA Goddard Space Flight Center
Manfred Bester	University of California at Berkeley
Raymond Bloomer	Air Force Academy
Bernard W. Bopp	University of Toledo
William J. Borucki	NASA/Ames Research Center
Louis J. Boyd	Fairborn Observatory
William Brown	Sand Point, Idaho
Jack O. Burns	New Mexico State University
Shane Burns	Harvey Mudd College
Murray Campbell	Colby College
Mark A. Caylor	Hughes Aircraft Company
Arthur D. Code	University of Wisconsin, Madison
Derick V. Deleo	University of Iowa
Robert J. Dukes, Jr.	The College of Charleston
David L. DuPuy	Virginia Military Institute
Thomas Edwards	Laramie, Wyoming
Alexei V. Filippenko	University of California at Berkeley
Michael Frazier	Ball Aerospace
David Fry	University of Calgary
Roy Garstang	University of Colorado
Russell M. Genet	AutoScope Corporation, and Fairborn Obs.
Roger F. Griffin	The Observatories, Cambridge, UK
Ranjan Gupta	IUCAA, Pune, India
Douglas S. Hall	Tennessee State University
Gordon Hammond	University of South Florida
Lee Hawkins	Wellesley College
R. Kent Honeycutt	Indiana University
Howard Hule	NCAR/High Altitude Observatory
MaryLou Jewett	Mill Valley, California
David Jewitt	University of Hawaii
Ronald H. Kaitchuck	Unified Software Systems
Dimitri Klebe	University of Denver
S. G. Kleinmann	University of Massachusetts
Karl K. Klett	Table Mountain Observatory
Dan Lester	University of Texas, Austin
Jeffrey F. Lockwood	University of Arizona
Frank Loob	Napa, California

Philip Lubin	University of California at Santa Barbara
William R. Luebke	Modesto Junior College
Matthew A. Malkan	University of California at Los Angeles
J. D. Mayfield	Ball Aerospace
Mike McCarthy	CE Technology
John T. McGraw	University of Arizona
Kim Mehlbach	Denver Astronomical Society
Nancy D. Morrison	University of Toledo
Casper Morsello	Bego Park, New York
Matthew J. Nelson	University of Wisconsin, Madison
Mark Nook	St. Cloud State University
Fritz Osell	Leeward College
Saul Perlmutter	University of California at Berkeley
Sig Peterson	Intel
Edward Quartemont	Eastbay Astronomical Society
Lawrence W. Ramsey	Pennsylvania State University
Michael W. Richmond	University of California at Berkeley
Edward G. Schmidt	University of Nebraska
Michael Shao	Jet Propulsion Laboratory
Earl J. Spillar	University of Wyoming
Jeff Stoner	Boulder, Colorado
Klaus G. Strassmeier	University of Vienna
Marc L. Tillman	Hughes Aircraft Company
Richard R. Treffers	University of California at Berkeley
Virginia Trimble	U. C. Irvine, and University of Maryland
Roland K. Vanderspek	Massachusetts Institute of Technology
Lee Youngblood	Loveland, Colorado

PART I.

RESULTS FROM AUTOMATIC PHOTOELECTRIC TELESCOPES

THE WISCONSIN APT: THE FIRST ROBOTIC TELESCOPE

ARTHUR D. CODE
Department of Astronomy, University of Wisconsin, 475 N. Charter St., Madison, WI 53706

ABSTRACT Around 1965 a small, computer controlled telescope was put into operation at the University of Wisconsin Pine Bluff Observatory in order to provide for real-time evaluation of extinction coefficients. This instrument represented a "spinoff" from the space program being actively pursued at Wisconsin at that time. The telescope could operate automatically from night to night without the need for human input, and it could carry out the basic logical decisions required for such an operation. It is the purpose of this paper to describe the history and functional performance of our early robotic telescope.

INTRODUCTION

The Wisconsin Automatic Photoelectric Telescope (APT) was a computer controlled photometric telescope developed in the mid 1960s to provide real-time computation of the atmospheric extinction coefficients in support of other photometric observations. The title "The First Robotic Telescope" was suggested to me at the time that I was invited to participate in this symposium. I have retained the title with some trepidation. Things are seldom the first. At about the same time as our APT was placed into operation, the 60-inch reflector on Kitt Peak was carrying out observations remotely from Tucson (Maran 1967). A few years earlier, we had installed an ultraviolet telescope on the X-15 rocket plane which necessarily operated automatically. Previously I had not considered the distinction between remote operation, automatic control, and robotic systems, but these three projects seemed to illustrate that distinction. By entering the proper commands at the computer terminal in Tucson and relaying that information to the mountain, the 60-inch telescope could be pointed to the correct position, the sequence of observations initiated, and the resulting data fed back to the remote observer. Today remote observing is becoming quite common. In the case of the X-15 observations, the pilot simply turned on the instrument and the telescope by appeal to a gyro-stabilized reference frame, and a star tracker centered the object and carried out the programmed sequence of observations. The observations were examined by the astronomer at a later time. The Wisconsin APT, however, did more than implement an automatic sequence of events. It performed a variety of logical decisions and could continue to operate for several days without human input.

In fact, as I started to review the operation of this instrument, I found that I was describing the APT as though it were a human observer.

This distinction between remote, automatic, and robotic has its analog in the modes of TV viewing. The remote control allows us to control the operation of the TV from a distance; we do not have to get up and go to the TV. The VCR can be set to automatically turn on in accordance with a preset program, and the recording may be viewed at a later time. As far as I know, however, only the parent exercises the necessary logic to tell the children to turn off the TV and go to bed, or a discriminating viewer chooses the right program. Robotic television is still something in our future, but the robotic telescope is with us today. It is the purpose of this paper to describe the history and evolution of Wisconsin's early robotic telescope.

EARLY HISTORY

In 1922, when Joel Stebbins arrived in Wisconsin with his newly developed photocell, the observatory moved directly from the era of visual astronomy to photoelectric photometry with not even a passing glance at photographic astronomy. The early photometry was done with the 15-inch refractor on the University campus where the quality of the sky and the scarcity of good nights required care and patience. Stebbins frequently remarked that one could make accurate measurements of atmospheric extinction or measure variable stars, but not both. Depending upon the nature of the program, astronomers have developed a variety of techniques to overcome this problem. At a good site it is often sufficient to employ average extinction corrections. A Polaris telescope has sometimes been used to monitor the sky conditions. The Wisconsin APT owes its origin to a desire to increase the efficiency and quality of observations obtained at our Pine Bluff Observatory west of Madison, Wisconsin.

Initially we had attempted to use a telescope, acquired from government surplus properties, employing an analog control system. This telescope had been part of the ground test system for an aircraft guidance system. If my memory serves me correctly, the guidance system was called STIBS, an acronym derived from stellar inertial bombing system. A large gyro-stabilized gimbal system employed a star tracker to update the gyros. The surplus telescope was slaved to the star tracker in order to provide for visual verification of the pointing. A photometer was fitted to this instrument, and it was installed in a roll-top roof housing some 100 meters NW of the main observatory.

The first tests were discouraging. The setting accuracy was poor due primarily to backlash, some gear teeth had been broken, and the servo units did not have sufficient resolution to drive smoothly at the sidereal rate. It looked as though we were going to have to do considerable machine shop work on the telescope mount if it were to perform reliably. It was at this stage that John McNall, who was leading the engineering activities at the Space Lab, stepped in. He had acquired a strong interest in artificial intelligence and robotics, and he thought that this would be an excellent application. He suggested that we digitize the control system by replacing the servo motors with stepping motors and shaft encoders. In addition, if we replaced the telescope itself by one of the 8-inch Orbiting Astronomical Observatory (OAO) filter photometer telescopes, we could use the same software for data acquisition that we were using for the satellite system. By providing a complete digital interface, the problems that we had encountered previously were now successfully addressed by software. This was probably the most important lesson that we learned in the execution of this project. A software solution that did not work could be easily removed, while hardware changes were often irreversible. The system that evolved is described below.

SYSTEM DESCRIPTION

The Wisconsin computer controlled telescope is described in detail by McNall *et al.* (1968). The telescope was similar to one of the instruments we had developed for the first OAO. It consisted of an 8-inch, $f/4$ off-axis parabolic reflector with the photometer and mechanism mounted on the side of the telescope tube. An entrance aperture of either 2 or 10 arc minutes in the focal plane was selected by a solenoid. Directly behind the entrance aperture was a five-position filter wheel operated by a stepper motor. The filter wheel contained standard U, B, V filters along with an interference filter that provided a U bandpass completely shortward of the Balmer jump. In addition, there was a dark position and a calibration source position consisting of a Čerenkov source. A Fabry lens imaged the telescope objective onto an end-on photomultiplier. The photomultiplier output was fed to both a pulse counter and a dc amplifier providing the redundancy of a digital signal and an analogue output. The gain of the dc amplifier and the integration time of the pulse counter could be controlled by the telescope computer and provided for extended dynamic range while maintaining digital accuracy.

This telescope was attached to the modified equatorial mount described above. The pointing of the instrument was accomplished by a motor control attached to the mount and commanded remotely by the telescope control computer. The shaft encoders provided position information to a second of time in hour angle and to six seconds of arc in declination; these correspond to one step of the stepper motors. A sidereal clock drove a counter and display that could be read by the computer to provide a continuous sidereal rate drive when actuated.

The telescope housing consisted of a small metal shed with electrically operated roll-back roof and hinged side walls. When the building was closed, the telescope monitored the light level. Additional sensors provided status data on the roof position and on rain, temperature, pressure, wind velocity, and the outdoor light level. The telescope control computer used these housekeeping data and the photometric data to execute a variety of logical decisions with regard to telescope operation and safety.

The telescope control system was located in the basement of the main observatory. The heart of the system was a DEC PDP-8 computer. The computer architecture was built around a 12-bit word length, and therefore the maximum addressable memory was 4096 12-bit words. In modern terms this was a remarkably small memory, and yet by clever programming it was possible to support essentially all the necessary control functions. The increase in computer memory is far more dramatic than the increase in computer speed. This 4 K magnetic core memory was approximately a 4-inch cube or a volume of 64 cubic inches (768 bits per cubic inch). Today one can buy 10 Mbyte SIMS package in a volume of approximately 1/8 cubic inch, or a density of 640 Mbits per cubic inch. The increase in onboard memory density has been nearly a factor of one million. Because of the limitations of the PDP-8, all programs were of minimum length with optimization in that direction at a sacrifice in speed if necessary. The library routines supplied with the computer were removed and replaced by more compact arithmetic subroutines for only those functions required for the task. The computer was modified by adding an analog-to-digital converter and an eight-channel multiplexer. Some functions, which in principle could have been carried out by the computer, were transferred to special purpose hardware located in a separate electronics rack. These included the photometer control, the housekeeping controller, and the required display functions. The interface with the user was provided by a punched paper tape star list input, while the output was provided in two forms via a Kleinschmidt printer and data storage on punched paper tape. A functional block diagram of the telescope control system is shown in Figure 1.

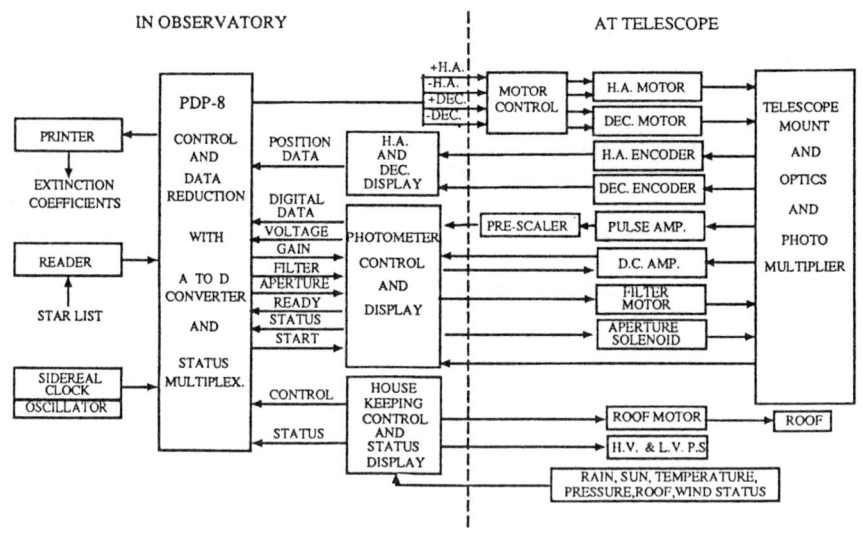

Fig. 1. The Wisconsin APT system functional block diagram.

TELESCOPE OPERATIONS

When the computer was first powered up, it was necessary to key in from switches on the front panel a primitive load program. This instruction set did one thing and one thing only: namely, it made it possible to read a more complete loader program from the punch paper tape reader. It is difficult to fully appreciate how intrinsically dumb a digital computer really is unless you have had this opportunity to teach it its first few words of machine code. Of course, despite this it often succeeded in outwitting us. After booting, the computer read the control programs and other software. The program execution began with reading the star list into core memory.

The high voltage was applied to the photomultiplier after activation and the light level inside the shed sampled. When it was sufficiently dark the rain sensor was interrogated. If it was not raining, and the wind and barometer readings were not too threatening, the roof was opened and the light level tested again. The first star was selected from the star list based on hour angle, and the telescope would slew to that position. If a star was found in the 10′ diaphragm, the telescope would center and switch to the 2′ aperture. Before final centering, the telescope would move 10′ north and measure the sky. After centering the star a check of the brightness was made, and if a magnitude had been given in the star list a verification of the star identification was performed. The gain and exposure time were then set from either the information provided or from the measured intensity. A symmetrical set of measurements was carried out, observing the star twice through each filter in a cyclical manner. The signal was sampled every 26 microseconds and averaged over a 12 second interval. The observation was completed by moving to a sky region 10 arc minutes south of the star.

The data from each observation sequence was reduced to a visual magnitude and three instrumental color indices using the current extinction coefficients. In the star list an object was identified as a standard or a program star. The standard stars were reobserved on a schedule determined by the zenith distance, and each reobservation was followed by a recalculation of the extinction coefficient. The magnitudes, colors, and extinction coefficients were printed out on the Kleinschmidt printer, and all data were put on paper tape for later processing. The telescope then moved to the next star and the sequence continued throughout the night. As dawn approached the Sun sensor and the photometer would signal the end of observing and the shed would close up.

Barring problems, the telescope could continue to operate in this manner night after night and did so successfully. If the telescope failed to find a star during set-up, a search routine was implemented which moved the star through a rectangular pattern about the center point. The default search pattern was set at 30′ × 30′. If it failed to find a star of the appropriate magnitude it would move on to the next star in the list. Three successive star-acquisition failures were taken to indicate that it was cloudy, in which case the system waited one hour and tried again. In the event of rain or an impending storm, the instrument shut down for one hour and then checked the status again. In the event of serious problems, a computer-controlled alarm sounded, and the system shut down and waited for servicing.

In developing the control program a number of difficult hardware problems were encountered. The gear backlash could amount to as much as 20′, but was effectively eliminated by ensuring that all final motions were made moving north or west. The motors did not have sufficient starting torque to operate at the nominal operating frequency of 400 Hz and so were ramped up starting at 200 Hz. There were also some resonances excited which could result in motor stall or in reversing the direction of drive. These frequencies were digitally filtered from the drive pulses. In addition to these problems, a defective Datex encoder on the declination axis gave incorrect readings at certain positions when moving south. By using a dynamic slewing and following the continuity of encoder readings that problem could be overcome. The procedure involved waiting many steps before concluding that the motion is incorrect.

SUMMARY

The Wisconsin Automatic Photoelectric Telescope, built in 1965, operated successfully for several years. During that time we were carrying out photometric programs on stellar energy distributions and on planetary and diffuse nebula line intensities. These programs required photometric nights which were a relatively precious commodity in Wisconsin. The use of an auxiliary extinction telescope increased the overall observing efficiency and quality of the data. The Wisconsin APT can quite properly be called a robotic telescope despite its limited functionality, since it was capable of making all the major logical functions required to fulfill its task. As indicated above, the computer control allowed us to program around a number of difficult hardware problems. Based upon our subsequent experience with space instrumentation, I believe that this lesson is not as true today. The reason is that as more and more dependence has been placed on software and less on the hardware components, the programs have grown very complex. There is no way that all branches of a large control program or applications software can be tested, and we are going to continue to be surprised by software bugs. Big systems are beyond human comprehension.

During the years we have moved away from programs of this nature at Pine Bluff and towards programs involving differential measurements. Today the observatory is used primarily for spectral polarization measurements which do not demand photometric skies, but do require a great deal of observing time and care. Our current control system activities are centered around the WIYN 3.5-m telescope control system. This will be a remote control system but not a robotic telescope. It will be remotely operated from the control building adjacent to the telescope housing and remotely from the campuses of the participating universities. And this is the final lesson I derive from the history of the Wisconsin APT; namely, that the purpose should not be to have a robotic or automatic telescope, but rather to do something better or something otherwise not possible. The extinction telescope freed up observing time so that continuous observation of the program stars was possible. Telescopes in space, at the south pole, or on the Moon allow us to explore different spectral regions and different dimensions of precision and time. The APTs and the newer Automatic Imaging Telescopes, pioneered by many of the participants of this symposium, have done just that — something not possible by traditional means.

ACKNOWLEDGEMENTS

The original idea to construct this automatic extinction telescope was due to the late Theodore E. Houck. The innovative implementation owes much to the late John F. McNall. Both Ted and John made invaluable contributions to space astronomy, and those of us who knew them were richer for their friendship.

REFERENCES

Maran, S. P. 1967, *Science*, **158**, 867.
McNall, J. F., Miedaner, T. L., and Code, A. D. 1968, *A.J.*, **73**, 756.

RESULTS FROM THE FOUR COLLEGE APT

ROBERT J. DUKES, JR.
Physics Department, The College of Charleston, Charleston, SC 29424

ABSTRACT The Four College Consortium Automatic Photoelectric Telescope (APT) has completed its first full year of operation. In this paper I describe the acquisition and testing of this APT. I also discuss preliminary results from the first year of operation. During this year we have paid particular attention to the problems of assessing the quality of the data. We find that it is possible to determine the sky quality from the combination of site logs and data, and to devise objective means of rejecting observations obtained on poor nights. When this is done we find that the standard deviation of a 10 second differential Strömgren measure of a pair of constant stars is better than 5 millimagnitudes. I also discuss some of the problems associated with scheduling a multiuser APT involving a variety of types of observing requests.

The Four College Consortium, consisting of The College of Charleston, The Citadel, Villanova University, and The University of Nevada (Las Vegas), is a group of primarily undergraduate colleges with a strong commitment to undergraduate research. These schools were funded by the National Science Foundation under the Research in Undergraduate Institutions (RUI) program for a three-year period for the acquisition and operation of an Automatic Photoelectric Telescope (APT). Data from this telescope would be used together with data gathered at campus observatories to permit undergraduates to perform astronomical research in the limited time available to them. The first part of this paper discusses the design and construction of our APT that was the first of a new generation of telescopes designed from the start for fully automatic operation.

Our original proposal called for us to duplicate the APT which Vanderbilt University had acquired with NSF funding. This is a 16-inch DFM Engineering telescope which has a control system and a photometer constructed by Louis Boyd of Fairborn Observatory. However, due to a change in policy by DFM, it was impossible for us to purchase this for the price originally quoted which was the amount we had budgeted. Fortunately, Fairborn Observatory offered to supply the prototype of a 30-inch automatic telescope for the budgeted amount. In doing this, they donated the primary mirror as well as engineering and design time. We requested and received approval from the NSF program officer for this change. We recognized that a new, as opposed to a tested, design involved increased risk, but also that getting a 0.75-m telescope for the price of a 0.4-m provided significant advantages.

Our proposal to the NSF outlined an observing program that was both more flexible and more sophisticated than those which had been used with previous APTs. Those then operating had yielded one differential measure of a variable, comparison, and check star group each night. The Four College Consortium's program required that our APT be capable of observing one star continuously for an extended period, or repeating observations of one group at fixed intervals during a single night. Additionally, we wanted to use many filter combinations. Some of our programs also involved transformations to a standard system and therefore required all-sky photometry.

To accommodate these requirements, the Automatic Telescope Instruction Set (ATIS) was developed by Russ Genet, Lou Boyd, Donald Hayes, and Diane Pyper-Smith. ATIS gives the flexibility that we needed and provides capabilities that we had not anticipated. For example, we now specify a list of groups of stars to the telescope. A group consists of one or more stars as well as one or more sky positions. For each group we specify an observing window in Julian Date and an observing window in Local Sidereal Time. The latter is used for specifying either the time or hour angle of observation.

A group is assigned a priority of 1 (highest) to 99 (lowest). This is used by the control program to sequence the groups. All groups of a given priority are observed in order from west to east. Then groups of the next lower priority are observed. In practice we have found that three levels of priority are all that are necessary for routine observations. We use priorities 5, 7, and 9 for groups that are to be observed over an extended period. Higher priorities are used for time critical observations. These, for example, might be observations coordinated with another observatory or continuous coverage of a star during a short eclipse. We reserve even-numbered priorities for extinction or standard stars.

There are two ways of obtaining more than one observation of a group during a night. We may request that a group be observed repeatedly by specifying more than one observation. In this case the group is placed repeatedly in the queue of equal priority groups until all the observations requested have been obtained or until the group is out of the observing window. On the other hand, if we are interested in obtaining several observations of a group but not continuous coverage, then we give the group a different name for each LST window desired. An example of this is a variable with a period of a few days which needs to be observed several times during a night to adequately cover the light curve.

The final control parameter is a probability. This allows us to request that a group be observed less frequently than once per night without having to specify multiple observing windows. If a request has a probability less than one, the control software generates a random number to decide whether to observe a group at all during the night. If the group is selected for observation, its priority is then compared with the priorities of all other groups and those with the highest priorities are observed first.

Scheduling an ATIS telescope requires programs which produce ATIS instruction files from simple requests submitted by the astronomer. Two such programs were developed by the Four College Consortium. The loader/ generator program designed by Diane Pyper-Smith of UNLV accesses separate telescope, star, group, and request files and produces an ATIS file for a night's observations. CREATE, developed by George McCook of Villanova, uses as a basis a group file containing all the information on a differential group and the stars in it. At present CREATE is more flexible but requires the entry of redundant information in some cases.

Programs to perform preliminary analyses of ATIS files were developed separately by André Hedrick, a College of Charleston undergraduate, and George McCook. A spreadsheet program for reducing standard star data to obtain extinction and transformation coefficients was coded by Don Hayes. ATIS files from an APT are prepared for use by this spreadsheet by a program that has many authors including Russ Genet, Don Hayes, Mike Seeds, and Diane Pyper-Smith. An additional spreadsheet program for reducing differential photometry as well as two programs to filter incomplete groups and to prepare ATIS files for use by it were written by Diane Pyper-Smith.

A mini-ATIS simulator was developed by George McCook, while the development of a full-scale ATIS Simulator was begun by André Hedrick. Finally, we have worked with Mike Seeds (Principal Astronomer of the Phoenix 10-inch APT) in developing procedures for determining the overall quality of a night.

We have had some problems with bringing an instrument of completely new design into operation. It became necessary to make a number of changes to the original design as the project progressed. For example, the smaller APTs find and center their targets by repeatedly moving in a square spiral pattern and using the photometer to determine when the target is in the field. The 0.75-m design had a moment of inertia great enough that this process was excessively time consuming. Thus, it was necessary to add a CCD camera for acquisition and centering. The initial telescope proved to have mechanical defects. A completely new one was constructed by a larger machine shop. The donated mirror was very thin and difficult to figure. It had to be refigured several times to achieve an acceptable figure. All of these changes and corrections were made by Fairborn Observatory at no cost to the Consortium.

Our time line given in the original proposal quickly became obsolete as a result of the change in telescope and the problems described above associated with perfecting and completing it.

While we were waiting for the Four College Consortium telescope to become operational, we observed a small number of stars on the Fairborn Observatory "Rent-a-Star" APT. These observations were provided by Fairborn Observatory at no cost to the consortium. This APT could only provide one observation per group per night and thus was not suitable for observing short period variables. Therefore, Dukes joined with Saul Adelman and Diane Pyper-Smith in work on some of the Ap stars described later.

The Four College Consortium APT was operating in a mode similar to the earlier APTs during the Spring of 1990 in that most groups were observed once per night. Pyper-Smith served as Principal Astronomer since Dukes had lost his residence in Hurricane Hugo and was spending most of his time dealing with insurance adjusters and contractors.

During this period, our operations suffered from a number of problems involving both the telescope and learning to program ATIS. These have been resolved. Operations since early November 1990 have been relatively routine. We have been affected by unforeseen events such as mice eating the cables, power failures on Mt. Hopkins, and difficulties in transmitting data between computers at the observatory. We have attempted to characterize the nights of the 1990-91 season according to whether we obtained a significant number of observations, whether the night was lost due to poor weather, or whether the night was lost due to equipment problems. The results of this are shown in Figure 1. To put the difficulties in perspective, equipment problems have caused us to lose only a small number of nights compared to those lost due to weather.

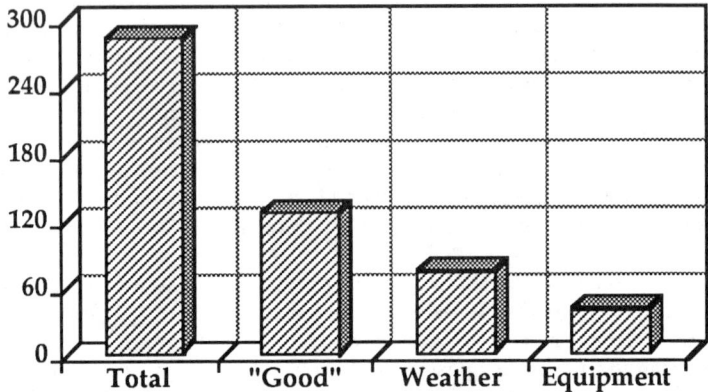

Fig. 1. The 284 nights between the beginning of the season in September 1990 and the end in June 1991, broken down into nights with significant data ("good"), nights with little data due to clouds or rain (weather), and nights lost due to equipment problems (equipment).

As of June 1991 the telescope has successfully obtained differential photometry in all the desired modes. We discovered that it was advisable to use a fourth or fifth magnitude star near a group as a navigation star. The telescope can easily acquire such a star and thus update the coordinates. By using such a navigation star we can routinely acquire ninth magnitude stars in all but the most crowded fields. With blind offsets from a brighter star we have been successful in observing the thirteenth magnitude quasar 3C 273. Recently we achieved the ability to change the ATIS observing program on a nightly basis. We have found that it is necessary to add neutral density filters to the alpha and beta wide filters to measure alpha and beta standards without fatiguing the tube. This will be done during the summer shutdown.

An objective of scheduling this telescope was to insure that each of the four institutions got the same allocation of telescope time. I began this process by asking each institution to submit the same number of high, medium, and low priority groups. I hoped that this would provide approximately equal time allocations. As can be seen from Figure 2, this didn't happen. There are a number of reasons for the great variation in time used. The two most significant are that while the number of groups was monitored, the time per group was not during the first part of the observing season. Since ATIS allows groups to be structured in a variety of ways, the time required to observe one group might be very different from the time for another group. Indeed, the users with the smallest total time were those whose groups were structured in the same manner as that used for traditional APTs. One of these groups might require about 15 minutes to observe. Other users modified this traditional structure to spend more time observing sky or to perform blind offsets to faint objects. Some of these groups required more than 30 minutes to observe. The other significant factor seemed to be how often observers would retrieve and analyze their data, since frequent retrieval permits rapid identification of groups that were aborting due to problems with the request file. Apparently, to most efficiently use an APT such as ours, it is necessary to monitor the operation carefully.

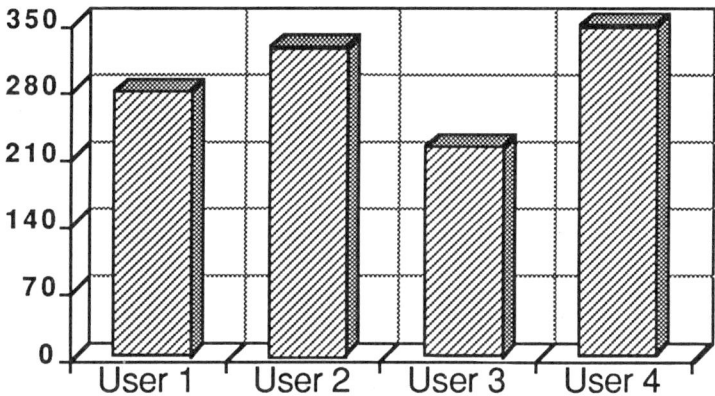

Fig. 2. A breakdown showing the number of observing hours accumulated by each user. The time recorded is the sum of time spent counting photons and in intragroup moves. It does not include intergroup moves.

One of the main objectives of the first year has been to develop techniques to evaluate the quality of the night from the data obtained during the night. One such way is to examine the consistency of the check minus comparison measures over a number of nights. Dukes has used this with the observations of 53 Persei discussed below. Figure 3 shows the Strömgren b magnitude difference between the check star (HR1261) and the comparison star (HR1482) for 17 nights during Fall 1990. This group was observed from 3 to 6 times per night during this period.

In panel (a) all the data are plotted. Several points are apparently discordant. The logs for these nights show that either the night had noticeable clouds present or the observation was obtained during twilight. After these data were removed the plot was repeated [panel (b)] with a magnified scale. Again a few points were discordant, and for the most part a check with the log revealed the reason. The exception are the observations made on JD 244874. These observations gave the appearance of the comparison rather than the variable varying. This night was eliminated from the analysis. Further investigation will attempt to determine the cause of this discrepancy. After these observations were eliminated, the standard deviation of a single 10 second check minus comparison measure was approximately 4 millimagnitudes.

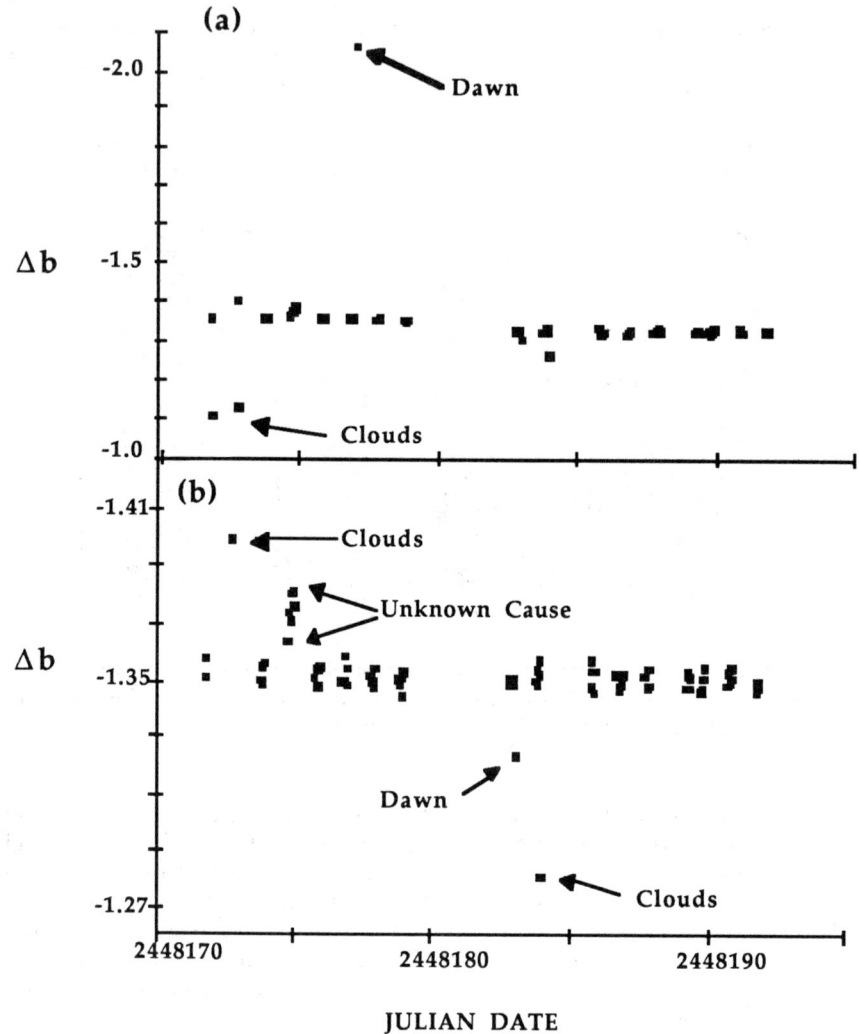

Fig. 3. The Fall 1990 check minus comparison data for the 53 Persei group. Panel (a) shows several discordant data points identified as discussed in the text. Panel (b) shows the data with these points removed and the scale expanded. Several more discordant data points with obvious explanations are shown. One night with an apparent variation of the comparison star is visible; this night was discarded.

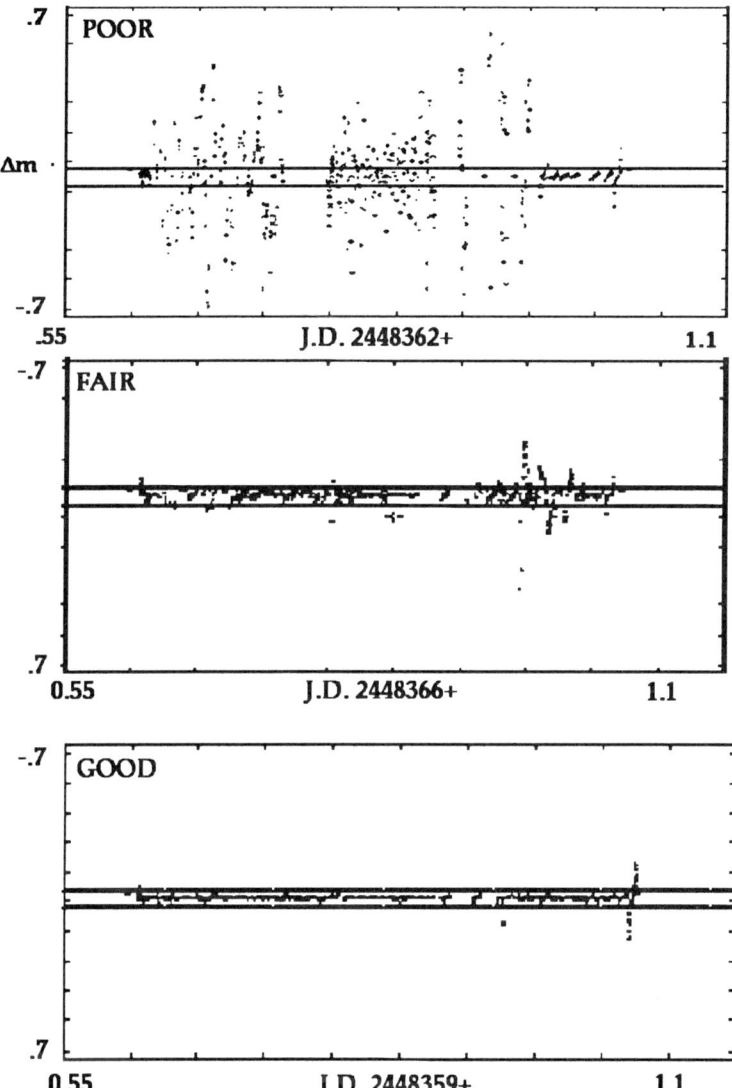

Fig. 4. Plots of the difference in magnitudes for each measure of a non-variable star in a group in a filter and the mean for all measures of the star in the filter that show the distinctions between good, fair, and poor nights. The horizontal lines across the center of each plot mark the 20 millimagnitude scatter about the mean.

McCook has been using CREATE to analyze the night quality using all the non-variable star data obtained by the telescope on a given night. The mean of the counts for each star in each group in each filter is calculated, and the difference from this mean of each of the measures is calculated. These differences are plotted against time. This is similar, but not identical, to the 20 millimagnitude check used for data from the non-ATIS APTs. Figure 4 shows what this plot should look like for good, fair, and poor nights. CREATE allows the user to edit the data, eliminating those measures with large standard deviations. Unfortunately, the file produced by this edit is no longer an ATIS file.

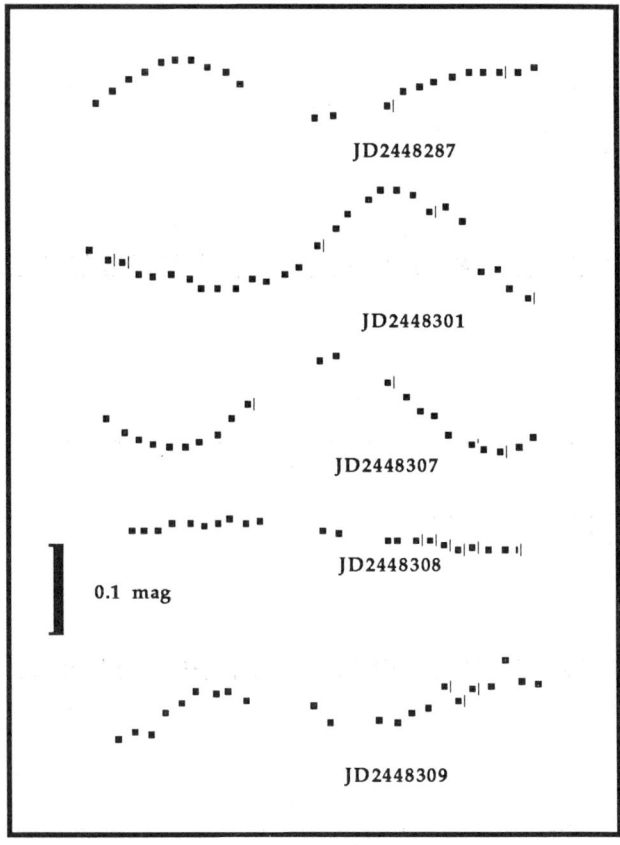

Fig. 5. Sample light curves of the Delta Scuti star 4 Canum Venaticorum obtained in the Spring of 1991. The check minus comparison observations indicated a standard deviation (of one point) of less than 4 millimagnitudes.

I have participated in a campaign involving coordinated ground-based and Voyager 2 observations of the non-radial pulsating B star 53 Persei. Nearly 200 Strömgren four-color ($uvby$) observations were obtained of this object from

October 1990 through January 1991. Preliminary analyses of these, done together with undergraduate student R. Martin, have shown a variation of approximately 0.1 mag in b.

I have obtained approximately 39 nights of observations of the Delta Scuti star 4 Canum Venaticorum. Sample light curves of the b magnitude difference of 4 CVn and the comparison star HD108100 are shown in Figure 5. Here, too, the standard deviation of a 10 second check minus comparison integration is better than 4 millimagnitudes.

I have also been observing several of the short period multimode Cepheids and have acquired a significant number of observations of BQ Serpens. A periodogram of the Johnson B data is shown in Figure 6. The two strong periods have been previously detected. I hope that the current observations will refine these and provide a baseline for monitoring period and amplitude changes of the two modes. The third period shown is interesting since it has a value close to that expected for the second overtone.

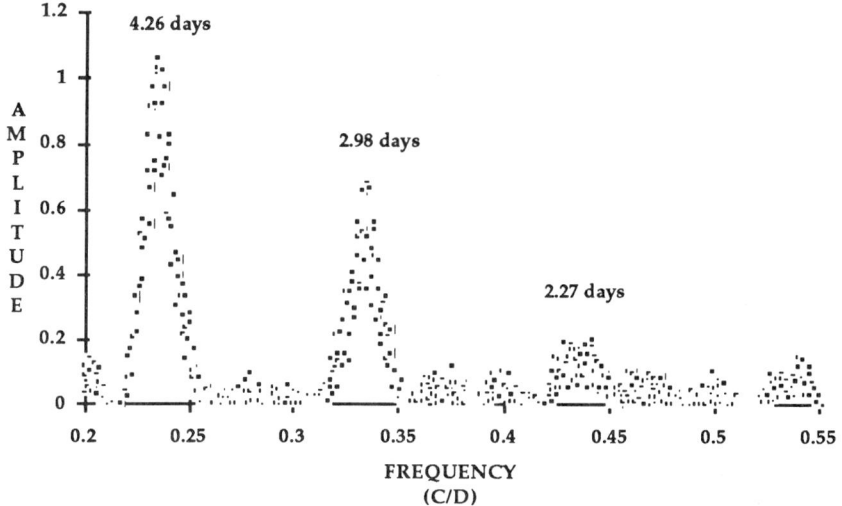

Fig. 6. A periodogram of 101 B observations of the double mode Cepheid BQ Serpens. The peak at 2.27 days is approximately the value of the second overtone, but is more likely to be an artifact of the observing window.

Adelman and Pyper-Smith are coordinating their observations of the magnetic Ap stars. They each selected some two dozen stars from the known chemically peculiar stars of the upper main sequence and are consulting each other in adding new stars to their programs. In this division of the Ap stars, Pyper-Smith selected many of the stars with the best known periods as she already had obtained some $uvby\beta$ photometry of them. Adelman chose many with relatively large amplitudes. At present Adelman and Pyper-Smith are systematically reducing and analyzing their Strömgren differential photometry of magnetic Ap variables and of a few Am and Be stars. The ability to easily add new stars to our program has gradually become known to other investigators of Ap stars. For example, Dr. David Bohlender, University of British Columbia, informed Adelman of a newly discovered magnetic star with a tentative period of 1.5 days. Although it was almost the end of its observing season, Adelman added it to his program and begun observations in less than two weeks.

To get a better feeling for the quality of data obtained with APTs, UBV observations for eight magnetic variables were obtained with the Phoenix 10-inch telescope. Adelman and I analyzed the check-comparison data and showed that none of the check and comparison stars are variable. Data for this telescope are distributed to users only if the rms deviation of the individual measurements (variable-check and check-comparison) is less than 0.02 mag. By accepting only data for which these values are given for U, B, and V, they found that the quality of the data was similar to that attained by astronomers observing with small telescopes. A paper on this work has been submitted to *The Astronomical Journal*. Besides refining the periods of many of these stars, the major utility of this exercise was to remind us that many of the published periods were based on rather small numbers of values. This, together, with the large separation between data sets, makes it very difficult to combine various sets of observations. After several seasons of APT data we will be better able to derive definitive periods. Such an observing strategy also will permit the derivation of the shape of the light curves and allow one to study possible changes in shape over periods of years.

In summary, we have been able to obtain observations of a number of interesting stars and prove the ability of the 30-inch APT to perform quality photometry. We also devised techniques to eliminate poor-quality observations from our data set, and we tested the ability of ATIS to permit a single telescope to perform a variety of types of observations for several observers on a single night. We expect that the coming year will see a number of observing projects be brought to fruition.

ACKNOWLEDGEMENTS

This work was supported in part by NSF Grant AST-8616362 to the College of Charleston and by a College of Charleston research award. I would like to thank my colleagues in the Four College Consortium, Saul Adelman, George McCook, and Diane Pyper-Smith, for their assistance in this project. I would also like to thank Mike Seeds, Bill Kubinec, and Rob Dillon for helpful discussions, and André Hedrick for aid with the programming and data reduction. Finally and most importantly, I would like to thank Lou Boyd and Russ Genet for designing, constructing, and operating our APT.

INTERROGATION OF DUPLICITOUS STARS WITH AN APT

BERNARD W. BOPP
Ritter Observatory, Department of Physics and Astronomy, University of Toledo, Toledo, OH 43606

ABSTRACT Preliminary results from intensive spectroscopic and APT monitoring of two interacting binary systems are presented. Both V644 Mon (Be + K:) and HD 37453 (F5 II + B) show complex, composite, and variable spectra. APT observations extending over three years show both stars to vary by 0.1–0.2 mag in V. The photometric variability of V644 Mon appears to be irregular, though there is some evidence for periodic behavior in the 50–60 day range. HD 37453 has an orbital period of 66.75 days; the best-fit photometric period is not quite half this value, indicating the star is an ellipsoidal variable.

INTRODUCTION

I have for a number of years entertained the fantasy that stars have personalities. There are certainly stars that I imagine to be honest, upstanding, solid-citizen types. A star like Arcturus might be counted among these "honest" stars, and I think I could imagine undertaking observations of Arcturus in a very civilized frame of mind: perhaps I would wear a coat and tie, and have a chilled bottle of wine at my side as I guided. But fate has decreed that most of my observing time would be devoted not to stars of high moral character, but to ones with *much* less credibility. In this case, the appropriate observing costume might be a trench coat, fedora, and a steely-eyed expression: these are stars that do not respond to occasional polite questions, but have to be **interrogated**.

The flavor that I am trying to convey by using "interrogate" in my title is that some stars need systematic, long-term observations before one can believe (or model) anything. Synoptic data of this sort can best be furnished by an APT. For the two examples that I will discuss, APT observations extending over months or years are providing information that describes or constrains models for what turn out to be complex, interacting binary systems. Given the timescale and amplitude of the light variations, it is hard to imagine any other effective way to photometrically observe these stars other than by using an APT. All the photometry discussed below was obtained with the Phoenix 10-inch APT on Mt. Hopkins.

V644 MONOCEROTIS

The interacting binary V644 Mon (HD 51480) has at least one claim to being one of the most remarkable stars in the sky. By an extraordinary coincidence, this sixth magnitude star was the subject of *two* independent spectroscopic and photometric studies, both of which were published (again, without prior planning) in the same issue of the same journal. The journal issue was (appropriately) the November 1989 *Publications of the Astronomical Society of the Pacific* (Bopp and Dempsey 1989; Halbedel 1989). Much to this author's relief, the observational data and conclusions reached in both papers were similar. A useful test of the accuracy and precision of APT observations was provided by partially overlapping photometry obtained by Halbedel at Corralitos and Kitt Peak Observatories: these measures agreed with the APT data to within 0.01 mag.

Spectroscopically V644 Mon appears to be, at least superficially, a Be star, showing an intense (and variable) Hα emission line. However, the CCD spectrum illustrated in Figure 1 also shows weak metallic absorption lines and very strong Na I D lines. A high-resolution observation of the D-line region (Fig. 2) is extraordinary. In addition to absorption lines from neutral metals (broadened to about 24 km s^{-1}) and a strong, broad He I 5876 Å absorption, the D-lines show extremely complex structure, including P Cygni profiles! Spectra in the region of the Ca II infrared triplet (Fig. 3) show these lines to be in emission; the spectrum in this region resembles that of the F + B binary HD 127208 (Dempsey *et al.* 1990). The spectrum of V644 Mon is both composite (Guinan, Koch, and Plavec 1984) and very complex.

APT observations during 1988 (Fig. 4) showed V644 Mon to vary by at least 0.2 mag in V. Standard period finding routines (e.g., PDM in the IRAF package) gave a weak indication of a period near 60 days, though this was not significant enough to mention in our 1989 paper. We resumed APT observations of V644 Mon in late 1989; a light curve extending over five months is presented as Figure 5. There is perhaps a stronger suspicion of periodicity in this data set, with minima separated by 50–60 days. While it is clearly premature to claim this as a possible photometric/orbital period for V644 Mon, continuing APT interrogations (and spectroscopy!) at least hold the promise of eventually untangling this duplicitous star.

INTERROGATION OF DUPLICITOUS STARS   21

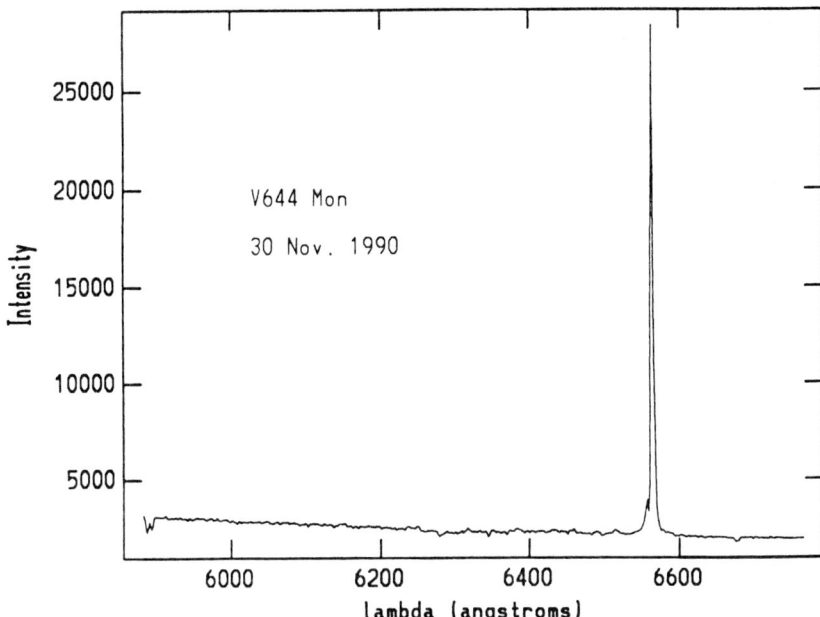

Fig. 1: CCD spectrum (2Å resolution) of the red region of V644 Mon. Note the weak metallic absorption features along with the strong Hα emission.

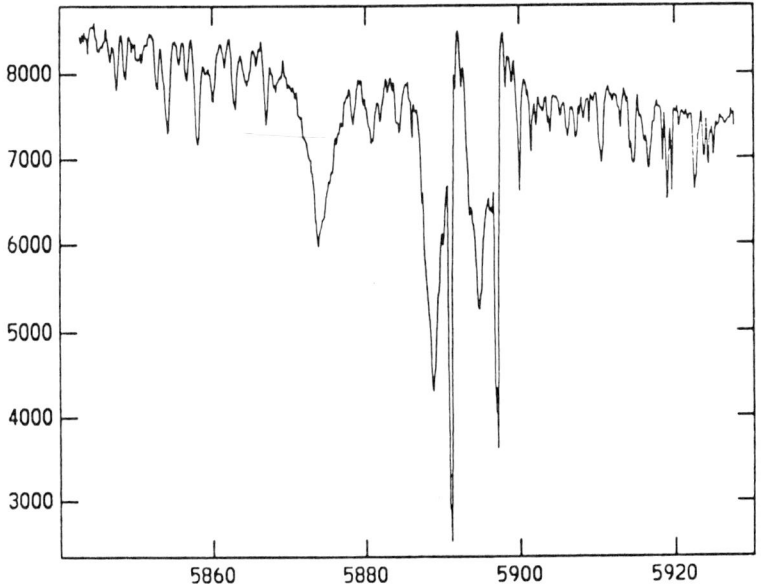

Fig. 2: High-resolution CCD spectrum of the Na I D-line region in V644 Mon. The strong, broad absorption feature near 5875Å is due to He I.

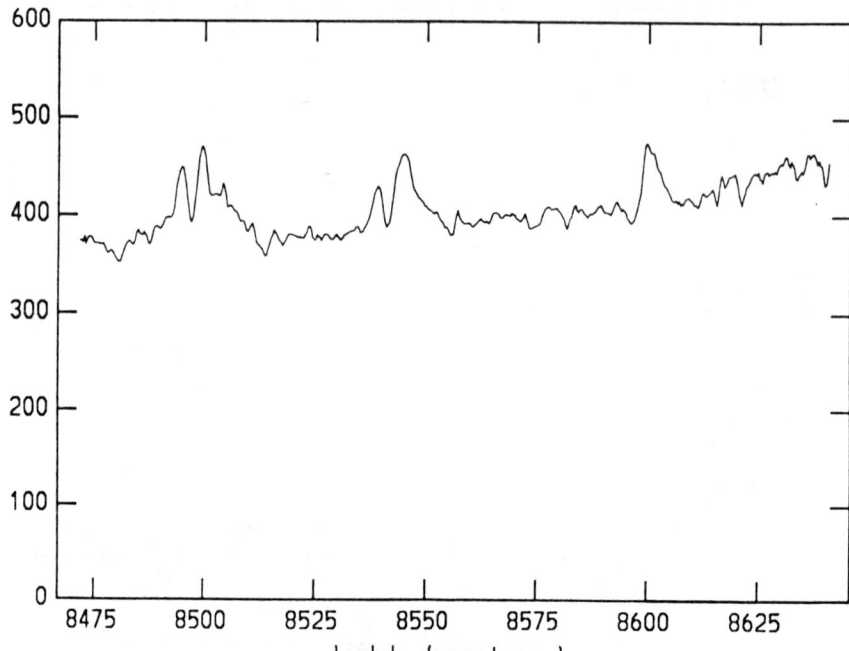

Fig. 3: The Ca II infrared triplet region in V644 Mon in October 1989.

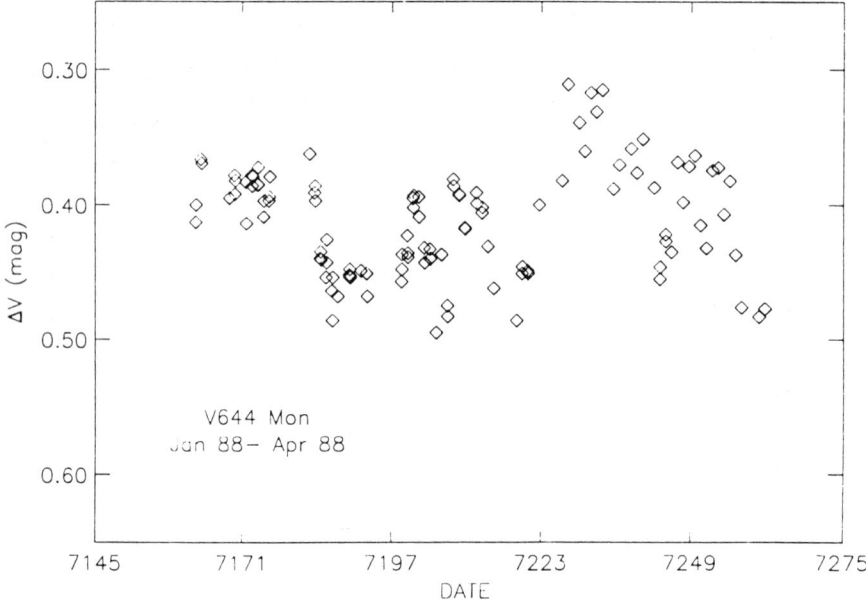

Fig. 4: APT observations of V644 Mon during 1988. The x-axis is Julian Date minus 2,440,000.

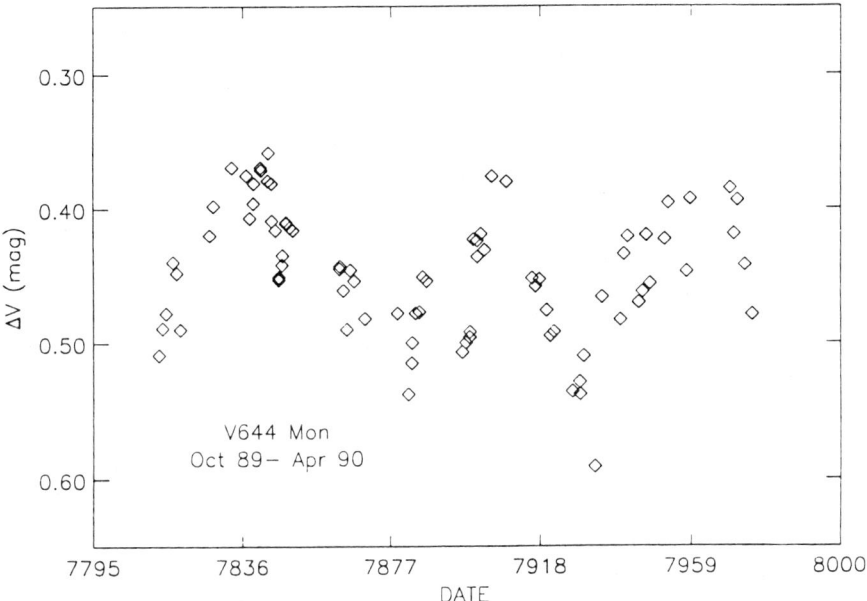

Fig. 5: Same as Fig. 4, for the Oct. 1989-April 1990 interval.

TABLE 1
Preliminary Orbital Elements for HD 37453

P = 66.75 ± 0.02 days
e = 0.04 ± 0.03
γ = 3.0 ± 1.3 km/s
K = 70.4 ± 1.7 km/s
T = JD 2,447,072.1 ± 5.5
ω = 232 ± 30 degrees
a sin i = 6.46 ± 0.16 x 10^7 km
f(m) = 2.39 ± 0.07 $M_\odot$

HD 37453

HD 37453 is a member of a group of binary systems which contain a main-sequence B star along with an evolved F, G, or K-type component. All these systems are characterized by strong Hα emission, which is produced by interaction between the components, rather than indicating that the early-type star is a "classical" Be (Fig. 6). Perhaps the most remarkable of these F + B systems is HD 207739 (Griffin *et al.* 1990), which shows [N II] 6548, 6584 Å emission, shell features in the Na I D lines, variable He I 5876 Å absorption, and a large mass function indicating that the cooler evolved component is the secondary in terms of mass.

Spectroscopically HD 37453 (F5 II + Be; Bidelman 1954, Table 9) is a very close match to HD 207739, exhibiting all the characteristics noted above. The orbital period of 66.75 days is well established by velocity measures made at Cambridge by Roger Griffin, as well as data from Kitt Peak. Preliminary orbital elements for HD 37453 are presented in Table 1. I caution that spectroscopic interrogation of HD 37453 has been arduous, extending over seven years, and there are still some puzzling aspects to the velocity measures, which during certain intervals exhibit systematic shifts from the Keplerian orbit by amounts of up to 10 km s^{-1}.

The APT program on HD 37453 has extended over "only" three years, during which nearly 200 V measures of the star have been obtained. In this case the variability with amplitude near 0.1 mag appears to be clearly periodic (Fig. 7). The small surprise is that the interval between minima is 33.24 days, not quite half the orbital period. When plotted with respect to the orbital period, a double sine curve results, indicating that HD 37453 is an ellipsoidal variable. It will be important to continue monitoring HD 37453 photometrically, since the photometric behavior of HD 207739 has been shown to be disturbingly variable (Bloomer 1985). Bloomer's photometry of HD 207739 reveals elliptical variations to be present, but there were season to season changes in the light curve, fluctuations (especially near the minima) on a short timescale, and even the suggestion of an "eclipse" (probably a brief occultation of a portion of an accretion disk or hot spot [Dempsey 1987]).

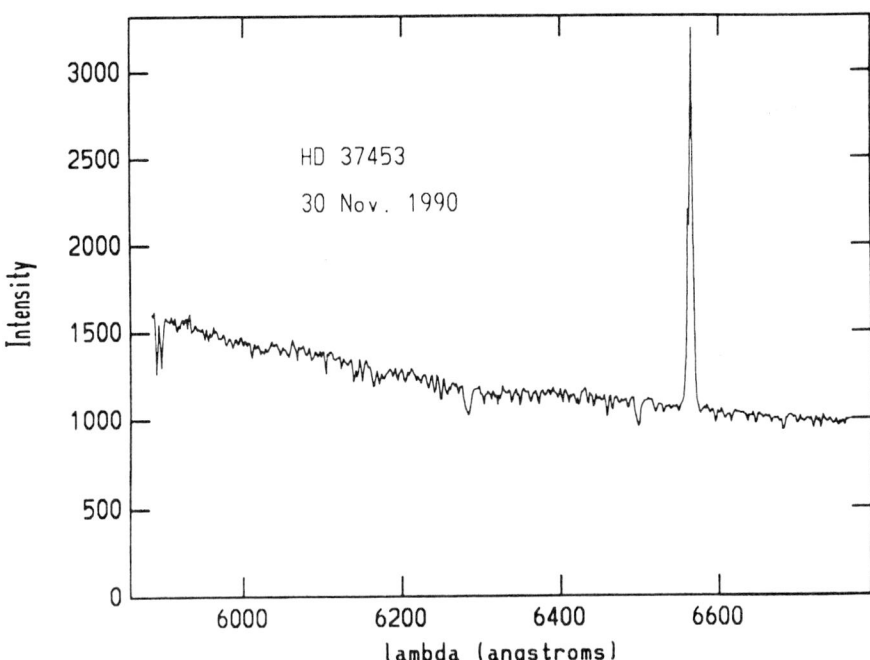

Fig. 6: CCD spectrum of the red region of HD 37453.

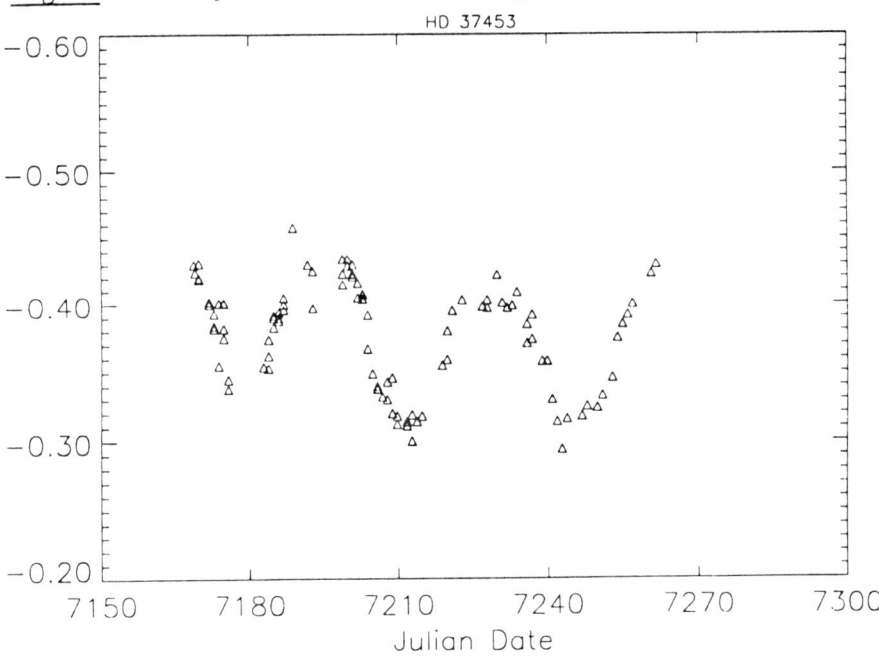

Fig. 7: A portion of the 1988 APT photometry of HD 37453, showing its periodic variability.

ACKNOWLEDGEMENTS

Many of the spectroscopic observations reported here were obtained and reduced by Robert Dempsey, whose collaborative effort and help I gratefully acknowledge. Roger Griffin contributed a number of radial velocity measurements of HD 37453, and made the first determination of its orbital period. I thank him for his contribution. Research on the interacting F + B systems is a collaborative effort by R. Dempsey, R. Griffin, S. Parsons, and myself, and is supported in part by NASA through grant NAG 5-954. Finally, I thank Virginia Trimble and Mel Dyck for noting that even Arcturus is not as honest as some naive astronomers might believe...

REFERENCES

Bidelman, W. P. 1954, *Ap. J. Suppl.*, **1**, 175.
Bloomer, R. H. 1985, *Bull. A.A.S.*, **16**, 913.
Bopp, B. W., and Dempsey, R. C. 1989, *Pub. A.S.P.*, **101**, 978.
Dempsey, R. C. 1987, *Master's Thesis*, University of Toledo.
Dempsey, R. C., Parsons, S. B., Bopp, B. W., and Fekel, F. C. 1990, *Pub. A.S.P.*, **102**, 312.
Griffin, R. F., Parsons, S. B., Dempsey, R. C., and Bopp, B. W. 1990, *Pub. A.S.P.*, **102**, 535.
Guinan, E. F., Koch, R. H., and Plavec, M. J. 1984, *Ap. J.*, **282**, 667.
Halbedel, E. M. 1989, *Pub. A.S.P.*, **101**, 995.

MULTIPERIODICITIES IN THE PHOTOMETRIC VARIABILITY OF
RS CVn BINARIES - WHAT WE KNOW AFTER 15 YEARS

DOUGLAS S. HALL

Center of Excellence in Information Systems, Tennessee
State University, Nashville, Tennessee

1. INTRODUCTION

Fifteen years ago (Hall 1976) the RS CVn binaries were defined, so now on the sesquidecennial is a good time to review what we have learned. Because the theme of this symposium was robotic telescopes, which to date have been used almost entirely for photometry of variable stars, this review will concentrate on the photometric behavior.

The group of RS CVn binaries is just one of ten groups of so-called chromospherically active stars, groups which include both single and binary stars, stars as small as dwarfs (V) and as large as bright giants (II), stars at different evolutionary stages (before, on, and after the main sequence), and stars with rotation periods shorter than a day to longer than a year. Such stars display all of the phenomena which make up solar surface activity, but they do so more vigorously, by orders of magnitude. The term "chromospherically active" is something of a misnomer in two respects: (1) chromospheric activity per se is just one in the panoply and (2) the sun is also "active" so one should distinguish the others as "very active".

The phenomena of chromospheric activity are best understood in the context of a stellar dynamo, and there has been some success in parameterizing the strength of the dynamo action with the Dynamo Number or with the Rossby Number. The second of these is the ratio between the rotation period and the convective turnover time, with a smaller Rossby Number associated with stronger dynamo action. To compress a lot of physics into four words one could say there are only two necessary ingredients for a strong stellar dynamo: "deep convection, rapid rotation". It is established now that binarity is a factor only indirectly, serving to make a star rotate faster than an otherwise similar star would if single. Likewise, evolutionary state or age is also only an indirect factor, among single stars, with the rapid pre-main-sequence rotation decaying once a star reaches the main sequence and with post-main-sequence expansion causing rotation to slow down.

Although much of my own photometry of chromospherically active stars has been done with two automatic telescopes, the prototype 10-inch and the Vanderbilt 16-inch, both on Mount Hopkins, much other photometry has helped in the difficult task of making sense of the multiperiodic photometric behavior of these peculiar stars. Other telescopes used by me have been the 24-inch at Dyer Observatory and the two 16-inch telescopes once at Kitt Peak. Moreover, I have collaborated with over a hundred different observers, each obtaining photometry with his own telescope. And, of course, scores of other observers not directly collaborating with me have observed these stars as well, a few of them with an automatic telescope.

The theme in this review paper will be that chromospherically active stars vary in brightness on a variety of different time scales, each involving very different physics. Some of them are literally periodic. Some are not strictly periodic but could be termed quasi-periodic or cyclical. And some, for now at least, can be described only in terms of a time scale. There are recent signs that we are indeed beginning to understand things: neat connections between some of the different periods, cycles, and time scales. These will be explained.

For background and additional detail, on the statements made above and on the material which follows, let me list several of my previously published papers. A paper on RS CVn itself (Hall 1972) got me started. Two review papers (Hall 1976, 1981) dealt with the RS CVn binaries as a group. The next (Hall and Kreiner 1981) dealt with orbital period changes and mass loss. The next review paper (Hall 1987) summarized what we knew then about the multi-faceted photometric behavior and raised many questions. A catalog of all known chromospherically active binaries (168 definite and 37 suspected) had me as one of the co-authors (Strassmeier et al. 1988). A four-paper series presented results of four years of continuous photometry of chromospherically stars with the proto-type 10-inch automatic telescope: 15 non-variable stars (Strassmeier and Hall 1988a), 5 variable single stars (Strassmeier and Hall 1988b), 49 variable binaries (Strassmeier et al. 1989), and the data base itself (Boyd et al. 1990). Hall (1989) made it clear that the classical Algols are also chromospherically active binaries, with much of the evidence of the activity buried by the dominant light of the much brighter early-type companion star. Three review papers presented at a N.A.T.O. Advanced Study Institute in Turkey dealt with period changes and magnetic cycles (Hall 1990a), starspot lifetimes (Hall and Busby 1990), and circularization, synchronization, and differential rotation (Hall and Henry 1990). Photometry of 50 suspected variables, most of them chromospherically active, revealed almost three dozen new variables (Hooten et al. 1989, Hooten and Hall 1990). A recent review placed photometric variability in the context of stellar dynamos (Hall 1991a), and a letter in press (Hall 1991b) deals with a newly discovered connection between orbital period changes and long-term variability in mean luminosity.

2. ORBITAL PERIOD

This time scale pertains only to those chromospherically stars which are in binaries.

Eclipses, if they occur, are strictly in phase with the orbital period and occur at the two conjunctions. A recent eclipse of RS CVn itself obtained with an automatic telescope is pictured on the cover of issue no. 45 (September 1991) of the I.A.P.P.P. Communications.

The ellipticity effect, which is measurable in most known eclipsing binaries, can occur in non-eclipsing binaries as well, which are then called ellipsoidal variables. Thirteen such binaries which are also chromospherically active were analyzed recently by Hall (1990b). To first order the ellipticity effect causes a sinusoidal variation with exactly half the orbital period, but limb-darkening effects on the pointed end of the tidally elongated star can cause the two minima to differ in depth by as much as $0^{m}.14$ (Hall 1990b). The amplitude of the light variability produced by ellipticity can be larger than many realize: if one star fills its Roche lobe, if the orbital inclination is 90°, and if there is no dilution by the light of the companion star, the full range can be $0^{m}.22$ measured from the shallower minimum and $0^{m}.36$ from the deeper (Hall 1990b). The light curve of the chromospherically active binary UV CrB = HD 136901 (Hall 1990b, figure 2), which coincidentally had no starspots at the time, provides a beautiful example of a large amplitude and markedly unequal minima.

The reflection effect, which is measurable in some eclipsing binaries, also can occur in non-eclipsing binaries, but there is no generally accepted name for such variables. The non-eclipsing chromospherically active binary HR 5110, which has the variable star designation BH CVn, varies only as a result of the reflection effect (Burke et al. 1980). The cooler star may have spots but its much more luminous hotter companion star dilutes any possible starspot variability.

3. ROTATION PERIOD

If a star has a longitudinally asymmetric surface brightness distribution and is not exactly pole-on, the brightness will vary as the star rotates. This variation is commonly called the "wave" and can appear in the light curve superimposed on additional variability which might be resulting from eclipses, ellipticity, or reflection. Photometry of the wave can yield remarkably accurate rotation periods, in general far more reliable than those derived from vsini measurements, especially so when the value of i is not known and/or when the absolute radius of the star is difficult to estimate and or/when the rotation itself is relatively slow.

On the chromospherically active stars one finds regions of lesser surface brightness, i.e., dark spots. There is no evidence to date that bright spots occur (Poe and Eaton 1985).

The range of possible light curve shapes and amplitudes produced by a dark spot was discussed by Friedemann and Gürtler (1975). If the inclination of the rotational axis is 90° and the spot is small, then the light curve will show constant light for almost half the rotation cycle and will have the approximate shape of the bottom half of a sine curve for the other half. Such was the basis of a model devised by Hall, Henry, and Sowell (1990) for first-order fits to light curves of spotted variables. More realistic models have been developed by other investigators. If two spots exist simultaneously, the light curve can take on quite a complex shape, depending on the latitude, longitude, and area of each spot and the inclination of the rotation axis. If the two spots happen to be separated by exactly 180° in longitude and be equal in area, the resulting light curve will approximate a sinusoid with half the rotation period. For a given constant dark area, maximum wave amplitude will result if there is one spot, a smaller amplitude if the same area is apportioned to two or more spots, and no amplitude at all if the dark area is in a longitudinally symmetric belt or polar cap.

Multi-bandpass photometry will show larger amplitudes at shorter wavelengths and the minimum amplitude around the I bandpass, where the temperature of a typical cool spot has its Planckian maximum. In a binary where the unspotted companion star is hotter, however, the amplitude of the wave in the composite light curve will be <u>less</u> at shorter wavelengths.

The wave in the W UMa binary 44 i Boo holds the record for the shortest period ever observed: $0^d.27$. The cool secondary stars in cataclysmic binaries probably are chromospherically active and hence probably spotted (Hall 1990a), although no wave has yet been detected observationally. If one were, it would have a period around $0^d.09$. The longest ever observed is the 385^d period in HD 181943 (Hooten and Hall 1990), although Strassmeier (1991) questions whether this really is the star's rotation period. The next longest, in another single star, is the 335^d period in HR 1362 = EK Eri (Strassmeier et al. 1990) and the longest in a binary is the 109^d period in HR 7428 = V1817 Cyg (Hall, Gessner, Lines, Lines 1990).

The largest wave ever detected is the one in HD 12545 (Nolthenius 1991). During the late 1990 observing season its amplitude was $0^m.8$, $0^m.6$, $0^m.5$, and $0^m.4$ in B, V, R, and I, respectively. The record holder before that was the $0^m.5$ wave in the late 1986 V-band light curve of II Peg (Doyle et al. 1988). A number of waves as small as $0\overset{m}{.}01$ in amplitude have been detected reliably. Two examples are the $0\overset{m}{.}013$ wave in HD 222317 = KT Peg (Strassmeier et al. 1989) and the $0\overset{m}{.}011 \pm 0\overset{m}{.}002$ wave in HD 17144 = UY For (Hooten and Hall 1990, figure 9). When a rather large or very large sunspot rotates into and out of view, the sun's bolometric luminosity dims by about $0\overset{m}{.}001$ or $0\overset{m}{.}002$ (Willson, Duncan, and Geist 1980), but even large sunspots are small compared to starspots which have been detected on chromospherically active stars.

Wave amplitude is perhaps the sharpest diagnostic of stellar dynamo strength. A survey of 277 stars with good photometric histories revealed (Hall (1991a) that the onset of heavy starspot coverage is abrupt, at a Rossby Number of 2/3. At larger Rossby Numbers (slower rotation) wave amplitudes are all less than about $0\overset{m}{.}01$, and all of the large wave amplitudes are found at smaller Rossby Numbers (faster rotation). Rossby Numbers were computed with the convective turnover times calculated by Gilliland (1985) for dwarfs (V) as a function of B-V color index, made 2.0 X longer for subgiants (IV) and 4.0 X longer for giants (III), with intermediate factors applied to intermediate luminosity classes.

4. MIGRATION

The 1963 and 1964 photoelectric light curves of RS CVn (Chisari and Lacona 1965) made quite an impact when first published. Each showed a well-defined wave of large amplitude (about $0\overset{m}{.}2$) but from one year to the next the wave had shifted about $0\overset{p}{.}1$ to the left in the light curve, i.e., towards smaller orbital phase. Subsequent photometry at Catania (Catalano, Frisina, Rodono 1980) showed that this trend had continued, with wave minimum falling at the same orbital phase 10 years later. This striking phenomenon has been termed "migration".

The basically correct explanation for wave migration was proposed by Hall (1972). The cooler star is spotted, rotates in approximate synchronism with the orbital period, but still has residual differential rotation akin to what is seen on the sun. Only one particular latitude (called the co-rotation latitude) rotates in exact synchronism, with lower and higher latitudes rotating faster and slower than synchronously. If the spot was anywhere other than exactly on the co-rotating latitude, then the rotation period traced by that spot would differ from the orbital period and wave migration would result.

The relation should be

$$1/P(migr) = 1/P(orb) - 1/P(rot) , \qquad (1)$$

where P(migr) is negative if P(rot) < P(orb), and positive if

P(rot) > P(orb). The former case was the one first seen in RS CVn, but both cases have been seen in other chromospherically active binaries, sometimes both in the same system at different epochs. Moreover, both cases occur in roughly equal number (Hall and Henry 1990, figure 10). It follows from equation (1) that

$$\Delta P/P(orb) = P(rot)/P(migr) , \qquad (2)$$

where $\Delta P = P(rot) - P(orb)$. Thus we see that migration period is not really a fundamental time scale. It is just a measure of the percentage difference between the rotation period and the orbital period.

Differential rotation on the sun is normally described with the equation

$$P(\emptyset) = P(eq) / (1 - k \sin^2\emptyset) , \qquad (3)$$

where $\emptyset$ is latitude. If P(rot) for a star is measured repeatedly by observing waves from spots occurring over the entire 90° range of latitude, then k can be determined very simply from

$$\delta P/\langle P \rangle = k , \qquad (4)$$

where here δP is P(max) - P(min) and $\langle P \rangle$ is the mean period. If the spots observed do not sample the full 90° latitude range, then

$$\delta P/\langle P \rangle = k f , \qquad (5)$$

where f is a factor which can be estimated (Hall and Busby 1990). Note that this approach to determining k works for single spotted stars as well as binaries. When k was determined in this way for a sample of 85 spotted stars, both single and binary, a dramatic result was found (Hall 1991a, figure 7). As P(rot) decreases, k decreases, i.e., rapidly spinning stars approach solid-body rotation. The result was expressed as

$$\log k = -2.0 + 0.8 \log P(rot) - 0.4 F , \qquad (6)$$

where P(rot) is in days, and F is Roche lobe filling factor. This explains in part a puzzling observation which few have realized was puzzling: virtually all migration periods are about the same, a few years in length, even though the rotation periods involved cover three orders of magnitude. Extremes of P(rot), $0^d.3$ and 300^d, inserted in equation (6) along with F = 1 and f = 0.3 yield remarkably similar migration periods: 2 yrs and 8 yrs, respectively.

Normally one thinks synchronous rotation is described by P(rot) = P(orb). This is true only in circular orbits. Synchronism in an eccentric orbit results in P(rot) < P(orb), the exact ratio P(rot)/P(orb) depending on the eccentricity in a relation given by Hall (1986) This situation is termed pseudosynchronization and has been found to occur in a number of chromospherically active binaries (Hall and Henry 1990). One of the clearest examples is BY Dra.

Some chromospherically active binaries, about 15%, simply do not rotate synchronously or pseudosynchronously (Hall and Henry 1990). Such systems are excluded when a relation such as that in equation (6) is developed. Remarkable examples of grossly asynchronous rotation are λ And, where P(rot)/P(orb) = $54^d.0/20^d.52$ = 2.6 (Hall et al. 1991), and HD 181809 = V4138 Sgr, where P(rot)/P(orb) = $60^d.23/13^d.05$ = 4.6 (Hooten and Hall 1990).

5. SPOT LIFETIMES

Hall and Busby (1990) calculated the time required for a spot to be disrupted by the shearing effect of differential rotation, t(c). Comparing this with spot lifetimes observed in several spotted stars, t(o), they found that t(o) = t(c) for spots larger than about 15° or 20° in diameter. Smaller spots died before their disruption times would dictate and, moreover, had lifetimes which were roughly proportional to spot area.

It is not certain that the shearing effect of differential rotation on the stellar surface is actually the physical mechanism for terminating the identity of a large spot. It is likely that differential rotation with radial depth occurs as well and has a gradient similar to k. Successive rotations might produce shearing which breaks the connection between spots on the surface and the underlying magnetic field in which the spot is rooted. Once connection is broken, a spot should disintegrate rather quickly.

It was realized early on that large spots on chromospherically active stars live quite long (years), considerably longer than do sunspots (months at most). Though these long lifetimes per se are perhaps partly understood now, it is a bit puzzling that they are all so similar, a few years. Hall and Busby (1990, equation 5) showed that large-spot lifetimes should be proportional to rotation period, which covers three orders of magnitude in the available sample. It seems that the relation in equation (6) again provides a partial explanation. If one considers a spot 20° in diameter and calculates k with equation (6) using F = 1 as before, one gets t(c) = 2 yrs for a rapidly rotating star, P(rot) = 1^d, and t(c) = 5 yrs for a slowly rotating star, P(rot) = 100^d. These are remarkably similar lifetimes for such a wide range in spin rate. This result is changed very little if somewhat smaller (15°) or larger

(25°) spots are considered.

6. MAGNETIC CYCLES

There seem to be magnetic cycles, akin to the 11-year sunspot cycle, operating in other stars as well. There are very many different observational manifestations of these cycles, and they are found in very many different types of stars. It is remarkable and surely significant that the cycle lengths observed all fall in a relatively narrow range: from somewhat shorter than 10 years to perhaps somewhat longer than 100 years. The most comprehensive review is by Hall (1990a), with an update by Hall (1991a). Cycles like this and their many manifestations cannot be regarded as understood; even the solar cycle is not understood well.

The manifestation which relates most closely to photometry is the cyclical variation in mean luminosity. The largest in amplitude observed to date is the 50-yr cycle in V833 Tau (Hartmann et al. 1981), which varies by $0^{m}5$. The smallest in amplitude must be that observed in the sun, which varies by $0^{m}001$ in phase with the solar cycle (Wolff and Hickey 1987). Two of the most recent ones established are the 11.4 ± 0.4 yr cycle in λ And, with a range of $0^{m}15$ (Hall et al. 1991), and the 45-year cycle in CG Cyg, with about the same range (Hall 1991b).

It should be stressed that the time scale of these magnetic cycles, though not very much longer than that of starspot lifetimes, must be a fundamentally different cycle. The proof is that starspot lifetimes have been determined in most of the stars for which the magnetic cycle length also has been determined, and the two are always different, with the latter always longer than the former by about a factor 5 for large spots and a factor 100 for small spots.

Another striking manifestation, which relates to photometry as well, is the cyclical variation in orbital period, which can be traced precisely by timing eclipses in eclipsing binaries. These changes are remarkably large. During the course of one cycle the period decreases by about one part in 10^5 and then, half way through the cycle, increases by about the same amount. It was clarified by Hall (1990a) that these are different from the truly periodic apparent orbital period changes caused by apsidal motion or orbital motion around a third body. Moreover, they are different from period changes caused by mass loss and/or transfer; those are monotonic in any one binary system, either increases or decreases, but not cyclical. Notable examples of such cycles in well known eclipsing binaries are the 9-year cycle in U Cep (Hall 1975), the 10-year cycle in V471 Tau (Skillman and Patterson 1988), the 25-year cycle in W UMa (Kreiner 1977), the 32-year cycle in

Algol (Soderhjelm 1980), the 50-year cycle in AR Lac (Hall and Kreiner 1981), and the 73-year cycle in SV Cam (Frieboes-Conde and Herczeg 1973).

The most recent and complete explanation of the period changes is that of Applegate (1991). The basically correct theory was first proposed by Matese and Whitmire (1984), although they had some of the details missing or not quite correct. All of the many theories before that were entirely wrong, including those suggested by me (Hall 1975, Hall and Kreiner 1981). The mechanism involves a waxing/waning magnetic field within or beneath the convective layer of one star. This exerts/releases a torque which increases/decreases the degree of differential rotation in the convective envelope. One consequence is an increase/decrease in the quadrupole moment, which is communicated to the orbit immediately and manifested as an orbital period change. Another consequence is a change in the star's luminosity. Since the star does not actually expand or contract in radius, the brightening or dimming should result from an increase or decrease in surface brightness. The phasing should be as follows. At maximum quadrupole moment the period should be shortest. That automatically tells us the maximum/minimum in the corresponding O-C curve should precede/follow it 90° in phase. Moreover, maximum luminosity should coincide with O-C curve minimum/maximum depending on whether the sense of the star's radial differential rotation gradient is such that it rotates faster/slower on the outside.

Recently Hall (1991b) showed that one particular eclipsing binary containing a chromospherically active star, CG Cyg, verifies the three most important predictions of the Applegate model. First, the variation in mean luminosity and the orbital period variation both had the same cycle length, in this case 45 years. Second, the epoch of maximum luminosity coincided with the epoch of one of the O-C curve's extrema (the minimum), namely, 1980.4 ± 0.4 <u>versus</u> 1980.1 ± 0.4, respectively. Third, the active star became bluer as it brightened, by the required amount. Because maximum luminosity coincided with the O-C curve's minimum rather than its maximum, we have learned as a bonus that differential rotation in CG Cyg has the outer layers rotating faster than the deeper layers.

7. FLARES

Until now in this paper, time scales have been introduced in order from shortest to longest. Pardon me for concluding with the shortest time scale of all. The most familiar flares, those on the sun, are very rapid, with rise times and decay times of tens of minutes. The interval between flares is an altogether different time scale, and it is definitely not periodic and probably should not be considered even cyclical.

An 8-year interval between maxima in flaring frequency has, however, been identified as a possible manifestation of a magnetic cycle in at least one star, AD Leo (Pettersen and Panov 1989).

Solar type stars experience similar flares, which can make the star brighten measurably if the star's absolute luminosity is much less than the sun's, as is the case with M-type dwarfs. Chromospherically active stars also experience flares, and theirs can be intrinsically more powerful than solar flares.

On the night of 14/15 December 1989 in the Arizona sky V711 Tau underwent the most powerful and energetic flare ever detected (Henry and Hall 1991). Compared to major solar white-light flares, it was 10^7 greater in bolometric luminosity and 10^8 greater in total energy release. Even though the host star was more luminous than the sun (K1 IV) and light from the companion star diluted the effect further, V711 Tau brightened $0\overset{m}{.}69$ in B and $0\overset{m}{.}42$ in V. It seems this was a double-barreled action. Another large flare, though not quite as large, fired about 12 hours earlier in the sky over China (Zhang et al. 1990). That precursor caused V711 Tau to brighten by $0\overset{m}{.}27$ in B and $0\overset{m}{.}18$ in V. The Chinese flare was observed continuously from start to finish and had a 4.5-hour duration. The Arizona flare was defined with fewer observations but probably had a similar duration.

ACKNOWLEDGEMENT

The writing of this paper and travel to the Symposium in Laramie were made possible by N.A.S.A. research grant N.A.G. 8-111.

REFERENCES

Applegate, J. H. 1991, Ap. J., in press.

Boyd, L. J., Genet, R. M., Hall, D. S., Busby, M. R., and Henry, G. W. 1990, I.A.P.P.P. Comm. no. 42, 44.

Burke, E. W., Eaton, J. A., Fisher, G. A., Hall, D. S., Heiser, A. M., Henry, G. W., Vaucher, C. A., Sabia, J. D., and Skillman, D. R. 1980, A.J. **85**, 244.

Catalano, S., Frisina, A., and Rodono, M. 1980, I.A.U. Symposium **88**, 405.

Chisari, D. and Lacona, G. 1965, Mem.Soc.Astr.Italiana **36**, 463.

Doyle, J. G., Butler, C. J., Morrison, L. V., and Gibbs, P. 1988, Astr. Astrophys. **192**, 275.

Frieboes-Conde, H. and Herczeg, T. 1973, Astr. Astrophys. Suppl. **12**, 1.

Friedemann, C. and Gürtler, J. 1975, Astr. Nachr. **296**, 125.

Gilliland, R. 1985, Ap. J. **299**, 286.

Hall, D. S. 1972, P.A.S.P. **84**, 323.

Hall, D. S. 1975, Acta Astr. **25**, 1.

Hall, D. S. 1976, I.A.U. Colloquium **29**, 287.

Hall, D. S. 1981, in Solar Activity in Stars and Stellar Systems, ed. R. M. Bonnet and A. K. Dupree (Dordrecht: Reidel), p. 43.

Hall, D. S. 1986, Ap. J. **309**, L83.

Hall, D. S. 1987, Publ. Astr. Inst. Czechoslovakia **70**, 77.

Hall, D. S. 1989, I.A.U. Colloquium **107**, 219.

Hall, D. S. 1990a, A.J. **100**, 554.

Hall, D. S. 1990b, in Active Close Binaries, ed. C. Ibanoglu (Dordrecht: Kluwer), p. 95.

Hall, D. S. 1991a, I.A.U. Symposium **130**, 353.

Hall, D. S. 1991b, Ap. J. Letters, in press.

Hall, D. S. and Busby, M. R., 1990 in Active Close Binaries, ed. C. Ibanoglu (Dordrecht: Kluwer), p. 377.

Hall, D. S., Gessner, S. E., Lines, H. C., and Lines, R. D. 1990, A.J. **100**, 2017.

Hall, D. S. and Henry, G. W., 1990 in Active Close Binaries, ed. C. Ibanoglu (Dordrecht: Kluwer), p. 287.

Hall, D. S., Henry, G. W., and Sowell, J. R. 1990, A.J. **99**, 396.

Hall, D. S. and Kreiner, J. M. 1981, Acta Astr. **31**, 387.

Hall et al. [30 authors] 1991, J. Astrophys. Astr., submitted.

Hartmann, L., Bopp, B. W., Dussault, M., Noah, P. V., and Klimke, A. 1981,Ap. J. **249**, 662.

Henry, G. W. and Hall, D. S. 1991, Ap. J. **373**, L9.

Hooten, J. T. and Hall, D. S. 1990, Ap.J. Suppl. **74**. 225.

Hooten, J. T. et al. [29 authors] 1989, I.A.P.P.P. Comm. no. 38, 19.

Kreiner, J. M. 1977, I.A.U. Colloquium **42**, 393.

Matese, J. J. and Whitmire, D. P. 1983, Astr.Astrophys. **17**, L7.

Nolthenius, R. 1991, I.B.V.S. no. 3589.

Pettersen, B. R. and Panov, K. P. 1989, in Cool Stars, Stellar Systems and the Sun, ed. M. Zeilik and D. M. Gibson (Berlin: Springer Verlag), p. 91.

Poe, C. H. and Eaton, J. A. 1985, Ap. J. **289**, 644.

Skillman, D. R. and Patterson, J. 1988, A.J. **96**, 976.

Soderhjelm, S. 1980, Astr. Astrophys. **89**, 100.

Strassmeier, K. G. 1991, I.B.V.S. no. 3618.

Strassmeier, K. G. and Hall, D. S. 1988a, Ap.J. Suppl. **67**, 439.

Strassmeier, K. G. and Hall, D. S. 1988b, Ap.J. Suppl. **67**, 453.

Strassmeier, K. G., Hall, D. S., Barksdale, W. S., Jusick, A. T., and Henry, G. W. 1990, Ap. J. **350**, 367.

Strassmeier, K. G., Hall, D. S., Boyd, L. J., and Genet, R. M. 1989, Ap. J. Suppl. **69**, 141.

Strassmeier, K. G., Hall, D. S., Zeilik, M., Nelson, E., Eker, Z., and Fekel, F. C. 1988, Astr. Astrophys. Suppl. **72**, 291.

Willson, R. C., Duncan, C. H., and Geist, J. 1980, Science **207**, 177.

Wolff, C. L. and Hickey, J. R. 1987, Science **235**, 1631.

Zhang, R., Zhai, D., Zhang, X., Zhang, J., and Li, Q. 1990, I.B.V.S. no. 3456.

STARSPOT PHOTOMETRY: OBSERVATIONAL REVIEW AND INTERPLAY WITH SPECTROSCOPY

KLAUS G. STRASSMEIER
Institut für Astronomie, Universität Wien,
Türkenschanzstraße 17, A-1180 Wien, Austria

ABSTRACT Most of what we know about starspots comes from time variations of broad-band lightcurves. A review is presented of the current observational knowledge of starspots on RS CVn-like F, G, and K stars, T Tauri, and W UMa-type stars. Recent Doppler maps are compared with photometric results.

INTRODUCTION

In this review I will focus on results obtained with automatic photoelectric telescopes (APTs) and will discuss the starspot phenomenon throughout the Hertzsprung-Russell diagram (HRD). Particular weight will be given to results obtained in connection with Doppler imaging. Progress in the study of spotted, late-type stars has come mostly through the analysis of their light curves. Their regular variations − modulations with the rotation period − and their quasi-regular long-term variations, such as the change of the mean brightness due to a spot cycle, have told us most of what we know about starspots.

OBSERVATIONAL EVIDENCE FOR THE EXISTENCE OF STARSPOTS

The solar analog

Sunspots are local magnetic fields with field strengths of ~ 1000 G which block the emerging flux from the interior and cool down the photosphere by about $2000-3000$ K. Typically, sunspots cover 10^{-4} to 10^{-5} of the solar surface and only during solar maximum reach about 10^{-3}. The ACRIM experiment, a radiometer on the SMM spacecraft, has measured dips of up to 0.2% of the solar irradiance when large spot groups where moving in and out of view (Foukal and Lean 1986). This verifies the assumption that light variations in active late-type stars are associated with the presence of surface magnetic fields. Magnetic activity is essentially found in regions of the HRD where stars with convective envelopes occur. Table 1 gives an overview of "spotted" stars throughout the HRD and their observed spot parameters. Despite their different evolutionary status, these stars have several things in common: convection, rapid rotation, they are generally not known as pulsators[1] and, they all show Ca II H and K emission due to an extended chromosphere.

---

[1] although, in principle, pulsation could also drive a dynamo

Table 1: STARSPOT MORPHOLOGY THROUGHOUT THE HRD

	spot occurence (spectral range)	spot temperatures $\Delta T = T_{phot} - T_{spot}$	spot coverage f (%)	activity components	differential rotation	lifetimes, variation timescales
T Tauri	G5 - M1	hot+cool spots hot:−1800...7450 cool:$\geq$200...1300	hot:0.07...3±1 cool:3...17±3	one/two-spot (prefer. one) mass accretion	0.1% relative (on V410 Tau) 20% (BP Tau)	1200 days (650 rot.) for V410 Tau, 2-8 days ($\leq$ 1 rot.)
BY Dra single	K5 - M5	cool spots 280 ... 850	2 ... 11	one/two-spot polar spot		years?(e.g.,AU Mic) $\leq$1 yr (e.g.,EV Lac)
BY Dra binaries	F8 - M5	cool spots $\geq$140 ... 1500	2 ... 23 (for BY Dra)	one/two-spot polar spot?		60yr cycle (V833 Tau) few days ($\leq$ 1 rot.)
lower m-s Hyades	F8 - K5-8	cool spots	< 1	$\leq$2% variabil. in b, y	5 ... 21%	$\leq$3yr (HD 206860) $\approx$10 days (e.g.,VB31)
Sun	(G2)	cool spots 1700 ... 3000 but $\Delta T(\bar{B}, age)$	$\leq$0.1 in max. 0.01 ... 0.001	spot groups isolated spots ±40°lat.	20%	several rotations for spot groups, hours for pores
RS CVn binaries	IV: F9 - K3-4 III: G5 - K3	cool spots 600 ... 1900	0.15 ... 16	one/two-spot polar spot	$\leq$3% −2% (UX Ari)	$\approx$1yr, maybe longer days ($\leq$ 1 rot.)
Algol binaries	G - K	hot and/or cool? ±500 (β Per)		one-spot		
W UMa binaries	A8 - K3.5	hot+cool spots hot:−600...1200 cool:200...1350	hot:$\approx$20 cool:0.6...4	mass transfer one/two-spot polar spot		8-9 yr cycle (VW Cep)
FK Comae type	G2-3 - K1	cool spots 400 ... 800	0 ... 7	one/two-spot polar spot	"very small"	phase coherence for decades, short-term var.

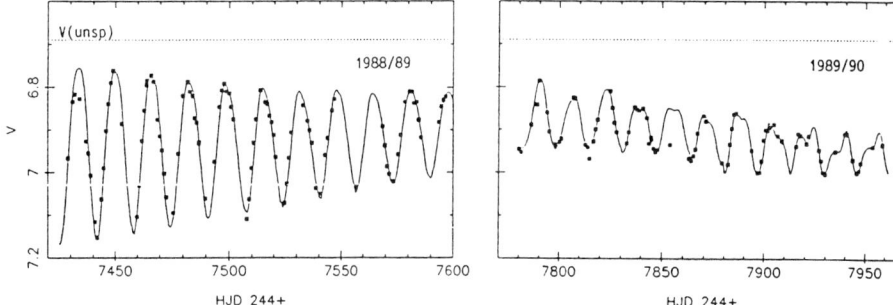

Figure 1: Light curve variations of HD 17433 due to starspots. The full line is a fit with an evolving spot model (adopted from Strassmeier and Bopp 1991).

Evidence from photometry

(1) The modulation of the continuum light accompanied by color variations in the sense that the star becomes redder when the light curve has a minimum. Increasing color amplitudes at longer wavelengths are also in agreement with the picture of *cool*, dark spots. Fig. 1 shows examples of V-band light variations typical for spotted stars.

(2) Photometric periods are in agreement with spectroscopically determined periods from rotational velocities but with the additional advantage that they are unequally more precise. This implies that the photometric period is the rotation period of the spotted star. A fact further strengthened by the long-term Mount Wilson Ca II H and K observations (Wilson 1978, Baliunas and Vaughan 1985) which showed, e.g., for δ CrB (Baliunas 1988), that the Ca II S-index varies with the *same* period as simultaneous broad-band photometry.

(3) Linear polarization appears to depend on the chromospheric activity level and is, when detectable, also modulated with the photometric period and its detection suggests a magnetic origin (Huovelin et al. 1988, Kemp et al. 1987).

(4) Infrared excess seems to be present in spotted stars, however, as pointed out by Busso et al. (1988), the excess is not correlated with the degree of activity nor with the evolutionary status, thus its origin is still not clear.

Evidence from spectroscopy

(1) The spectral flux ratio of II Peg, taken as the ratio of spectra when the (presumably) most spotted and the least-spotted hemispheres were in view, showed a steep rise to the red (Vogt 1981a). This flux ratio, actually the relative energy distribution of the spotted region itself, showed pronounced molecular absorption features of TiO and VO, characteristic of very cool stars.

(2) The existence of "bright" bumps in the profile of a rotationally-broadened absorption line which are correlated with phase (Fekel 1980). A cool spot emits less light than the surrounding photosphere and produces, at the rotation velocity

of the spot on the stellar surface, a lack of photons absorbed by the line and thus an apparent emission bump (Vogt and Penrod 1983). This led to the application of the Doppler-imaging technique to spotted, late-type stars.

(3) The direct detection of surface magnetic fields with different techniques, e.g., Fourier transform methods (Robinson et al. 1980), line profile modeling (Marcy and Bruning 1984, Saar 1988), Stokes V profile modeling or Zeemann-Doppler imaging (Donati and Semel 1991).

Using the method of *speckle imaging*, Lynds et al. (1976) were able to produce a direct image of the red supergiant α Ori – the first of its kind – which showed two large "spots" on the surface. However, using better data, Wilkerson and Worden (1977) could not confirm any structure on the stellar disk. Thus the most convincing proof for the existence of starspots, a direct stellar image, is still needed.

MODELING THE PHOTOMETRIC VARIATIONS
Geometric spot models

Basically there are two different modeling procedures, a more physical one which includes effects such as gravitationally deformed binary components, reflection effect, secondary contribution, a.s.o., and the classical light curve rectification technique. A variety of computer spot modeling programs with different degrees of sophistication were developed (Torres and Ferraz-Mello 1973, Bopp and Evans 1973, Friedemann and Gürtler 1975, Budding 1977, Eaton and Hall 1979, Bopp and Noah 1980, Vogt 1981b, Poe and Eaton 1985, Rodonó et al. 1986, Dorren 1987, Strassmeier 1988, Kang and Wilson 1989). The basic set of integral equations were outlined, e.g., by Strassmeier (1988). The most serious problem is the ambiguity of the light curve fit as demonstrated in the review by Vogt (1983) and others.

Assumptions and *a priori* knowledge

The following four points briefly discuss the necessary observations and *a priori* knowledge before one can obtain a reasonably constrained solution with a geometric spot model.

(1) The first step is to determine the photometric period and assume it to be the rotation period of the spotted star. Periods obtained from data taken at different epochs can vary by several percent (see the summary in Hall and Busby 1990). Hall (1972) [see also Hall and Henry 1990] attributed these differences to the existence of differential rotation on the stellar surface in accordance with the solar behaviour.

(2) The next step is to estimate the orbital inclination, i, and to assume that the rotation axis of the spotted star is perpendicular to the orbital plane. For example, this was carried out for the non-eclipsing, SB1 binary HK Lac by Olah et al. (1985), using the observed mass function, the spectral classification of the spotted star, the invisibility of the secondary star in the spectrum, and the absence of eclipses. This yields inclinations with uncertainties of about $\pm 10°$ to

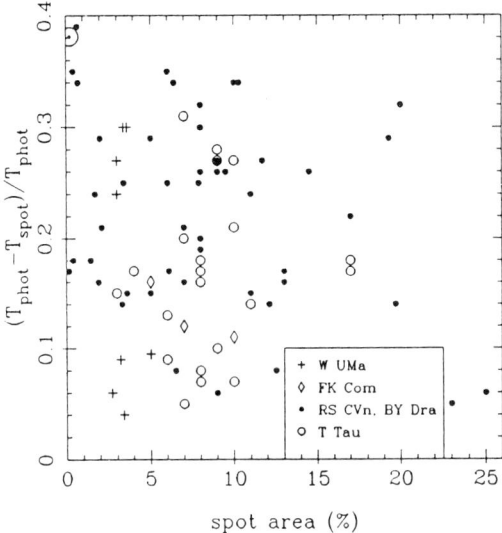

Figure 2: Spot temperatures and coverages (in % of the stellar surface) from modeling the broad-band light curve variations.

20°.

(3) Find the brightest magnitude ever observed and assume it to be the "unspotted" or "least spotted" brightness level.

(4) Determine the temperature *difference*, photosphere minus spot (ΔT), from the V-I color amplitude — actually the difference least-spotted hemisphere minus most-spotted hemisphere — which ranges from "undetectable" (say, ≤ 0.01 mag) to about 0.05 mag with uncertainties in the spot temperature of about 100 to 200 K. Furthermore, assume ΔT is the same for each individual spot or spot group and is constant within the time of observation.

Spot temperatures and coverages

Table 2 and Fig. 2 summarize the results from modeling the light and color curves of 28 late-type stars, where ΔT is the spot temperature difference, n_{spots} is the number of spots, and f is the spot covering fraction in percents of the entire stellar surface. This database is supplemented with spot modeling results for eleven T Tauri stars taken from Bouvier and Bertout (1989) and five W UMa binaries and is plotted in Fig. 2. We arrive at the same (though preliminary) conclusion as Bouvier and Bertout — there is no evident difference between spots on pre-main sequence and spots on post-main sequence stars nor between singles and binaries. There also seems to be no obvious correlation between temperature and size for any of the plotted types of variable stars. The large *range* of spot temperatures and coverages is probably the only significant feature in this figure. For sunspots, Parker (1979) showed that young spots must emit MHD waves which cool them within about 3 minutes (the time for the energy to

Table 2: STARSPOT TEMPERATURES AND COVERAGES

star	ΔT (K)	$\Delta T/T_{phot}$ (-)	n_{spots} (-)	f (%)	ref.
RS CVn	650-1580	0.14-0.34	2	10.0-19.7	(1)
	1300,1600	0.28-0.34	2	...	(2)
BY Dra	≥140	0.04	1	2-23	(3)
	600	0.15	2	3.6-11	(4)
VY Ari	1200	0.26	2,3	9.5-14.5	(5)
EI Eri	1850	0.34	2,3	6.4-10.3	(6)
V711 Tau	1800	0.32	2	8-20	(7)
	1400	0.29	2	5-19.3	(1)
	1200	0.25	2	6-7.9	(4)
σ Gem	620	0.14	2	3.3-12.1	(8)
	570	0.13	2	6.4	(3)
BF Lyn	1500	0.32	2	...	(9)
AR Lac	230-1320	0.05-0.29	2	2-9	(1)
	1200	0.25	2	3.4	(4)
SV Cam	2020	0.35	1	0.4-6	(10)
	1550	0.29	1	5	(11)
RT And	1070	0.17	1	0.15-6.1	(12)
BH Vir	980-1290	0.16-0.21	2	1.9-2.1	(15)
	2010-2300	0.34-0.39	2	0.62-0.66	(15)
	1100	0.18	1	0.4-1.44	(16)
	1300-1400	0.22	1	1.5	(17)
II Peg	730-1020	0.17-0.23	1,2	9-13	(3)
	1200	0.26	1	10	(18)
	1200	0.27	2	11.7	(4)
	950	0.20	2	16	(25)
HK Lac	950-1200	0.2-0.26	2	8	(3)
HR 7275	1200	0.26	2	9	(3)
IM Peg	920	0.21	2	7	(3)
λ And	800	0.16	2	2-5.5	(19)
	1020	0.22	2	17	(3)
UX Ari	1420	0.30	2	8	(3)
LX Per	730	0.15	2	5	(3)
SZ Psc	1200	0.25	1,2	...	(20)
UZ Lib	few100	...	2	...	(21)
CC Eri	...	1	1	3.3	(22)
YY Gem	...	1	1	2.5	(22)
AU Mic	850	0.24	2	1.7-11	(4)
ER Vul	1000	0.16	1,2	13	(23)
RT CrB	340-1000	0.06-0.19	2	8-25	(24)
EV Lac	280	0.08	1	6.5-12.5	(26)
FK Com	500-800	0.10-0.16		5-7	(14)
	600	0.12		13	(13)
HD 199178	580	0.11	1	10	(27)

References:
(1) Kang and Wilson (1989)
(2) Eaton (1991)
(3) Poe and Eaton (1985)
(4) Rodono et al. (1986)
(5) Strassmeier and Bopp (1991)
(6) Strassmeier (1990)
(7) Dorren and Guinan (1982)
(8) Strassmeier et al. (1988b)
(9) Strassmeier et al. (1989b)
(10) Zeilik et al. (1988)
(11) Cellino et al. (1985)
(12) Zeilik et al. (1989)
(13) Holtzman and Nations (1984)
(14) Rucinski (1981)
(15) Zhai et al. (1990)
(16) Zeilik et al. (1990)
(17) Scaltriti et al. (1985)
(18) Vogt (1981b)
(19) Bopp and Noah (1980)
(20) Eaton and Hall (1979)
(21) Bopp et al. (1984)
(22) Budding (1977)
(23) Hill et al. (1990)
(24) Zhai and Chen (1989)
(25) Byrne (1987)
(26) Gershberg et al. (1991)
(27) Jetsu et al. (1990)

cross the spot) while observations (Chou 1987) reveal cooling times between 0.5 to 9 hours, i.e., a factor of 10 to 100 longer than predicted. From Tables 1 and 2 we see that starspots are larger than sunspots by about a factor of 100 to 1000, therefore their respective cooling times should be substantially longer and one and the same starspot could have been observed at quite different temperatures which would explain the observed large range of ΔT in Fig. 2. For example, one of the best studied stars in the sample, AR Lac, was reported with two spots of $\Delta T = 1320\pm590$ K and 770 ± 700 K in 1978, and with 550 ± 450 K and 230 ± 180 K in 1981 (Kang and Wilson 1989). Note that the large uncertainties came mostly from the availability of only B and V data instead of red and near-infrared light curves which would constrain the temperatures much better. A similar example is RT CrB (Zhai and Chen 1989) with $\Delta T \approx 340$ to 1000 K. High time-resolution monitoring of V-I color variations with an APT could give us some first estimates of cooling times of starspots and, moreover, provide some clues to explain the obvious discrepancy between theory and observation in the case of sunspots.

A spectroscopic technique to derive starspot temperatures and coverages is the spectrum synthesis of TiO-band observations at 8860 Å (Ramsey and Nations 1980), who applied it to V711 Tau and found $\Delta T \approx 1000$ K. Contemporaneous and subsequent work (Vogt 1981a, Huenemoerder and Ramsey 1987, Huenemoerder et al. 1989a,b, Saar and Neff 1990) yielded spot temperatures and coverages for II Peg, UX Ari, V833 Tau, and HD 82558, in agreement with photometric results. Spot temperatures from spectral line profile modeling are still too uncertain and sparse to be taken into account, e.g., for HD 32918, $\Delta T \approx 500$ K (Piskunov et al. 1990).

MODELING THE SPECTROSCOPIC VARIATIONS

Doppler imaging is a technique to derive a resolved image of a star by using the relation between wavelength position across a spectral line and spatial position across the stellar disk (Struve 1930, Deutsch 1970). Cool spots on the surface produce distortions in the line profiles which can be followed throughout a rotation cycle of the star (for references on this technique I'd like to refer the reader to the recent review of Collier-Cameron 1991).

A comparison between results from Doppler imaging and photometric spot modeling can be made in the case of V711 Tau (HR 1099). In Fig. 3 we compare a Doppler map (Vogt 1988) from 1981 with an independently derived map from contemporaneous photometry (Rodonó et al. 1986). [Note to Fig. 3: the upper two images in the first column of the Doppler maps are to be compared with the photometric maps]. The agreement is very encouraging since both techniques required a big spot at or close to the rotation pole as well as a second spot at the equator (a third, very small spot was missed by the photometric technique).

Four teams in Strassmeier et al. (1991) applied their Doppler-imaging versions to data of the RS CVn binary EI Eri (HD 26337). Fig. 4 shows maps for the 1988.85 epoch obtained with three different techniques (only one rotational phase is shown). The lower panel in Fig. 4 compares their theoretical V-band light

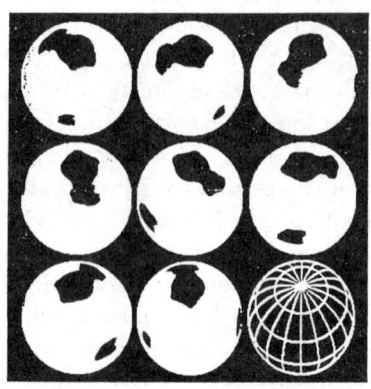 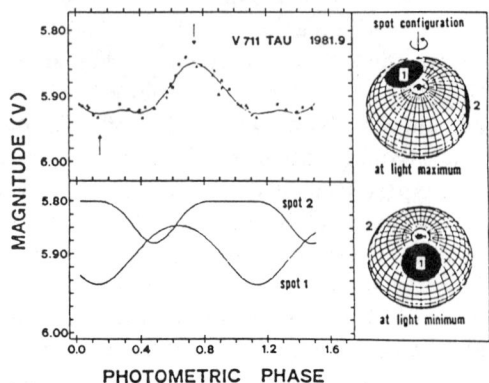

Figure 3: A comparison of a Doppler map (left; Vogt 1988) and a photometrically derived map (right; Rodonó et al. 1986) for V711 Tau in 1981.

curves with contemporaneous APT photometry (the shown maps correspond, from left to right, to the dotted, full, and dashed light curves, respectively). All four imaging techniques (one has been published earlier and is not repeated here) yielded similar images with spots at or close to the rotation pole – a feature not known from the solar paradigm. So far, Doppler maps have been derived for six late-type stars. Large polar spots were seen on V711 Tau (Vogt and Penrod 1983, Vogt 1988), HD 199178 (Vogt 1988), UX Ari (Vogt and Hatzes 1991), and EI Eri (Strassmeier 1990). Middle and high-latitude spots (even touching the pole) were seen on AB Dor (Kürster and Schmitt 1991). An equatorial belt of spots and an appendage to middle latitudes were determined for HD 32918 (Piskunov et al. 1990).

STARSPOTS THROUGHOUT THE H-R DIAGRAM

Photometric starspot models of RS CVn stars have recently been reviewed by Eaton (1991). Reviews of evidence of magnetic and related cycles can be found in, e.g., Maceroni et al. (1990a), Hall (1990), Catalano (1990), and Baliunas and Vaughan (1985). I will therefore concentrate on some aspects not, or only marginally, covered by these authors. These are mainly the questions of morphology and onset of spot activity throughout the Hertzsprung-Russell diagram.

Spots on F, G, and K stars

Convection zones are thought to appear at a spectral type near F0. From observations of CIV and HeI line strengths in ∼80 late A and F stars, Wolff et al. (1986) found the onset of *chromospheric* activity (CA) near $B-V=0.28$, i.e., near spectral type F0. Starspots, i.e. *photospheric* activity, seem to occur at somewhat later spectral types. Differential photometry of 24 Hyades main-sequence stars at Lowell Observatory (Radick et al. 1987) confirmed 18 to be variable – all later than spectral type F8. Another study, using over 12,000 differential UBV measures of 49 late-type CA binaries made with the prototype APT (Strassmeier *et al.* 1989a), yielded 54 Cam (F9IV+F9IV) as the "earliest" subgiant with the characteristic starspot "wave". Zeilik *et al.* (1983) report a 0.04 mag modulation in HD 108102, a double-lined (F8V+F8V) binary. Another RS CVn-type system, σ^2 CrB (F6V+G0V), shows strong chromospheric Ca II H and K emission from *both* components but only the G0 star seems to have also starspots. The earliest class III giants in the "Chromospherically Active Binary Star" catalog (Strassmeier *et al.* 1988a) are α Aur (G1III+K0III; Strassmeier and Fekel 1990), AY Cet (wd+G5III), 93 Leo (A6V+G5III-IV), and ϵ UMi ($\sim$F0V+G5III), but only AY Cet and 93 Leo show signs of a starspot wave. No giant in a detached binary system earlier than spectral type G5 is known to have starspots. The most "prominent" example is Capella's G1III secondary, which is otherwise known to be chromospherically active (e.g., Ayres and Linsky 1980) but no conclusive evidence for photometric variability exists (see Jackisch 1963). Of course, the G5 onset might be a selection effect since there is at least one *single* class III giant with starspot activity: FK Comae (G2-3 III; see, e.g., Jetsu *et al.* 1991). Another early giant with starspot activity is δ CrB (Baliunas 1988). However, note that δ CrB was originally classified as G3.5III-IV, but has recently been reclassified as G5III-IV by Keenan and McNeil (1989). Most single, early-G giants do *not* show photometric variability (Strassmeier and Hall 1988) such as ψ^3 Psc (G0III), 31 Com (G0III), 42 Cap (G2IV), or HR 1023 (G5III).

Spots on pre-main sequence stars

The T Tauri stars are long known for their *irregular* brightness variations. Some stars, most noticably V410 Tau, are also *periodic* variables (Rydgren and Vrba 1983) and the idea that these periodic or quasi-periodic variations might be due to rotational modulation by dark starspots goes back to Hoffmeister (1965). In the meantime, a growing number of spotted T Tauri's is known. Phenomenologically, three groups could be identified. Strictly periodic variables such as V410 Tau (Vrba *et al.* 1988, Herbst 1989, a.o.), DN Tau (Bouvier *et al.* 1986), DH Tau, DI Tau, GG Tau, AA Tau, HP Tau (Vrba *et al.* 1989), DF Tau, FK1, FK2, WK2, GW Ori, LH$_\alpha$332-20/1, CoD-33°10685, RY Lup, SR12, SR9 (Bouvier and Bertout 1989), stars with a periodic component superimposed on an irregular light variation such as T Tau (Herbst *et al.* 1986), SY Cha (Schaefer 1983), UX TauA (Bouvier and Bertout 1989), BP Tau (Simon *et al.* 1990), TW Cha (Bouvier *et al.* 1988), and stars where no periodicity has been found such as RY Tau, SU Aur, CO Ori, RU Aur (Herbst and Levreault 1990, Herbst *et al.* 1987). Bouvier (1990) showed that the correlation between x-ray fluxes and rotation of T Tauri stars is the same as that found for cool dwarfs and active binaries in agreement with the likely analogy between spots on pre-main sequence and post-main sequence stars obvious from Fig. 2.

It is generally believed that the irregular light variations are due to a circumstellar disk left over from an earlier epoch of star formation and that a hot spot results from steady mass accretion onto the surface. Herbst and Levreault (1990) rule out the possibility that cool spots are also responsible for the irregular light variations. It is no surprise, therefore, that mostly weak-emission

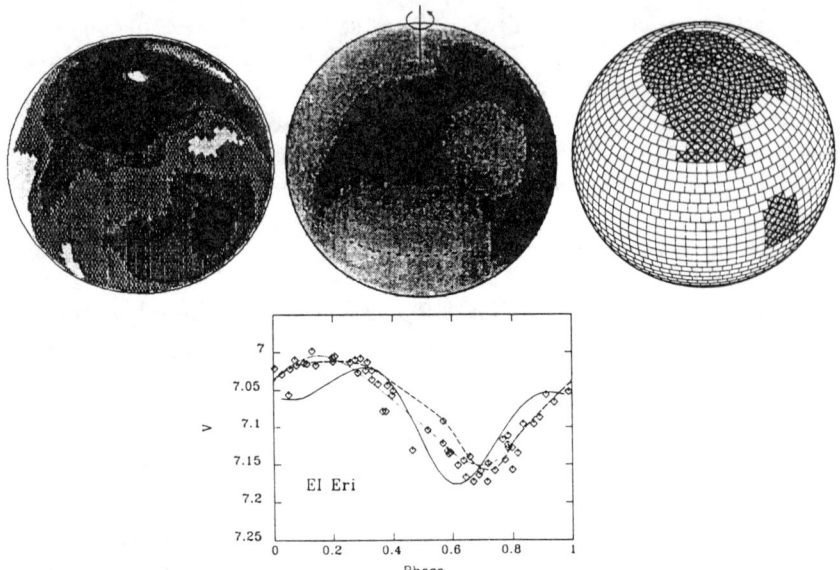

Figure 4: A comparison of Doppler maps of EI Eri from three different mapping techniques (upper panels) and their respective theoretical light curves (lower panel). See also the text (adopted from Strassmeier *et al.* 1991).

line stars (the so-called naked T Tauri's) show periodic light variations, while the classical, strong-emission stars tend to also show irregular variations, and extreme T Tauri's do not show periodicities at all. In at least one case, DN Tau (Vrba *et al.* 1986), a cool *and* a hot spot are found to be spatially associated, while on other stars, e.g., GW Ori (Bouvier and Bertout 1989), the light and color variations would be consistent with a hot *or* a cool spot solution.

Spots on W UMa-type contact binaries

Dark spots on W UMa systems were first proposed by Binnendijk (1970) to explain the asymmetric eclipses in some systems and Mullan (1975) followed up with a more detailed investigation. In his original list of RS CVn and related binaries, Hall (1976) included W UMa binaries as stars with an "uneven surface brightness distribution". Of course, there are surface brightness variations *not* caused by starspots such as polar brightening due to $T \propto g^{0.08}$ (effect of the gravitationally deformed component), ellipticity effect, a "bright" and a "dark" hemisphere on the cooler component due to irradiation (reflection effect), a hot spot on the mass-gaining component due to mass transfer, common envelope oscillations, and others. To gain some information on local temperature variations solely due to starspots, all the mentioned effects must be isolated and subtracted. It is therefore not surprising that only a few systems have been studied with modern light curve synthesis programs allowing for starspots. Table 1 summarizes some starspot parameters of W UMa stars as determined with modern synthesis programs. Note that the entry "hot spot" refers to solutions with a hot, circular region at the *substellar* point of the secondary component which is not magnetic in origin, e.g., VW Boo (Rainger et al. 1990), or BX And (Bell et al. 1990a). Spot coverage factors for W UMa binaries are hard to be put on a common scale with other stars (Table 1) due to their contact, common-envelope configuration and must be looked at with care when intercompared (Fig. 2).

Using short-wavelength IUE spectra, Eaton (1983) demonstrated the existence of active chromospheres on several W UMa's having more active primaries but less active secondaries. From photometry the situation is not as clear. In the case of AG Vir (A2+A8), Bell et al. (1990b) demonstrate the ease with which the spot phenomenon can be invoked to explain the light curve variations *and* to provide conflicting results. Their study (see their Fig. 9) shows the ambiguity between a deep-contact plus cool-spot model and a marginal-contact plus warm-spot[2] model. A similar ambiguity was encountered for VZ Psc (Maceroni et al. 1990b): a near-contact plus cool-spot model or a deep-contact model with no spots (Hrivnak and Milone 1989). The fact that the absolute dimensions from the near-contact plus cool-spot solution are in better agreement with the late spectral types, and that the fits are generally better, argues in favor of the existence of spots on VZ Psc, but the evidence remains inconclusive. In the case of AC Boo (Linnell 1991), the light curve modulation persists into totality, when one component is eclipsed, and the spots can be unequivocally placed onto the foreground component. A similar case is VW Cep (G5+K0). Bradstreet and Guinan (1990) presented near-simultaneous photometry and UV spectroscopy. Their photometry showed sometimes a secondary minimum deeper than the primary minimum and eclipse asymmetries appeared to be greatest at times when the system is *least* luminous. This suggests the existence of starspots cooler by about 500 K. Furthermore, Bradstreet and Guinan present evidence for spatial connection between these starspots and chromospheric and TR-emission regions. There is also good evidence that locations of starspots change with time. In U Peg (Zhai et al. 1988) the spot was shifting in longitude and latitude by an amount much too large to be caused by observational and model uncertainties. The temperature difference (photosphere minus spot) had also changed by ≈ 60 % from 1961 to 1970. Spot migration periods were derived from light curve variations, e.g., for W UMa ≈ 500 days (Rigterink 1972), and VW Cep ≈ 720 days (Leung and Jurkevich 1969) while Bradstreet and Guinan (1990) observed a cyclic behaviour with 155 and 320 days (they interpreted these periods as the beat periods between the orbital and the rotational period).

The onset of spot activity in contact or semi-detached systems is not clear since the sample of spotted components is too limited but Olson (1987) reports spot activity on the G4III-IV component in the Algol binary U Sagittae. The "earliest" W UMa system with cool starspots seems to be the A7-9 component of AG Vir (Bell et al. 1990b). However, as discussed by these authors, the detection

---

[2]Not to be confused with the *hot* spot due to mass transfer or irradiation

was not conclusive. Another early W UMa-type component with spot activity is U Peg (spectral type G2, Lu 1985).

This review is dedicated to my friend and mentor "Doug" Hall whom I thank for all his guidance and patience during the past years. It is, as always, a pleasure to thank my financial sources: the University of Vienna, and especially my wife *Elfi* who gave up some personal wishes so that I could attend this meeting.

REFERENCES

Ayres, T. R., Linsky, J. L. 1980, *Ap. J.*, **241**, 279.
Baliunas, S. L. 1988, in *Automatic Small Telescopes*, eds. D. S. Hayes, R. M. Genet, Fairborn Press, Mesa, p. 83.
Baliunas, S. L., Vaughan, A. H. 1985, *Ann. Rev. Astr. Ap.*, **23**, 379.
Bell, S. A., Rainger, P. P., Hilditch, R. W. 1990b, *M.N.R.A.S.*, **247**, 632.
Bell, S. A., Rainger, P. P., Hill, G., Hilditch, R. W. 1990a, *M.N.R.A.S.*, **244**, 328.
Binnendijk, L. 1970, *Vistas in Astr.*, **12**, 217.
Bopp, B. W., Evans, D. S. 1973, *M.N.R.A.S.*, **164**, 343.
Bopp, B. W., Noah, P. V. 1980, *Pub. A.S.P.*, **92**, 717.
Bopp, B. W. et al. 1984, *Ap. J.*, **285**, 202.
Bouvier, J. 1990, *A. J.*, **99**, 946.
Bouvier, J., Bertout, C. 1989, *Astr. Ap.*, **211**, 99.
Bouvier, J., Bertout, C., Bouchet, P. 1986, *Astr. Ap.*, **158**, 149.
Bouvier, J., Bertout, C., Bouchet, P. 1988, *Astr. Ap. Suppl.*, **75**, 1.
Bradstreet, D. H., Guinan, E. F. 1990, in NATO-ASI, *Active Close Binaries*, ed. C. Ibanoglu, Kluwer, Dordrecht, p. 467.
Budding, E. 1977, *Ap. Space Sci.*, **48**, 207.
Busso, M., Scaltriti, F., Persi, P., Ferrari-Toniolo, M., Origlia, L. 1988, *M.N.R.A.S.*, **234**, 445.
Byrne, P. B. 1987, *Irish Astr. J.*, **18**, 84.
Catalano, S. 1990, in NATO-ASI, *Active Close Binaries*, ed. C. Ibanoglu, Kluwer, Dordrecht, p. 347.
Cellino, A., Scaltriti, F., Busso, M. 1985, *Astr. Ap.*, **144**, 315.
Chou, D.-Y. 1987, *Ap. J.*, **312**, 955.
Collier-Cameron, A. 1991, in Armagh Obs. Colloq., *Surface Inhomogeneities in Late-type Stars*, ed. P. B. Byrne, D. J. Mullan, Springer, in press.
Deutsch, A. 1970, *Ap. J.*, **159**, 895.
Donati, J.-F., Semel, M. 1991, in IAU Colloq. 130, *The Sun and Cool Stars: activity, magnetism, dynamos*, eds. I. Tuominen, D. Moss, G. Rüdiger, Springer, Berlin Heidelberg New York, p. 326.
Dorren, J. D. 1987, *Ap. J.*, **320**, 756.
Dorren, J. D., Guinan, E. F. 1982, *Ap. J.*, **252**, 296.
Eaton, J. A. 1991, in Armagh Obs. Colloq., *Surface Inhomogeneities in Late-type Stars*, ed. P. B. Byrne, D. J. Mullan, Springer, Berlin Heidelberg New York, in press.
Eaton, J. A. 1983, *Ap. J.*, **268**, 800.
Eaton, J. A., Hall, D. S. 1979, *Ap. J.*, **227**, 907.
Fekel, F. C. 1980, *Bull. A.A.S.*, **12**, 500.
Foukal, P. V., Lean, J. 1986, *Ap. J.*, **302**, 826.
Friedemann, C., Gürtler, J. 1975, *Astron. Nachr.*, **296**, 125.
Gershberg, R. E., Ilyin, I. V., Shakhovskaya, N. I. 1991, in IAU Colloq. 130, *The Sun and Cool Stars: activity, magnetism, dynamos*, eds. I. Tuominen, D. Moss, G. Rüdiger, Springer, Berlin Heidelberg New York, p. 373.
Hall, D. S. 1972, *Pub. A.S.P.*, **84**, 323.
Hall, D. S. 1976, in IAU Colloq. No. 29, *Multiple Periodic Variable Stars*, ed. W. S. Fitch, Reidel, Dordrecht, p. 287.

Hall, D. S., 1990, in NATO-ASI, *Active Close Binaries*, ed. C. Ibanoglu, Kluwer, Dordrecht, p. 95.
Hall, D. S., Busby, M. R. 1990, in NATO-ASI, *Active Close Binaries*, ed. C. Ibanoglu, Kluwer, Dordrecht, p. 377.
Hall, D. S., Henry, G. W. 1990, in NATO-ASI, *Active Close Binaries*, ed. C. Ibanoglu, Kluwer, Dordrecht, p. 287.
Herbst, W. 1989, *A. J.*, **98**, 2268.
Herbst, W., Levreault, R. M. 1990, *A. J.*, **100**, 1951.
Herbst, W. et al. 1987, *A. J.*, **94**, 137.
Herbst, W. et al. 1986, *Ap. J. (Letters)*, **310**, L71.
Hill, G., Fisher, W. A., Holmgren, D. 1990, *Astr. Ap.*, **238**, 145.
Hoffmeister, C. 1965, *Veröff. Sternwarte Sonneberg*, **6**, 97.
Holtzmann, J. A., Nations, H. L. 1984, *A. J.*, **89**, 391.
Hrivnak, B. J., Milone, E. F. 1989, *A. J.*, **97**, 532.
Huenemoerder, D. P., Ramsey, L. W. 1987, *Ap. J.*, **319**, 392.
Huenemoerder, D. P., Buzasi, D. L., Ramsey, L. W. 1989a, *A. J.*, **98**, 1398.
Huenemoerder, D. P., Ramsey, L. W., Buzasi, D. L., 1989b, *A. J.*, **98**, 2264.
Huovelin, J., Saar, S. H., Tuominen, I. 1988, *Ap. J.*, **329**, 882.
Jackisch, G. 1963, *Veröff. Sternwarte Sonneberg*, **5**, 231.
Jetsu, L., Huovelin, J., Tuominen, I., Vilhu, O., Bopp, B. W., Piirola, V. 1990, *Astr. Ap.*, **236**, 423.
Jetsu, L., Pelt, J., Tuominen, I., Nations, H. 1991, in IAU Colloq. 130, *The Sun and Cool Stars: activity, magnetism, dynamos*, eds. I. Tuominen, D. Moss, G. Rüdiger, Springer, Berlin Heidelberg New York, p. 381.
Kang, Y. W., Wilson, R. E. 1989, *A. J.*, **97**, 848.
Keenan, P. C., McNeil, R. C. 1989, *Ap. J. Suppl.*, **71**, 245.
Kemp, J. C. et al. 1987, *Ap. J. (Letters)*, **317**, L29.
Kürster, M., Schmitt, J. H. M. M. 1991, in Armagh Obs. Colloq., *Surface Inhomogeneities in Late-type Stars*, ed. P. B. Byrne, D. J. Mullan, Springer, Berlin Heidelberg New York, in press.
Leung, K. C., Jurkevich, I. 1969, *Bull. A.A.S.*, **1**, 251.
Lu, W. X. 1985, *Pub. A.S.P.*, **97**, 1086.
Linnell, A. P. 1991, in IAU Colloq. 130, *The Sun and Cool Stars: activity, magnetism, dynamos*, eds. I. Tuominen, D. Moss, G. Rüdiger, Springer, Berlin Heidelberg New York, p. 377.
Lynds, C. R., Worden, R. D., Harvey, J. W. 1976, *Ap. J.*, **207**, 174.
Maceroni, C., Bianchini, A., Rodonó, M., Van't Veer, F., Vio, R. 1990a, *Astr. Ap*, **237**, 395.
Maceroni, C., van Hamme, W., van't Veer, F. 1990b, *Astr. Ap.*, **234**, 177.
Marcy, G. W., Bruning, D. H. 1984, *Ap. J.*, **281**, 286.
Mullan, D. J. 1975, *Ap. J.*, **198**, 563.
Olah, K. et al. 1985, *Ap. Space Sci.*, **108**, 137.
Olson, E. C. 1987, *A. J.*, **94**, 1043.
Parker, E. N. 1979, *Ap. J.*, **230**, 905.
Piskunov, N. E., Tuominen, I., Vilhu, O. 1990, *Astr. Ap.*, **230**, 363.
Poe, C. H., Eaton, J. A. 1985, *Ap. J.*, **289**, 644.
Radick, R., R., Thompson, D. T., Lockwood, G. W., Duncan, D. K., Baggett, W. E. 1987, *Ap. J.*, **321**, 459.
Rainger, P. P., Bell, S. A., Hilditch, R. W. 1990, *M.N.R.A.S.*, **246**, 47.
Ramsey, L. W., Nations, H. L. 1980, *Ap. J. (Letters)*, **239**, L121.
Rigterink, P. V. 1972, *A. J.*, **77**, 230.
Robinson, R. D., Worden, S. P., Harvey, J. W. 1980, *Ap. J. (Letters)*, **236**, L155.
Rodonó, M. et al. 1986, *Astr. Ap.*, **165**, 135.

Rucinski, S. M. 1981, *Astr. Ap.*, **104**, 260.
Rydgren, A. E., Vrba, F. J. 1983, *Ap. J.*, **267**, 191.
Saar, S. H. 1988, *Ap. J.*, **324**, 441.
Saar, S. H., Neff, J. E. 1990, in 6th Cambridge workshop on *Cool Stars, Stellar Systems, and the Sun*, ed. G. Wallerstein, ASP Conf. Series, **9**, p. 171.
Scaltriti, F., Cellino, A., Busso, M. 1985, *Astr. Ap.*, **149**, 11.
Schaefer, B. E. 1983, *Ap. J. (Letters)*, **266**, L44.
Simon, T., Vrba, F. J., Herbst, W. 1990, *A. J.*, **100**, 1957.
Strassmeier, K. G. 1988, *Ap. Space Sci.*, **140**, 223.
Strassmeier, K. G. 1990, *Ap. J.*, **348**, 682.
Strassmeier, K. G., Bopp, B. W. 1991, *Astr. Ap.*, submitted.
Strassmeier, K. G., Fekel, F. C. 1990, *Astr. Ap.*, **230**, 389.
Strassmeier, K. G., Hall, D. S. 1988, *Ap. J. Suppl.*, **67**, 439.
Strassmeier, K. G., Hall, D. S., Boyd, L. J., Genet, R. M. 1989a, *Ap. J. Suppl.*, **69**, 141.
Strassmeier, K. G., Hall, D. S., Zeilik, M., Nelson, E., Eker, Z., Fekel, F. C. 1988a, *Astr. Ap. Suppl.*, **72**, 291.
Strassmeier, K. G., Hooten, J. T., Hall, D. S., Fekel, F. C. 1989b, *Pub. A.S.P.*, **101**, 107.
Strassmeier, K. G. et al. 1988b, *Astr. Ap.*, **192**, 135.
Strassmeier, K. G. et al. 1991, *Astr. Ap.*, in press.
Struve, O. 1930, *Ap. J.*, **72**, 1.
Torres, C. A. O., Ferraz-Mello, S. 1973, *Astr. Ap.*, **27**, 231.
Vogt, S. S. 1981a, *Ap. J.*, **247**, 975.
Vogt, S. S. 1981b, *Ap. J.*, **250**, 327.
Vogt, S. S. 1983, in IAU Colloq. 71, *Activity in Red-Dwarf Stars*, eds. P. B. Byrne, M. Rodonó, Reidel, Dordrecht, p. 137.
Vogt, S. S. 1988, in IAU Symp. 132, *The Impact of Very High S/N Spectroscopy on Stellar Physics*, eds. G. Cayrel de Strobel, M. Spite, Kluwer, Dordrecht, p. 253.
Vogt, S. S., Hatzes, A. P. 1991, in IAU Colloq. 130, *The Sun and Cool Stars: activity, magnetism, dynamos*, eds. I. Tuominen, D. Moss, G. Rüdiger, Springer, Berlin Heidelberg New York, p. 297.
Vogt, S. S., Penrod, G. D. 1983, *Pub. A.S.P.*, **95**, 565.
Vrba, F. J., Herbst, W., Booth, J. F. 1988, *A. J.*, **96**, 1032.
Vrba, F. J., Rydgren, A. E., Chugainov, P. F., Shakovskaya, N. I., Weaver, W. B. 1989, *A. J.*, **97**, 483.
Vrba, F. J., Rydgren, A. E., Chugainov, P. F., Shakovskaya, N. I., Zak, D. S. 1986, *Ap. J.*, **306**, 199.
Wilkerson, M. S., Worden, S. P. 1977, *A. J.*, **82**, 642.
Wilson, O. C. 1978, *Ap. J.*, **226**, 379.
Wolff, S. C., Boesgaard, A. M., Simon, T. 1986, *Ap. J.*, **310**, 360.
Zeilik, M., Batuski, D., Burke, S., Elston, R., Smith, P. 1983, *I.B.V.S.*, No. 2257.
Zeilik, M., Cox, D. A., De Blasi, C., Rhodes, M., Budding, E. 1989, *Ap. J.*, **345**, 991.
Zeilik, M., De Blasi, C., Rhodes, M., Budding, E. 1988, *Ap. J.*, **332**, 293.
Zeilik, M., Ledlow, M., Rhodes, M., Arévalo, M. J., Budding, E. 1990, *Ap. J.*, **354**, 352.
Zhai, D. S., Chen, H. 1989, *Chinese Astr. Ap.*, **13**, 9.
Zhai, D. S., Lu, W.-X., Zhang, X.-Y. 1988, *Ap. Space Sci.*, **146**, 1.
Zhai, D. S., Qiao, G. J., Zhang, X. Y. 1990, *Astr. Ap.*, **237**, 148.

PART II.

CCD IMAGING WITH ROBOTIC TELESCOPES

THE SCIENTIFIC POTENTIAL OF AUTOMATIC CCD IMAGING TELESCOPES

ALEXEI V. FILIPPENKO
Department of Astronomy, University of California, Berkeley CA 94720

ABSTRACT   I discuss the scientific potential of moderate-aperture (~ 0.8 m) automatic imaging telescopes (AITs) equipped with CCDs. Such instruments are relatively inexpensive to construct, yet they have a wide range of uses for projects that cannot otherwise be conducted at most private or national observatories. They also have many important advantages over conventional automatic photoelectric telescopes. For illustrative purposes, I concentrate on three scientific areas being pursued by the Berkeley AIT: light curves of supernovae and novae, active galactic nuclei, and solar system objects. AITs are also useful for essentially all types of variable stars. There are enormous gains to be made by using AITs in coordinated ground-based and space-based observations. Finally, AITs have great potential in undergraduate laboratory astrophysics classes.

INTRODUCTION

As described elsewhere in this volume (Richmond, Treffers, and Filippenko 1992; Treffers, Richmond, and Filippenko 1992; see also Richmond and Filippenko 1991), we are in the process of establishing a fully-automated, computer-controlled, 0.8-m reflecting telescope that will be equipped with a charge-coupled device (CCD) imaging camera and an autoguider. A 0.5-m prototype is already being tested at Leuschner Observatory, near the U. C. Berkeley campus. This automatic imaging telescope (AIT) is capable of long, autoguided exposures that enable us to study faint objects. The basic goal is to conduct long-term studies of variable and ephemeral objects over a range of time scales.

   Area photometry with CCDs offers many advantages over conventional photometry through large apertures. The high quantum efficiency of CCDs relative to photomultiplier tubes (PMTs) means that fainter objects can be observed with a given telescope. The study of faint stars is also enhanced by the fact that the effective entrance aperture is defined on the image itself, and can be made very small if the seeing was good. This decreases the sky contribution, and also allows a local background subtraction. Thus, while a 0.8-m automatic photoelectric telescope (APT) can reach $m \approx 13$ mag, an AIT of comparable diameter is able to detect stars with $m \approx 21$ mag in long, guided exposures. In addition, one can obtain accurate measurements of objects in crowded fields, especially if a two-dimensional point spread function is used to fit the profile of each star. With an area detector, useful data can be taken through thin clouds, since accurate differential photometry is performed between objects in the same CCD frame. (The differential photometry is converted to an absolute scale by measuring standard stars on photometric nights.) Another very important point for ephemeral objects such as novae and supernovae (SNe) is that the underlying background can be determined from a "template" image obtained after the object has faded. By carefully scaling all images to match the template, accurate magnitudes of ephemeral objects can be measured even when they are quite faint (e.g., Filippenko et al. 1986; Porter 1989). Finally, CCD imaging can be used to monitor the positions and brightnesses of moving objects such as asteroids and comets.

Here I describe some of the projects we are currently conducting, or plan to pursue, with the Berkeley AIT (see also Filippenko and Richmond 1991). Our team is concentrating on long-term photometry of *known* variable or ephemeral objects, since other groups are busy searching for new objects (e.g., SNe, massive compact halo objects, comets in the Oort cloud).

LIGHT CURVES OF SUPERNOVAE AND NOVAE

Supernovae

The Type II SN 1987A has shown, more than any other supernova, how much can be learned about the death of stars when high quality, frequently sampled data are obtained over very long time intervals (e.g., Arnett et al. 1989). The sharp rise in brightness and the initially rapid color evolution, for example, supported the identification of a *blue* supergiant, rather than a much larger red supergiant, as the progenitor star. The luminosity of the supernova and the rate of its exponential decline implied that 0.07–0.08 $M_\odot$ of radioactive $_{28}Ni^{56}$ had been produced during the explosion, a result that was later confirmed through infrared spectroscopy and measurements of gamma rays. Subsequent deviations from the exponential decline, which coincided with the detection of X-rays, showed that not all of the radioactive decay energy was being trapped by the ejecta, and led to the ideas of significant clumping and mixing. The light curve may someday exhibit conclusive evidence for a significant additional energy source, perhaps a pulsar or an accreting neutron star.

Especially important in the physical study of SNe is the *bolometric* light curve, which consists primarily of near-ultraviolet to near-infrared radiation during the first one or two years. It is almost impossible, however, to actually procure such data for most SNe with existing telescopes at private and national observatories; too little time is available for a given project, and there is insufficient quality control and uniformity when many different observers are involved. This is indeed unfortunate, given that SNe in the Virgo cluster have $m \approx 11\text{–}13$ mag at maximum (depending on their type) and can, in principle, provide us with 1–2 year light curves (down to $m \approx 21$ mag) nearly as good as that of SN 1987A.

Precise light curves are essential not only for a more thorough understanding of the physics of SNe, but for cosmological studies as well (e.g., Kirshner 1990). For example, most Type Ia SNe appear to have almost the same light curves and maximum luminosities (± 0.3 mag), and there are good theoretical reasons for believing that this should be the case. Since SNe Ia are extremely luminous ($M_V \approx -19$ to -20 mag), they can be seen at very great distances, and a simple comparison of apparent and absolute magnitudes at maximum yields a distance modulus. If the redshifts of their parent galaxies are measured with spectrographs on large telescopes, both the Hubble constant (H_0) and the deceleration parameter (q_0) can be determined. An accurate calibration of the maximum luminosity (L_{max}) is necessary for the former, while the latter only requires L_{max} to be constant over all cosmic epochs under consideration. Detailed studies of nearby SNe Ia, however, are needed to establish the degree to which different SNe Ia have similar observed properties, and to calibrate L_{max}. If it is found, for example, that L_{max} depends on the metallicity of the parent galaxy, then the assumption of constancy of L_{max} over cosmic time will have to be questioned.

An independent method to determine cosmologically interesting distances with SNe is through the expanding photosphere method (Kirshner 1990, and references therein). The velocity of the photosphere can be estimated from the blueshifted absorption components of the P Cygni line profiles. Together with the measured time since the explosion, this gives a radius for the photosphere. Its temperature can be determined from the spectral energy distribution; hence, we can derive the luminosity of the supernova. Comparison with the measured brightness consequently gives the distance. This technique can even be applied if the explosion date is unknown; one simply compares the measurements taken on several sufficiently well-separated dates.

There are currently several successful searches for supernovae being conducted throughout the world. The Berkeley Automated Supernova Search, for example, has discovered about 20 SNe (Perlmutter *et al.* 1992; Muller *et al.* 1992). Similarly, Rev. Robert Evans has found over 20 bright SNe. Followup observations of these objects, on the other hand, could certainly be much improved. We plan to obtain images of galaxies containing known SNe at short time intervals in at least the V, R, and I filters, with occasional observations of bright objects in U and B. (Our current CCD system is not very sensitive at short wavelengths.) Data will be obtained on a nightly basis at early times, when the SNe are bright and changing rapidly, and less frequently at late times, when they are dim and changing slowly.

Novae

Our fully automated telescope will also be of great use in monitoring novae. Although novae have been known and well studied for many decades, there are still numerous surprises and poorly understood phenomena associated with them. An excellent example is V404 Cygni, which in 1989 achieved an X-ray flux 17 times that of the Crab nebula and exhibited extremely rapid variability of flux and emission-line profiles. Detailed spectroscopic studies strongly support the hypothesis that V404 Cyg contains a black hole rather than a neutron star (Casares, Charles, and Naylor 1992).

Multi-band observations of novae can be used to construct approximate bolometric light curves, as with SNe. For the best results, the sampling must be done over a wide range of time scales. This is difficult to accomplish at most observatories, and data sets produced with different telescopes suffer from discrepancies between the exact bandpasses provided by various CCD/filter combinations. With a fully automated telescope, we will be able to obtain accurate, well-sampled bolometric and color curves of several novae each year, following them to the late stages which have been less thoroughly studied. This may also advance our understanding of the precursors of SNe believed to arise in binary systems. An intriguing, but admittedly remote, additional possibility is that we will identify a highly-obscured Galactic SN from the light curve (and subsequent spectroscopy with other telescopes) of a "nova."

Besides being interesting objects in and of themselves, novae are potentially useful distance indicators, at least out to the Virgo cluster, because their light-curve shapes appear to be correlated with L_{max} (e.g., McLaughlin 1960): the rate of decline is rapid for luminous novae, and slower for less luminous novae. To strengthen the use of novae for cosmological studies, we should more accurately calibrate this correlation with Galactic novae. Moreover, we must verify that novae in other nearby galaxies behave in exactly the same manner. We plan to do this at least for novae in M31, the brightest of which reach $m \approx$ 15-16 mag at maximum. Amateur astronomer James Bryan in Texas is quite successful at finding novae in M31, and he hopes to accelerate his efforts during the next few years. Having located a nova, he will contact us to initiate high quality CCD photometry with the Berkeley AIT.

LIGHT CURVES OF QSOS AND AGNS

One of the most important characteristics of quasi-stellar objects (QSOs) and active galactic nuclei (AGNs) is their rapid variability, and this makes them very attractive for study with robotic telescopes (e.g., Keel 1986). Excluding relativistic effects, which are certainly present in some of these objects, the variability time scale gives us a rough idea of the maximum size of the emitting region. For example, if all stars in our Milky Way Galaxy simultaneously brightened by a factor of two, an extragalactic observer located in the Galactic plane would see light from the near side about 100,000 years sooner than light from the much more distant far side (ignoring extinction by dust). Thus, the intrinsically instantaneous brightening would *appear* to be spread out over a very long time interval.

Significant variations are observed in many QSOs and AGNs over time scales as short as years, months, weeks, or (in some cases) even days. Hence, the emitting regions of these objects are at most a few light years (months, weeks, days) in size. When combined with the stupendous luminosities of QSOs and AGNs, this fact leads to the conclusion that the "central engine" is probably a supermassive black hole ($M \lesssim 10^{10}\ M_\odot$) accreting material at a rate of up to 10 $M_\odot$ yr^{-1} (e.g., Rees 1984). Since the size of the emitting region is generally determined from the shortest time scale over which high-amplitude variability is observed, it is important to continue frequent, yet long term, monitoring of QSOs and AGNs. Extremely rapid variability may reveal the presence of relativistic beaming, such as in the highly-polarized BL Lac objects, or of instabilities in the accretion disk thought to surround the black hole. Furthermore, most low-luminosity AGNs seem to have shorter minimum time scales than very luminous objects, implying that their black holes may be less massive, but this correlation must be verified to eliminate the possibility that it results from selection effects. Of course, the scientific utility of all these data will be enhanced when analyzed in conjunction with other ground-based and space-based observations of the same objects.

Another potentially very fruitful area is the study of gravitationally lensed QSOs (e.g., Canizares 1987). The most famous example is Q0957+561, which consists of two bright ($m \approx 17$ mag), well-separated ($\sim 6''$) QSO images. The lensing object is a massive cluster of galaxies whose redshift z is 0.36. The QSO images are also detected at radio wavelengths, even with the technique of very long baseline interferometry. Because of these extensive data, the geometry of the lens is reasonably well understood, and one can calculate the difference in the light travel time to each of the QSO images. This difference is inversely proportional to H_0, the Hubble constant. If the luminosity of the QSO suddenly increases, we will see the two images brighten, but with a relative time delay that can therefore be used to determine H_0. A heroic study of this type is already being conducted elsewhere (Schild 1990, and references therein), but it relies on a human operator for the telescope. With the Berkeley AIT, we will be able to conduct similar projects over longer time intervals and with better temporal sampling. It will be important, for example, to confirm the suspected gravitational microlensing of Q0957+561A,B by halo stars in the lensing galaxies (Schild and Smith 1991).

Although the studies described above are primarily based on broad-band photometry, very interesting results can also be obtained with images taken through intermediate-band filters. For instance, we plan to monitor the total flux of the broad ($\lesssim 400$ Å) Hα and Hβ emission lines in the nuclei of low-redshift Seyfert 1 galaxies. The time delay between changes in the continuum and emission-line fluxes is a measure of the size of the region containing the clouds of gas that produce the broad emission lines. Moreover, geometrical properties of the "broad-line region" can be determined from the relative shapes of the continuum and emission-line light curves. This has been a very active area of extragalactic research during the past few years, and already there are strong indications that the broad-line region is considerably smaller than was previously believed; see, for example, Peterson et al. (1991) and references therein. With our telescope, we will be able to monitor many Seyfert galaxies on an almost nightly basis. Unlike the case with a large entrance aperture, careful analysis of the CCD images will allow us to exclude much of the starlight that contaminates the near-nuclear regions and dilutes the intrinsic variability of the active nucleus.

An extra benefit of long-term monitoring of QSOs and AGNs is that we will be able to notify other observers of their current brightnesses. This information could be valuable to astronomers scheduled to use the highly oversubscribed Hubble Space Telescope, since exposure times can sometimes be adjusted to achieve the desired signal-to-noise ratios. Some observers wish to obtain very high resolution spectra of QSOs in order to study the absorption lines produced by intervening galaxies and clouds, and this can be done most efficiently when the QSOs are bright. Others desire spectra when the active nuclei are faint, so as to maximize the relative signal from the host galaxy. This has especially been the case in studies of BL Lac objects, whose nuclei can completely obliterate light from the surrounding "fuzz" when they are undergoing an outburst.

LIGHT CURVES OF SOLAR SYSTEM OBJECTS

Photometry has long been established as one of the fundamentally most important tools for gaining physical information on small and distant solar system bodies such as asteroids and Pluto. Here I mention some applications of AITs in this area. For a more complete discussion of comets, see Jewitt (1992).

Asteroids

Photometry of asteroids reveals their rotational light curves, from which their rotation rates and approximate shapes can be derived. Statistical distributions of these properties can be used to constrain models for the collisional evolution of the asteroids (e.g., Binzel et al. 1989). When composite light curves from different apparitions in different parts of the sky are combined, it is sometimes possible to determine the orientations of spin axes (Magnusson et al. 1989). Light curve observations are also useful for high-precision radiometric determinations of albedo, and for interpreting radar observations.

If the light curve is sampled over more than a few days (preferably 2-3 months for a main belt asteroid), so that the solar phase angle varies over a substantial range, the phase relation, or absolute light level as a function of solar phase angle, can be defined. This relation yields clues to the surface physical structure (roughness, albedo, porosity) of the asteroid, and allows accurate comparison of the intrinsic light levels at different apparitions, even if the exact same phase angles are not observed (Bowell et al. 1989). For asteroid rotational light curve work, the typical amplitude is at least 0.2 mag. Therefore, even moderate precision differential photometry ($\pm 0.01 - 0.02$ mag) is sufficient to fully resolve the light curve.

Although previous observational surveys of asteroid rotational properties have been made, these have principally been performed using small telescopes with low quantum efficiency detectors. The resulting observational bias has been toward large, inner-belt asteroids. The statistical properties of small and distant asteroids remain poorly known. Data on these bodies are needed in order to obtain a complete picture of the processes of collisional evolution in the asteroid belt. Such knowledge is required before we can fully understand the processes involved in the formation and evolution of our solar system.

2060 Chiron

The minor planet (comet?) 2060 Chiron is an ideal target for an AIT. Over several years, it has been seen to fade and brighten by about 1 mag, too large to be explained by inverse-square effects of Earth and Sun distances. On the other hand, the light curve was well observed in 1986, and found to have a full range amplitude of only 0.09 mag (Bus et al. 1989). Two obvious explanations of this peculiar behavior are that we see variable light from a coma, or that outbursts occasionally "paint" the surface brighter or darker. The only way to understand in detail what is happening is to monitor Chiron extensively. The rotation period is known ($P = 5.9178$ hr), so a well designed program could be accommodated with only about one hour per night, for about 3 months a year, while Chiron is well placed near opposition.

The controversy regarding the nature of Chiron is related to the question of which, if any, other asteroids are actually "dead" or dying comets. Our AIT might monitor other "asteroids" that are in eccentric (comet-like) orbits for comatic activity (dust) near perihelion. Asteroids other than Chiron do not seem to show comae, but deeper broad-band red CCD images of asteroids closely approaching Earth are needed to more definitively settle the issue.

Pluto

Photometry of Pluto reveals its 6.38 day rotational light curve, which is well known to display secular variations in shape and amplitude over time scales of years. These variations may be due to Pluto's changing aspect angle, seasonal changes in its surface/atmospheric structure, or both. A few years ago Pluto passed perihelion, and continued observations of its rotational light curve at the current epoch are critically needed to investigate the physical properties of Pluto's surface and atmosphere and their possible interaction. Results from the once-per-century series of Pluto-Charon mutual events reveal a very high average albedo (~ 0.5) for the planet and an even brighter south polar cap. Such high albedos are difficult to understand for a methane-covered surface since bombardment by solar UV photons and Galactic cosmic rays tends to darken such a surface on timescales of order 100 years. Thus, the surface and atmosphere of Pluto must be undergoing significant interactions over the course of its orbit.

Important constraints can be placed on Pluto's atmospheric structure through monitoring if and when the recondensation process of methane begins. Such monitoring can only be performed through routine Earth-based observations of Pluto's 6.38 day rotational light curve. Pluto's rotation rate has been known since the 1950s, when photometry revealed a periodic variation with an amplitude of ~ 0.1 mag. Since that time Pluto's rotational light curve amplitude has increased to ~ 0.3 mag, while its overall brightness has decreased. This may be a geometric effect as the sub-Earth latitude on Pluto has moved from the south polar region northward through the equator. Decoupling geometric effects from atmospheric-surface interactions requires accurate albedo mapping of Pluto's surface. Although such maps are being derived from the Pluto-Charon mutual events, these are restricted to only one hemisphere of the planet (the sub-Charon hemisphere) owing to the synchronous orbit of the satellite. Thus, the modeling of one-half of Pluto's surface at the current epoch remains dependent on observations of Pluto's rotational light curve. Continued observations of Pluto are needed to define its rotational light curve at the current epoch so as to refine and improve models for the albedo distribution of its surface. Such models will then be able to serve as a baseline for detecting significant albedo changes due to the condensation of atmospheric methane on Pluto's surface.

Comets

Our knowledge of the physics of comets is based on a large but eclectic set of observations taken using visual, photographic, and electronic detectors, both ground-based and space-based. The observational sample is highly biased towards comets which are close to the Sun and/or unusually active. Low-activity comets are rarely studied in any detail; neither are active comets once they have receded from the Sun to distances greater than a few AU. Even the brightest comets generally attract only intermittent attention and, with the notable exception of P/Halley, very few quantitative data are available for comets at heliocentric distances greater than 3-4 AU. It is likely that our perception of comets is at least partly biased by the selective nature of the currently available information.

In addition to the gradual changes in the morphology and total activity which occur as a comet approaches and recedes from the Sun, many comets show stochastic variations (often called "outbursts"). The most famous case is that of P/Schwassmann-Wachmann 1 (SW1), in which outbursts of $\Delta m = 3-5$ mag are common, but outbursts occur in many comets. The few studies which have had sufficient temporal resolution invariably yield results of considerable interest (e.g., McFadden et al. 1987). The properties (frequency of occurrence, magnitude, mass loss rate, composition, dust size) of the outbursts are very poorly known, even in SW1. This is unfortunate because the outburst properties may depend on structures in the upper few meters of nuclei, and a physical understanding of the outburst mechanisms might give valuable information about these structures. For instance, viable hypotheses include reaction of unstable radicals, phase transition in amorphous ice, and rupture of a surface crust or "mantle," among others. A physical understanding of the outbursts, founded on a strong observational base, might reveal a great deal about near-surface structures on nuclei.

One general objective of an AIT could be to obtain a set of homogeneous photometric images of a moderate sample ($\sim$ 10) of comets over many years, with carefully selected time resolution. While an underlying interest is to obtain the light curves of comets over a wide range of heliocentric distances, a more immediate need is to understand the (sometimes dramatic) variations which are known to occur in many or most comets on short timescales at fixed distance. AITs would rectify the chronic lack of time series observations of comets. The data could be used to (a) produce time-resolved continuum and gas band light curves for comets having a variety of dynamical ages, heliocentric distances, and apparent gas/dust ratios, (b) interpret these light curves using a sublimation algorithm already applied to P/Halley with some success, (c) characterize outbursts in the same sample of comets by means of the photometric and morphological information in CCD images obtained with high temporal resolution, and (d) compare the statistical properties of outbursts with models. Moreover, rotation periods could be obtained for comets whose non-uniform nuclei induce a periodic variation in the rate of mass loss and, consequently, in the strength of the coma. Reliable periods are currently available for only a few comets.

Guiding and Photometry

Old periodic comets, as well as known asteroids, Pluto, and Chiron, have well determined orbital elements, and their location and non-sidereal rates can be predicted with sufficient accuracy to allow fairly long integrations. A different guide star must be found each night (or every few nights), but this is not a problem. It is important to note that offset guiding should be possible with our AIT even at non-sidereal rates. The centroid algorithm in the guider CCD can be modified to compute offsets with respect to a moving center (the guide star), instead of with respect to a fixed center. The direction and rate of the moving center are entered as part of the observational parameter file associated with each object on the observing list. (To improve the photometric quality of results obtained from images having trailed stars, it may sometimes be necessary to acquire an additional image, guided at the sidereal rate, so that the stars appear circular.) Initially, the rates of new comets may not be very well known, but in this case the comets usually are brighter, and many short exposures can be added to achieve whatever integration time is desired.

It will, of course, be difficult to obtain photometrically accurate data for moving objects on cloudy nights, since the field stars will not generally be the same in different images of a given object. However, *some* of the same field stars will be visible on overlapping frames taken over the course of several nights when imaging sufficiently slowly-moving objects. This makes it possible to calibrate cloudy nights if photometric nights are not infrequent. One can also reobserve, during photometric nights, those fields through which the moving object previously passed on cloudy nights.

Advantages of CCD AITs
It is important to note that CCD photometry of solar system objects can be obtained with significantly greater precision than for most stationary celestial sources by utilizing a novel reduction technique. The technique is essentially the same as using archival images to subtract the background level from an image of a supernova, as described previously. The rapid apparent motion of asteroids allows acquisition of a source image and a second exposure of the same background field in the absence of the source within a time span of minutes to hours. Registration of the two images and subtraction of the background frame permits a photometric analysis of asteroids to faint levels rarely attempted in the past. At $m \approx 17$ mag, the sky is often "lumpy" at the several percent level, even in the smallest aperture size (or synthetic aperture on an image) practical under the observing conditions. Near the Milky Way, this "lumpiness" makes traditional photometry very difficult below 13 mag, and leads to substantial data loss due to confusion even for CCD images. The method proposed here should allow proper background subtraction except when the target body is swamped by a confusing source many times brighter than itself. Reduction and analysis of the images should be possible with existing software developed for supernova photometry.

OTHER IMPORTANT USES OF AITs

Variable Stars
Many variable stars are much too faint to study with conventional APTs, yet are astrophysically important. Examples can be found in the literature among classes such as dwarf novae, T Tauri stars, and Cepheid variables. A representative sample of these could easily be accommodated in the observing schedule of the Berkeley AIT. RR Lyrae variables in globular clusters, used to determine distances and to study stellar evolution in the instability strip, could also be monitored on a regular basis with the CCD area photometer. Since very bright objects ($m \lesssim 10$ mag) are probably best observed with conventional photometers, the study of large-amplitude variables (e.g., Mira stars) could benefit from a program involving the use of both APTs and AITs.

As just one example of a fascinating star that deserves further study with an AIT, I mention SS 433. This is perhaps the only known object in the Universe that, so far, is truly "unique" in its major observed properties — yet it may represent a short-lived stage of evolution through which many massive binary stars pass, and it is certainly related to the more general class of X-ray binaries. In addition, the relativistic two-sided jet emerging from SS 433 may be a small-scale version of the phenomenon occurring in many AGNs and QSOs, and could therefore teach us much about the physics of relativistic jets. SS 433 deserves to be intensively monitored until it is understood. There is evidence for variability of SS 433 on almost all time scales, but it has traditionally been difficult to study both the short-term and long-term behavior during the course of one observational program. We plan to make SS 433 a high-priority object for the Berkeley AIT.

Coordinated Observations
Many astronomical programs require, or are greatly enhanced by, simultaneous or quasi-simultaneous photometric observations. For example, emission-line variability detected during rapid spectroscopy of cataclysmic variable stars can be more thoroughly interpreted if good photometry over the same time interval is available, but this generally cannot be achieved through a narrow slit. We plan to coordinate our efforts with many other astronomers to ensure that the best possible science is done in the available time.

Nearly simultaneous optical photometry will be particularly desirable during some of the space-based observations planned for the next decade. For example, we hope to monitor bright QSOs and other interesting objects in the field of view of the Gamma Ray Observatory. Similarly, when ROSAT begins its pointed observations, we plan to complement the X-ray observations with optical photometry. Coordinated optical and ultraviolet observations of stars will be conducted after the Extreme Ultraviolet Explorer is launched in 1992. In addition, we anticipate that many observations to be made with the Hubble Space Telescope will benefit from ground-based photometry, sometimes over long time scales. These include multi-wavelength studies of BL Lac objects, as well as rapid monitoring of changes in the ultraviolet spectra of variable objects.

EDUCATIONAL PURPOSES

We will incorporate our AIT into several undergraduate laboratory astrophysics courses at U. C. Berkeley. Students will learn how to take and reduce data with modern solid-state instruments, thereby providing excellent preparation for graduate work. One specific project we plan to pursue in our first-semester course is the determination of the mass of Jupiter through systematic observations of its Galilean satellites. We might also break the class into several groups, with each group determining the light curve of a particular variable star, QSO, or supernova. Advanced techniques of image processing, data manipulation, and statistical analysis will be taught in upper-division courses, and in some cases it is expected that the work will be published. A great advantage of our scheme will be that students can accomplish most of the work *on campus*, during the day; we avoid all the problems associated with transporting a class to the observatory and back at night. Real-time access, if needed occasionally, will be possible through an Ethernet link to campus.

This kind of early exposure to *real* science is very important for student development. It encourages creativity, illustrates the thrill of scientific discovery, and provides a sense of participation and accomplishment at a time when many students feel overwhelmed by formal coursework involving dry textbooks and lectures. It also makes science seem much more of an active, exciting field to non-science students, and it can therefore significantly increase the appreciation and understanding of science among the general public.

CONCLUSIONS

It is clear that there is much science to be done with automatic imaging telescopes. AITs are the logical next step for robotic observatories, whose "existence proof" has so successfully been provided by the APTs on Mt. Hopkins and elsewhere. As is readily apparent from the talks given at this symposium, APTs have made a significant impact on astronomical research. Given the much fainter limiting magnitudes achievable with autoguided CCDs, I anticipate that even greater progress will be made with AITs. It will certainly take time to convince the skeptics, but we are already well on our way. The 1990s are the decade in which robotic telescopes will come of age.

ACKNOWLEDGEMENTS

My work on automatic imaging telescopes has been supported largely by the National Science Foundation in the form of a Presidential Young Investigator Award (NSF grant AST-8957063). Additional funds have been provided by NSF Cooperative Agreement AST-8809616 (Center for Particle Astrophysics, U. C. Berkeley), NSF grant AST-9115174, CalSpace grant CS-96-91, and the University of California at Berkeley. Donations made by AutoScope Corporation, IBM, Photometrics Ltd., and Sun Microsystems are much appreciated. I have received useful advice from many individuals, especially Russ Genet, Jack Borde, Joe Miller, and Joe Shields. Much of the information on observations of solar system objects was provided by Rick Binzel, Alan Harris, David Jewitt, and Hyron Spinrad. Last, but certainly not least, I am very grateful to Michael Richmond and Dick Treffers for their phenomenal contributions to the success of the Berkeley Automatic Imaging Telescope.

REFERENCES

Arnett, W. D., Bahcall, J. N., Kirshner, R. P., and Woosley, S. E. 1989, *Ann. Rev. Astr. Ap.*, **27**, 629.
Binzel, R. P., Farinella, P., Zappala, V., and Cellino, A. 1989, in *Asteroids II*, ed. R. P. Binzel, T. Gehrels, and M. S. Matthews (Tucson: Univ. of Arizona Press), p. 416.
Bowell, E., *et al.* 1989, in *Asteroids II*, ed. R. P. Binzel, T. Gehrels, and M. S. Matthews (Tucson: Univ. of Arizona Press), p. 524.
Bus, S. J., Bowell, E., Harris, A. W., and Hewett, A. V. 1989, *Icarus*, **77**, 223.
Canizares, C. R. 1987, in *Observational Cosmology*, ed. A. Hewitt, G. Burbidge, and L.-Z. Fang (Dordrecht: Reidel), p. 729.
Casares, J., Charles, P. A., and Naylor, T. 1992, *Nature*, **355**, 614.

Filippenko, A. V., Porter, A. C., Sargent, W. L. W., and Schneider, D. P. 1986, *A.J.*, **92**, 1341.
Filippenko, A. V., and Richmond, M. W. 1991, in *Advances in Robotic Telescopes*, ed. M. A. Seeds and J. L. Richard (Mesa: Fairborn Press), p. 89.
Jewitt, D. 1992, in *Robotic Telescopes in the 1990s*, ed. A. V. Filippenko (San Francisco: Astron. Soc. Pacific), p. 183.
Keel, W. C. 1986, in *Automatic Photoelectric Telescopes*, ed. D. S. Hall, R. M. Genet, and B. L. Thurston (Mesa: Fairborn Press), p. 176.
Kirshner, R. P. 1990, in *Supernovae*, ed. A. G. Petschek (New York: Springer-Verlag), p. 59.
Magnusson, P., et al. 1989, in *Asteroids II*, ed. R. P. Binzel, T. Gehrels, and M. S. Matthews (Tucson: Univ. of Arizona Press), p. 66.
McFadden, L., A'Hearn, M. F., Feldman, P., Roettger, E., Edsell, D., and Butterworth, P. 1987, *Astr. Ap.*, **183**, 333.
McLaughlin, D. B. 1960, in *Stars and Stellar Systems*, Vol. 6, ed. J. L. Greenstein (Chicago: Univ. of Chicago Press), p. 585.
Muller, R. A., Newberg, H. J. M., Pennypacker, C. R., Perlmutter, S., Sasseen, T. P., and Smith, C. K. 1992, *Ap. J. (Letters)*, **384**, L9.
Perlmutter, S., Muller, R. A., Newburg, H. J. M., Pennypacker, C. R., Sasseen, T. P., and Smith, C. K. 1992, in *Robotic Telescopes in the 1990s*, ed. A. V. Filippenko (San Francisco: Astron. Soc. Pacific), p. 67.
Peterson, B. M., et al. 1991, *Ap. J.*, **368**, 119.
Porter, A. C. 1989, in *Remote Access Automatic Telescopes*, ed. D. S. Hayes and R. M. Genet (Mesa: Fairborn Press), p. 281.
Rees, M. J. 1984, *Ann. Rev. Astr. Ap.*, **22**, 471.
Richmond, M. W., and Filippenko, A. V. 1991, in *Advances in Robotic Telescopes*, ed. M. A. Seeds and J. L. Richard (Mesa: Fairborn Press), p. 81.
Richmond, M. W., Treffers, R. R., and Filippenko, A. V. 1992, in *Robotic Telescopes in the 1990s*, ed. A. V. Filippenko (San Francisco: Astron. Soc. Pacific), p. 105.
Schild, R. E. 1990, *A.J.*, **100**, 1771.
Schild, R. E., and Smith, R. C. 1991, *A.J.*, **101**, 813.
Treffers, R. R., Richmond, M. W., and Filippenko, A. V. 1992, in *Robotic Telescopes in the 1990s*, ed. A. V. Filippenko (San Francisco: Astron. Soc. Pacific), p. 115.

A DOUBLY ROBOTIC TELESCOPE: THE BERKELEY AUTOMATED SUPERNOVA SEARCH

SAUL PERLMUTTER, RICHARD A. MULLER, HEIDI J. M. NEWBERG, CARLTON R. PENNYPACKER, TIMOTHY P. SASSEEN, and CRAIG K. SMITH
Lawrence Berkeley Laboratory, Space Sciences Laboratory, and Department of Physics, University of California, Berkeley, CA 94720

ABSTRACT   We have designed, built, and are successfully using a completely robotic supernova search, with an automated observatory and automated real-time analysis and scheduling. This system has detected 20 supernovae so far, resulting in early supernova observations, surprising supernova rates, and new evidence against a true "inclination effect" in galaxies.

INTRODUCTION

The Berkeley Automated Supernova Search is a working research tool that is robotic in two different ways. First, the observatory is completely automated, with computer controlled weather station, dome, telescope, and CCD camera. Second, and perhaps more unusual, the image analysis and real-time scheduling of observations are also automated (Perlmutter et al. 1988). This makes it possible to identify and reobserve supernova candidates in real time. The result of all this automation is that for the past three years we have been observing and analyzing between 400 and 600 galaxies every successful clear night, and have in the process detected 20 supernovae.

SEARCH SYSTEM DESIGN

The search uses a 30-inch telescope at Leuschner Observatory, located about 10 miles from the U. C. Berkeley campus. (This was intended as our prototype system, so we were willing to put up with the bright skies and poor seeing — typically 5″.) An IBM XT points the telescope and opens the shutter, an IBM AT reads out the CCD camera, and a microVax analyzes the images, chooses the next observation, and directs the other two computers via ethernet connections. The microVax also polls the various weather-station indicators, and opens or closes the dome slit depending on wind and moisture conditions. If in the course of the night the image analysis routines cannot recognize ten consecutive galaxy images, the dome is closed for an hour on the assumption that there is cloud cover. This observatory automation can allow night after night of successful observations even when there are only a few hours of clear skies in a night. Skilled technicians are only required for maintenance and repairs, and a caretaker who lives nearby fills the liquid nitrogen dewar every other day.

During the course of each clear night the telescope observes galaxy after galaxy, one per minute. The microVax analyzes each image, matching it to an older reference image of the same galaxy and then subtracting the images to find a supernova. While this image is being analyzed on the microVax, the next image is "flat-fielded" on the AT, and the image after that is observed; this parallel operation saves time. If a supernova candidate is discovered on the subtracted image, then the microVax schedules a follow-up observation to be obtained five minutes later. This eliminates cosmic ray false alarms. If a supernova candidate is found at the same sky location on the second image, a third observation is scheduled for one hour later to rule out false alarms due to asteroids.

Finally, if this third image confirms the supernova candidate, then all of the images are shipped back over a dedicated phone connection to the Lawrence Berkeley Laboratory computers for the scientists to examine in the morning. We report the supernova discovery that same day using the IAU telegram system. This series of checks has proven very reliable: we have never falsely reported a supernova.

TABLE 1
Supernovae Detected by the Berkeley Automated Supernova Search

SN	Galaxy	Galaxy Type	Date	SN Type	mag (CCD)	cos(i)
1986I	NGC 4254	Sc	5/17	II	14	0.91
1986N	NGC 1667	Sc	12/11	Ia	15	0.75
1986O	NGC 2227	Scd	12/24	Ia	14	0.73
1987K	NGC 4651	Sc	7/28	II,Ic	15	0.68
1988H	NGC 5878	Sb	3/3	II	15.5	0.42
1988L	NGC 5480	Sc	5/3	Ic	16.5	0.78
1989A	NGC 3687	Sbc	1/19	Ia	15.3	1.00
1989B*	NGC 3627	Sb	1/30	Ia	12	0.47
1989L	NGC 7339	Sbc	6/1	II	16	0.24
1990B	NGC 4568	Sbc	1/20	Ic	16	0.42
1990E	NGC 1035	Sc	2/15	II	16.7	0.36
1990H	NGC 3294	Sc	4/9	II	16	0.52
1990U	NGC 7479	Sc	7/28	Ic	16.0	0.75
1990aa	UGC 540	Sc	9/4	Ic	17.0	0.5
1991A	UGC 6872	Sc	1/2	Ic	17	0.53
1991B	NGC 5426	Sc	1/12	Ia	16	0.56
1991G*	NGC 4088	Sbc	3/11	II	17	0.41
1991M	IC 1151	Sc	3/13	Ia	15	0.28
1991N	NGC 3310	Sbc	3/30	Ibc	15	0.92
1991T*	NGC 4527	Sbc	4/16	Ia	11.5	0.34

RESULTS AND DISCUSSION

Table 1 lists the 20 supernovae detected by this search through June 1991. The asterisks indicate those objects discovered first by other observers, and later automatically found by our search. In response to our announcements, spectra were obtained for all of these supernovae. In cases where a supernova looked particularly interesting — an early discovery or an unusual pre-discovery light curve — we called observatories around the world to organize immediate follow-up observations. For example, we were able to see hour by hour spectral evolution in SN 1990E, which we detected within five days of the explosion.

A crucial aspect of this project is the automated record keeping and archiving. Previous supernova searches often have been unable to measure the rates of supernovae because they did not keep detailed records of the many observations that do *not* yield supernovae. The Berkeley search automatically archives every image it observes on 8 mm magnetic tape, and saves a detailed description of each observation and the results of analysis of each image. These observation and analysis records are sent back to the laboratory computers where they are automatically loaded into a database (currently Oracle database software is used).

One can use such records to determine the total integrated time that a given galaxy has been under surveillance. Since supernovae in a reasonably nearby galaxy can be detected for a few months, this surveillance can be intermittent and still catch a supernova. Therefore the total surveillance time, denoted by "galaxy-years," must be estimated separately for each class of supernova that exhibits a different characteristic light curve. We can then divide the number of discoveries of a given supernova type by the number of galaxy-years surveyed for that type to get the supernova rate for that type. Table 2 from Muller *et al.* (1992) shows this result for the three-year period beginning January 1988. (The supernovae from this period are shown between the two horizontal lines in Table 1.) The surprising result here is the high rate for Type Ic supernovae, particularly in the late-type spiral galaxies (Sbc, Sc, Scd, and Sd). It appears likely that these supernovae were previously missed because they are somewhat less luminous, and our search has a more sensitive detection threshold than most previous searches. In fact, it is possible that there are even more supernovae to be discovered at still fainter magnitudes.

Along with the complete observation record, the excellent dynamic range of a CCD camera gives this supernova search an additional advantage over most previous searches. The supernova-rate literature discusses a bias against discovering supernovae in edge-on galaxies. Van den Bergh (1990) pointed out that the 17% of galaxies with inclination angle greater than 0.88 (almost face-on) yielded 45 of the 95 supernovae in Sc galaxies with multiple supernovae. We find that the 17% of galaxies with the highest inclination angle yield about 15% of our supernovae, consistent with no inclination bias at all. Since most of these previous searches were photographic, the poor dynamic range is a likely explanation for their results.

TABLE 2
Supernova Rates for January, 1988 through January, 1991

	All galaxies		
	Ia	Ib and Ic	II
No. supernovae	3	5	4
Galaxy years*	1447	747	587
Average rate*	$0.21^{+0.12}_{-0.07}$	$0.67^{+0.25}_{-0.17}$	$0.68^{+0.31}_{-0.20}$

	Late spiral galaxies (Sbc to Sd)		
	Ia	Ib and Ic	II
No. supernovae	2	5	3
Galaxy years*	546	319	265
Average rate*	$0.37^{+0.29}_{-0.16}$	$1.57^{+0.59}_{-0.39}$	$1.13^{+0.62}_{-0.38}$

*Galaxy years should be multiplied by a factor of h^{-2}. Rates are in supernovae per 100 years per 10^{10} $L_\odot$ and should be multiplied by a factor of h^2.

FUTURE DEVELOPMENTS

We are now preparing to move the search from this prototype telescope to a remote site with dark skies, good weather, and good seeing. We have purchased a 30-inch telescope from AutoScope that is now being modified in the lab for accurate pointing and tracking. We hope to move it to a mountaintop site in the next year or so. We are also now switching to a new CCD camera at the present site, with a greater number of (smaller) pixels and significantly better quantum efficiency. These improvements will allow us to find out if we truly are just beginning to dip into a population of faint Type Ic supernovae.

ACKNOWLEDGEMENTS

Valuable contributions to this project were made by Ned Hamilton, Deepto Chakrabarty, Shawn Carlson, Frank Crawford, Robert Smits, and Richard Treffers. This work was supported by the Department of Energy under Contract No. DEAC03-76SF000098. The automated telescope was developed, in part, with funding from the National Science Foundation, the California Space Institute, the MacArthur Foundation, NASA, and the Research Corporation.

REFERENCES

Muller, R. A., Newberg, H. J. M., Pennypacker, C. R., Perlmutter, S., Sasseen, T. P., and Smith, C. K. 1992, *Ap. J. (Letters)*, **384**, L9.

Perlmutter, S., Crawford, F. S., Muller, R. A., Pennypacker, C. R., Sasseen, T. P., Smith, C. K., Treffers, R., and Williams, R. 1988, in *Instrumentation for Ground-Based Optical Astronomy*, ed. L. B. Robinson (New York: Springer-Verlag), p. 674.

van den Bergh, S. 1990, *Astr. Ap.*, **231**, L27.

AUTOMATED CCD PHOTOMETRY AT BEHLEN OBSERVATORY

EDWARD G. SCHMIDT
Behlen Observatory, Department of Physics and Astronomy, University of Nebraska, Lincoln, NE 68588

ABSTRACT The telescope and CCD camera system at the Behlen Observatory have been increasingly automated over the past four years. This automation allows us to acquire at least 100 light curve points for variable stars per night. A large survey of relatively faint stars from the General Catalogue of Variable Stars has been undertaken. Some early results are discussed. These include the reclassification of six supposed RR Lyrae stars as binaries, the identification of four new s-Cepheids, the identification of a possible new beat Cepheid, and the discovery of two peculiar stars which may be RR Lyrae stars. Additionally, significant period corrections are needed for 14 stars.

INTRODUCTION

The advantages offered by both CCDs and automation for stellar photometry are especially applicable to the study of variable stars. In addition to the high quantum efficiency, simultaneous sky measurement, and ability to discriminate optimally against the background, the CCD offers the advantage of recording the comparison stars on the same exposure as the variable. This gives high efficiency and good photometric accuracy even under poor sky conditions. On the other hand, the application of computer control and data handling allows full utilization of the capabilities of the CCD. In the context of variable star observations, automation allows the rapid observation of variables over the whole sky at the times dictated by the phases of the stars. It also permits the observation of adequate numbers of standard and extinction stars without seriously reducing the number of variables which can be observed.

INSTRUMENTATION

The CCD camera and telescope at Behlen Observatory have been automated to an increasing degree over the past four years. At the present time, the observing process itself only requires the observer to verify the identification of star fields prior to recording the data. This will soon be automated and the observer will only need to handle unforeseen circumstances such as threatening weather or instrumental failures. The observing program is prepared before each night, taking into account the phases available and those already observed for each variable. Since between 150 and 200 entries are needed in the observing list, this is done interactively with the computer. The reduction and archiving of the data only require intervention by the astronomer to assess the quality of the data and to make scientific decisions.

Currently the CCD photometry system is capable of observing between eight and eleven variable star fields in two colors (generally V and R) per hour. Extinction and standard stars are observed at a rate of about fifteen per hour and represent a small overhead. Thus, we are able to obtain about one hundred light curve points in an average night. We expect some improvement to this efficiency as the automation is completed. In any event, this system is capable of producing many thousand light curve points per year.

THE VARIABLE STAR SURVEY

Variable stars are important to many interesting astrophysical problems (e.g., the cosmic distance scale, stellar evolution, stellar pulsation, galactic structure). Some types of variables relevant to these areas are not numerous (e.g., long period RR Lyrae stars, Schmidt et al. 1990; s-Cepheids, Antonello et al. 1990; beat Cepheids, Balona 1985), and studies are sometimes limited by the available sample. There are also many stars with unique properties that have not been explained satisfactorily (e.g., V473 Lyrae, a classical Cepheid with a long term light curve modulation, Burki et al. 1986; RU Cam, a star which stopped pulsating for a protracted period, Demers and Fernie, 1966; XZ Cet, a long period RR Lyrae star with a peculiar light curve, Teays and Simon 1985). Since these samples are drawn for the most part from the brighter variables, there is considerable potential to improve this situation through an examination of fainter stars which are poorly studied.

An additional motivation for observing faint variable stars can be appreciated through examination of the General Catalogue of Variable Stars (GCVS, Kholopov 1985, 1987). For stars fainter than tenth magnitude at maximum light, 97 percent lack photoelectric photometry, about two-thirds lack the light curve asymmetry parameter, $m - M$, and more than a third lack a period. Thus, our knowledge of these faint stars is poor indeed.

Based on these considerations and the capabilities of the Behlen Observatory automated CCD photometry system, a survey of poorly studied variable stars has been undertaken. It will include northern stars from the GCVS which are classed as pulsating variables but will exclude the long period, semi-regular and irregular variables. Initially we will observe stars between tenth and fifteenth magnitude, but the CCD is capable of reaching much fainter limits and we will extend the program to fainter stars at a later stage. For each star a sparsely sampled light curve will be obtained. This will allow us to determine mean magnitudes on a modern photometric system, to check the periods and coordinates of the stars, and to derive various parameters descriptive of the light curves. It will also allow the identification of interesting stars for further study.

SOME RESULTS

Up to now, light curves have been obtained for 116 stars. The following discussion will consider 93 of these which have been fully analyzed.

For 14 stars the GCVS periods are in error by more than one percent. Of these, four are in error due to the misclassification of binary stars (discussed below), and three are in error due to aliasing with one day. Both effects produced the erroneous GCVS period for one star. The six remaining period errors cannot be easily explained.

To assess the classification of the variables, the appearance of the light curves, the various light curve shape parameters, the relative amplitudes in V and R, and the color-magnitude plots for all the stars were examined. A pulsating star should exhibit a strong correlation between color and magnitude as it goes through its cycle. For this reason, the color-magnitude plots proved especially valuable in distinguishing pulsating stars from binary variables (eclipsing binaries and ellipsoidal variables). From among the 93 variables, there are 26 stars for which changes from the classifications given in the GCVS were indicated. Of these, 13 represent minor improvements to the classifications.

Six stars (DN Aur, V508 Cyg, V830 Cyg, KV Gem, V719 Her, and KN Per) were classified in the GCVS as RR Lyrae stars but turn out to be eclipsing binaries (or possibly ellipsoidal variables). These represent more than seven percent of the 79 RR Lyraes in the sample. If this percentage is valid for the 6000 RR Lyrae stars in the GCVS, it implies that there are 400 to 500 stars in the catalogue which are classified as RR Lyrae stars but are, in fact, binaries. It should be noted that most of these are likely to be classed as Bailey type c RR Lyraes. Since there are only about 400 type c stars in the catalogue, the survey may result in a significant decrease in the number of such objects. On the other hand, the newly identified binaries are all of short period and generally have rounded maxima (otherwise they would not have been mistaken for RR Lyrae stars). Thus, they are potentially interesting stars and many of them will reward further study. There are only about 1200 binary stars in the GCVS with periods less than one day; the reclassification of RR Lyraes as binaries will cause a significant increase in the number of such stars known.

Two stars with periods appropriate to RR Lyrae stars were found to have light curve peculiarities. V442 Her (period = 0.442 days) has a still-stand on rising light which lasts about one-third of the cycle. The light curves in V and R for BK UMa (period = 0.636 days) appear normal for a Bailey type ab RR Lyrae star. However, the amplitudes in the two colors are equal, which cannot be the case for a pulsating star. This might be explained by a suitable blue companion. However, that explanation is not supported by the normal color of the star.

V428 Per has a GCVS period of 0.664 days and is classified as an RR Lyrae star. However, a period of about 4.13 days is required to fit our photometry, and even then there is considerable scatter in the light curve. Given the period, we conclude that this star is a classical Cepheid. The scatter in the light curve suggests it might be a beat Cepheid, and further observations are needed to confirm this possibility.

Finally, among our sample there are four stars (BD Cas, NO Cas, CI Per, and CN Tau) which appear from their periods and light curves to be s-Cepheids. Two of these (NO Cas and CN Tau) were previously classed as RR Lyrae stars due to period errors. Since the s-Cepheids are attracting considerable attention (cf. Antonello et al. 1990), the discovery of new examples is welcome.

With over one hundred stars finished, the survey is just getting underway. Eventually it will encompass in excess of 1800 stars. Data will accumulate much faster in the future due to improvements to the automation now in place or soon to be completed. However, even with this small beginning, nearly fifteen percent of the stars studied here have proven to be of special interest. The survey promises to lay the foundations for further detailed studies in the future.

ACKNOWLEDGEMENTS

The facilities of the Minnich Astronomical Computing Center were used for the work described here. This support from Commander Charles B. Minnich is gratefully acknowledged. The observatory instrumentation was purchased with funds from the National Science Foundation under grant AST–8504072 and the observational program is supported by NSF grant AST–8815806. The author wishes to thank A. Groebner, C. Loomis, C. Potter, K. Wiese, and G. Wolf for making some of the observations reported here.

REFERENCES

Antonello, E., Poretti, E., and Reduzzi, L. 1990, *Astr. Ap.*, **236**, 138.
Balona, L. A. 1985, in *Cepheids: Theory and Observations*, ed. B. F. Madore (Cambridge: Cambridge University Press), p. 17.
Burki, G., Schmidt, E. G., Arellano Ferro, A., Fernie, J. D., Sasselov, D., Simon, N. R., Percy, J. R., and Szabados, L. 1986, *Astr. Ap.*, **168**, 139.
Demers, S., and Fernie, J. D. 1966, *Ap. J.*, **144**, 439.
Kholopov, P. N. 1985, *General Catalogue of Variable Stars, Fourth Edition, Vols. 1 and 2* (Moscow: Nauka Publishing House) (GCVS).
Kholopov, P. N. 1987, *General Catalogue of Variable Stars, Fourth Edition, Vol. 3* (Moscow: Nauka Publishing House) (GCVS).
Schmidt, E. G., Loomis, C. G., Groebner, A. T., and Potter, C. T. 1990, *Ap. J.*, **360**, 604.
Simon, N. R., and Teays, T. J. 1982, *Ap. J.*, **261**, 586.

ARCHITECTURE OF THE SOFTWARE FOR THE INDIANA CCD AUTOMATED TELESCOPE

R. KENT HONEYCUTT and GEORGE W. TURNER
Department of Astronomy, Indiana University, Bloomington, IN 47405

ABSTRACT The structure of the software which controls the CCD automated imaging telescope at Indiana University is described. The order in which particular tasks are performed as well as the nesting and interactions of the various modules determine many of the overall characteristics of the software. Particular attention is paid to the ways in which the system can be monitored and, as necessary, controlled by the astronomer. Software techniques for error handling are also discussed.

INTRODUCTION

Conversion of the 16-inch telescope of the Goethe link Observatory to an automated system capable of unattended CCD stellar photometry began in 1988. The system is now completed and produces photometry on several types of interacting binary stars in order to monitor modulations in mass transfer rate. Some aspects of the project have been treated briefly in other papers (Honeycutt et al. 1989; Honeycutt et al. 1990; Honeycutt 1992). This paper presents some details on the organization of the software for the system. Information on other software components, on the hardware, and on the system performance will appear elsewhere. Descriptions of software for controlling the automated telescopes of the Fairborn Observatory, and those of the AutoScope Corporation, can be found in Genet and Hayes (1989).

The telescope/software system is optimized to obtain differential broadband photometry for stars in the magnitude range 12–17. Scheduling software selects the next object from among about 120 program stars on the basis of the orbital ephemeris of each star, the position in the sky of each star, and the elapsed time since the last observation of each star. The telescope, dome, and filter are positioned and a single unguided CCD exposure is obtained of the desired field. A typical exposure time is 240 seconds. The telescope sets to within about 15"; the program star and the comparison stars are therefore assured of falling within the 8' CCD field. The CCD image is flat-fielded on-line and all stars which exceed a specified threshold above the sky noise are located and centroided. The mean full-width at half-maximum (FWHM) of the centroided stars fixes the size of the numerical diaphragm which is used to extract instrumental magnitudes from the image. The position of each stellar image, its FWHM, and its instrumental magnitude are retained for later analysis, and the original image is normally discarded.

A typical exposure contains between 5 and 200 stars, and the data saved from each image typically amounts to a few thousands bytes. The original 800 × 800 pixel image occupies 1.3 Mbytes. The resulting data compression of nearly 1000× makes it practical to store and analyze the data from the 100 to 200 exposures per clear night that the system produces.

The system receives miscellaneous human interaction and attention at least every few days. Otherwise the observatory is designed to operate totally unattended, and all necessary daily functions have been successfully automated. These functions include filling the liquid nitrogen detector dewar, taking flat fields, monitoring precipitation and Polaris for open/close decisions, plus observing operations. These observing operations include automation of the telescope motion and focus, telescope mirror cover, filter selection, dome rotation and dome shutter.

These observatory functions as well as data acquisition, data reduction, and data archiving have been integrated into a single program which we call UNATTEND. This paper describes how the various modules of UNATTEND have been assembled to arrive at a system that is not only fully automated, but also allows the kind of routine human interaction that we feel is necessary for quality assurance and reliability. Some of the architecture of UNATTEND evolved because of the special capabilities and/or limitations of our particular hardware/software environment (namely, Fortran running under VMS on a MicroVAX II). But many of the design features of UNATTEND are broadly applicable to APTs operating with a variety of detectors, a variety of scientific programs, and different kinds of hardware and software. We concentrate here on those broad features of UNATTEND, and on those aspects of UNATTEND which address reliability. We start our description with the module OBSERVE which is embedded in UNATTEND and performs the actual nighttime observing. Next we describe UNATTEND itself, followed by a discussion of the way in which UNATTEND is supported by a suite of auxiliary programs that run concurrently. Finally, we describe how UNATTEND handles errors and other problems that sometimes occur during the night.

THE "OBSERVE" MODULE

Figure 1 is a diagram of the module OBSERVE, which is used by UNATTEND to acquire and reduce images when conditions allow. Star selection is made from a list of binary stars which must be observed at particular orbital phases, called "ephemeris" stars, and "interval" stars which are to be observed at any orbital phase but at specified average intervals of time. After star selection, OBSERVE performs a number of checks to make sure that the observation can actually begin. These checks include whether the star will remain in the observing window set by the hour angle and altitude limits of the telescope for the duration of the exposure, whether the weather sensors indicate that it is safe to proceed, whether liquid nitrogen or dome flats are needed, and whether dawn has arrived. Also, on each pass through these "exposure checks," the system clock of the computer is set from a WWV receiver and the status of the uninterruptible power supplies is queried. These checks are made inside loop A over the various filters and exposures for each star because some sequences of exposures on a given star (particularly an ephemeris star) can last an hour or more. Loop A is cycled each five minutes or less, ensuring that the checks are performed at a short enough interval to be effective and safe.

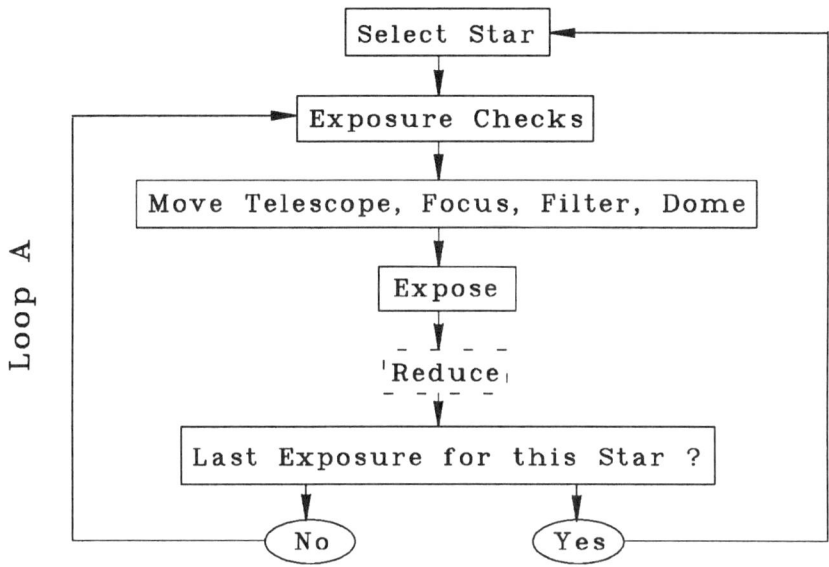

Fig. 1. Organization of the software module OBSERVE, which controls the observatory when the dome slit is open.

OBSERVE then calculates the apparent position of the star using corrections for precession, aberration, refraction, and flexure. The focus is calculated from the telescope temperature using a predetermined relationship. Then the telescope, filter, focus, and dome are positioned. These motions are for the most part multiplexed using intellegent motor controllers, thereby relieving VMS of any critical real-time control tasks. The time to perform them is the same as the most time consuming component, which is nearly always the dome rotation. Having the checks inside loop A takes care of dome tracking for a long series of observations on the same star as well as coordinate changes that might occur because of flexure and refraction.

Next the CCD image is exposed and read out. The reductions consist of making the flat field correction, finding the stars, photometering the stars, and storing the data. These reductions can take place either (1) in UNATTEND directly following the exposure or (2) via a separate program which runs concurrently. The advantage of method (1) is that UNATTEND knows how many stars were found in the previous exposure before proceeding to next exposure. This information can be used to aid in the close-down decision; i.e., when greater than five exposures in a row have less than two stars, UNATTEND closes the observatory. The advantage of method (2) is that the duty cycle is more favorable because reductions take place mostly during the exposure of the next star when the computer is otherwise relatively idle. We now mostly use procedure (1) but we expect to eventually use mostly procedure (2) because more stars can be observed in a night. Furthermore procedure (2) can still provide close-down information based on the number of stars found in each image, albeit with a time lag that must be accounted for.

THE "UNATTEND" PROGRAM

OBSERVE makes up one of the primary modules of UNATTEND, as shown in Figure 2. A companion module called STANDBY controls all observatory functions when the dome slit is closed, while OBSERVE controls all functions when the dome slit is open. The UNATTEND program itself simply cycles continuously between STANDBY and OBSERVE using the modules OPENUP and CLOSEDOWN as paths between those two routines. A hysteresis of 20 minutes is incorporated to avoid rapid opening and closing under marginal sky conditions.

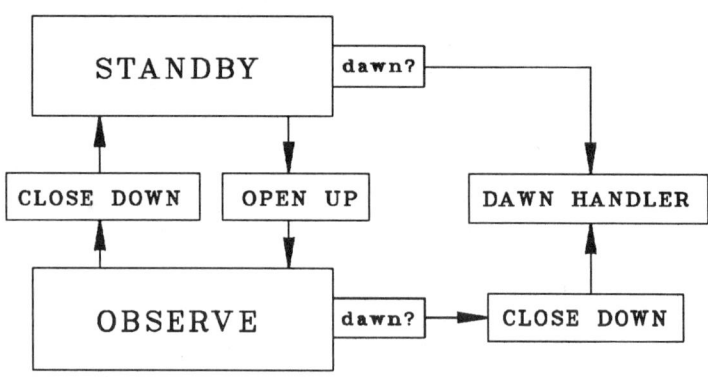

Fig. 2. Organization of the program UNATTEND.

STANDBY performs many of the same checks that were described in OBSERVE. Whereas OBSERVE performed these checks just before each exposure (at an interval that varies between one and six minutes depending on the exposure time), STANDBY makes these checks each five minutes and hibernates the rest of the time. The two checks of greatest importance to STANDBY are liquid nitrogen autofill and the dome flats. If twilight flats are needed a flag is set for OBSERVE to take them and UNATTEND goes to OBSERVE through OPENUP. STANDBY also calls OPENUP when dusk has arrived and the weather and transparency sensors indicate that observing is appropriate. Note that UNATTEND will stay in STANDBY all night if it is cloudy.

CLOSEDOWN first positions the telescope and dome to the home position. In this position the liquid nitrogen probe can be inserted without moving the telescope, and the telescope is pointed at the dome-mounted flat-field screen. The mirror cover mechanism is mounted to the non-moving dome wall. This mirror cover extends itself upward and then down onto the top of the closed-tube telescope when the telescope is in the home position. CLOSEDOWN also "unwinds" the dome cable. We have no dome commutator, so power is supplied to the dome slit motor via a tensioned power cable which is fixed to the top of the moving dome on one end and to a boom on the other. Continuous dome rotation in one direction is not allowed because the power cable would be twisted beyond its limits. To minimize long dome moves during the night, up to ± 2 dome rotations from the home position are allowed. Unwinding the dome at the end of the night repositions the dome to its central rotational position.

Dawn arrival (or more precisely the arrival of a solar altitude limit beyond which the sky is too bright for effective exposures; this is currently set to $-10^\circ.3$) is a special time in UNATTEND. At dawn a dawn-handler routine performs a number of housekeeping chores. These are performed once every 24 hours at dawn regardless of whether any nighttime data has been obtained.

The dawn-handler performs the following tasks. First UNATTEND is exited; it is resubmitted after the housekeeping. This frees up the master log file which UNATTEND was keeping. This master log file is a relatively detailed record of the activities of UNATTEND for the last 24 hours. Each mechanical action and each data reduction step is documented as it proceeds by recording information about calculated, observed, and derived quantities, and about the start and completion of various steps. This master log file is too large to routinely keep more than a few days, or even to regularly inspect, but it has proven invaluable for diagnosing problems that occur infrequently and which may be difficult to reproduce.

The dawn-handler strips 3 smaller files from the UNATTEND master log. This is accomplished using the VMS command SEARCH, operating on key ASCII sequences that have been embedded in the master log file for that purpose. The first such file is called a "stamp" log, and consists of all master log entries that contain time-stamped information. These entries usually denote the beginning and the successful completion of various steps; this stamp file is identical to the EAVESDROP display described in the next section. A portion of a stamp log file is displayed in Figure 3. A "mini" log is also extracted which consists of a series of one line summaries of the completion of each major task. This mini log occupies only 2 or 3 pages for a full night of observing. An example of a portion of a mini log is shown in Figure 4. For each completed exposure, a one line summary lists the star, the filter, the exposure time, the number of stars found in the image, and the mean FWMH of the stellar images in both RA and DEC. If the object is an ephemeris star the phase is also noted. For an interval star, the elapsed days since the last observation is listed in that field. A graphics log is also extracted which contains the Tektronix 4014 ASCII graphics commands which produce a plot of the position of each star in the sky at the time it was observed. This graphics log is quite useful for fine-tuning the star selection criteria for the interval stars, in order to ensure that most stars

```
22:58:52EST Star chosen: Obs.#= 112 SCO V0818      ExpTp=   1
22:58:56EST START  HA MOVE
22:58:56EST START DEC MOVE
22:58:57EST START FILTER MOVE
22:58:57EST Already at requested filter, move not made
22:58:57EST Beg.&target dome rev,tol   62.625 62.472   0.005 ;go left
22:59:09EST End dome rev,error   62.473  -0.001 #read,rej   60504     0
22:59:09EST HA MOVE COMPLETED
22:59:09EST DEC MOVE COMPLETED
22:59:09EST Begin focusing telescope
22:59:16EST Ccd expose      60.00
23:00:37EST Flattening V w/norm    15135.00
23:00:37EST Data dc        4134.14
23:01:15EST Data sky,sig     4287.754     20.223
23:01:15EST Proc sky,sig     4292.757     10.121
23:01:27EST Bias=    4134.14  Sky+bias=    4292.76  Sky sigma =      10.12
23:01:40EST Mean X,Y fwhm,sigmas,diap. dia:  4.88   0.78  6.02  1.41  7.22
23:01:44EST #@ SCO V0818 V 0060S Alt=34.9D Stars= 11 JDdf= 2.903 Sz 5. 6.
```

Fig. 3. A sample portion of a "time-stamp" file which has been extracted from the master log file.

are observed near the meridian except for catching rising or setting objects that would otherwise be missed. A graphics log output is shown in Figure 5.

The observatory MicroVAX is networked to the Astronomy Department's VAX cluster on campus using a point-to-point DECnet connection over a normal voice telephone line. The two telephone circuits involved are dedicated to this use and the phone connection typically remains intact for several days at a time.

```
23:57:49EST #@ Exited STANDBY,    call OPEN_UP: 26-JUN-91,civil
23:57:49EST #@ Exited OPEN_UP, call OBSERVE, episode  1 26-JUN-91,civil
00:00:44EST #@ CYG SW R 0030S Alt=61.0D Stars= 33 Ph=0.99059  Sz 5. 6.
00:02:43EST #@ CYG SW R 0030S Alt=61.5D Stars= 31 Ph=0.99103  Sz 6. 5.
00:04:47EST #@ CYG SW R 0030S Alt=61.9D Stars= 29 Ph=0.99133  Sz 5. 6.
............
03:00:58EST #@ AQR FO V 0060S Alt=36.4D Stars=   4 JDdf= 8.872 Sz 7. 6.
03:03:42EST #@ CYG SS V 0060S Alt=77.0D Stars=  47 JDdf= 2.042 Sz 5. 5.
03:09:45EST #@ LAC DI V 0240S Alt=65.8D Stars=124 JDdf= 9.092 Sz 5. 4.
03:15:34EST #@ BOO TT V 0240S Alt=26.7D Stars=   5 JDdf= 2.020 Sz 5. 6.
03:22:14EST #@ LAC DK V 0240S Alt=65.7D Stars=152 JDdf= 9.093 Sz 5. 6.
............
03:31:53EST #@ LN2 Regular scheduled fill
03:31:54EST #@ Move tel to LN2 fill position
03:32:18EST #@ Liquid nitrogen autofill sequence
03:51:03EST #@ SGE U  V 0012S Alt=56.6D Stars=  5 Ph=0.98791  Sz 6. 7.
03:52:34EST #@ SGE U  V 0012S Alt=56.3D Stars=  6 Ph=0.98830  Sz 6. 6.
03:54:05EST #@ SGE U  V 0012S Alt=56.0D Stars=  6 Ph=0.98861  Sz 5. 6.
............
04:16:57EST #@ Leave OBSERVE at dawn
04:16:58EST #@ Exited OBSERVE,    call CLOSE_DOWN: 27-JUN-91,civil
04:20:19EST #@ Exited CLOSE_DOWN, call EXIT@DAWN: 27-JUN-91,civil
```

Fig. 4. Sample portions of a "mini" log file which has been extracted from the master log file.

As part of the dawn housekeeping, the mini log file is sent by e-mail to a mailing list on campus. The graphics log is copied to a campus computer as well as all data files (i.e., "photometry" files in the terminology of the next section) from the previous night. A mail message is sent to the distribution list upon completion of these copying tasks.

The information which is distributed to the "observer" is limited to the minimum amount that is judged necessary to monitor the system performance and the data quality. In the event of problems the astronomer can examine progressively more detailed logs of the night's work using the stamp log and the master log as well as the environmental log (which is described in the next section). After completion of these housekeeping chores, the UNATTEND program is automatically resubmitted.

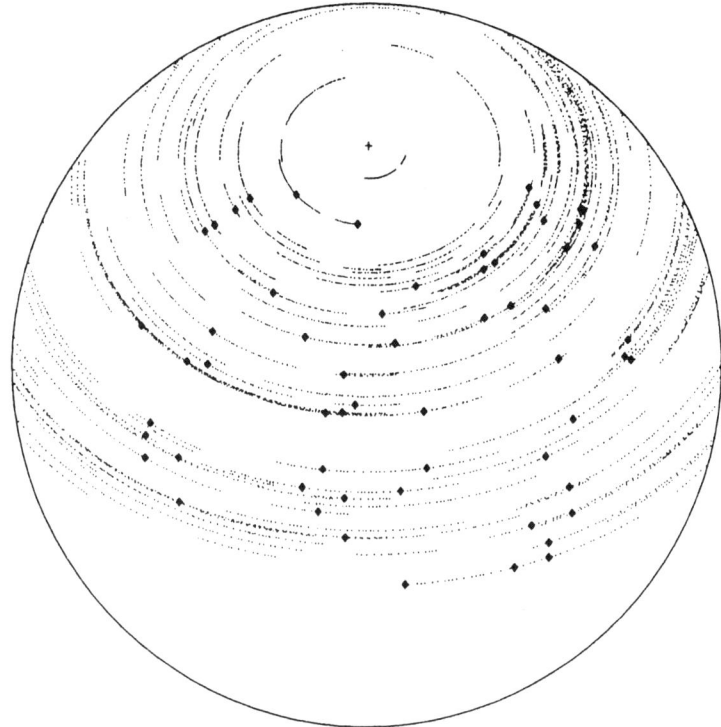

Fig. 5. An example of the graphics output from UNATTEND for a night's observations, showing the loction on the sky where each program star was observed and the diurnal track of each program star.

AUXILIARY ROUTINES

UNATTEND communicates with related programs and with the astronomer through a section of shared memory containing a set of status and control variables as shown in Figure 6. This shared memory is "installed". It is therefore permanent and the parameters are retained as various programs are stopped and started or the computer is re-booted. A number of other programs run concurrently with UNATTEND, most accessing these shared variables. Some programs such as ENVIRON run continuously; others such EAVESDROP and INTERCEDE are run only as needed.

The EAVESDROP program continuously updates a scrolling display of the last 22 lines of the time-stamped component of the master log as UNATTEND runs. This EAVESDROP display is therefore equivalent to the most recent 22 lines of the stamp log, as shown in Figure 3. Using EAVESDROP the operation of UNATTEND can be monitored in real time from the observatory, from the office, or from home. In STANDBY, UNATTEND issues regular reports to the stamp log (and to EAVESDROP) to confirm its continued normal operation.

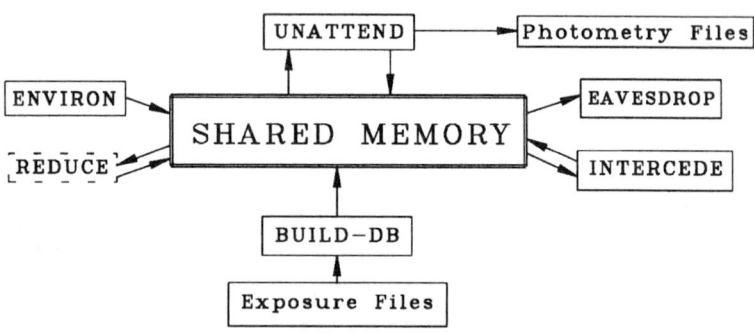

Fig. 6. The program UNATTEND communicates with a number of concurrent programs using shared memory.

The ENVIRON program samples each five minutes a number of analog inputs containing environmental information. Examples include telescope temperature, rain and snow monitor, the level of the liquid nitrogen storage dewer, and the Polaris transparency monitor. A list of these signals is shown in Figure 9. The instantaneous values are recorded to an environmental log each 5 minutes and a 20-minute running mean is written to shared memory for use by UNATTEND. (UNATTEND also accesses these analog signals directly when it needs instantaneous values for periodic checks in OBSERVE and in STANDBY.)

The INTERCEDE program serves to display the status of the variables in shared memory as well as change these variables and flags. Figure 7 is an example of the screen display from INTERCEDE. The environmental override tag allows the user to instruct UNATTEND to ignore the Polaris monitor and open up regardless, or close down regardless. Analogous override functions are provided for the type of data recording and the style of data reduction. Time records are kept in INTERCEDE for the last occurrence of particular operations such flat field exposures, liquid nitrogen fills, etc. Permission flags allow the astronomer to turn off various modules in UNATTEND without stopping and resubmitting the UNATTEND program. This feature is most often used for testing and for diagnostics. Some flags can be used to bypass noncritical modules in UNATTEND if desired. The dome slit can be turned off if one wishes to

```
OVERRIDE TAGS: Environ. =    0 Record =    0 Reduce =    0
ENCODER CONTROL: ha_enc = 1 dec_enc = 1
ha,dec    encoder zeropts =    250253   249801  6-JUL-91 22:09:29
ha,dec motor step zeropts =         0         0 00-JAN-00 00:00:00
  |--------1 DOME FLATS-------------|------2 TWILIGHT FLATS--------|CHOICE
  |Permit      Last          Counts |Permit      Last        Counts |
U|   0     00-JAN-00 00:00:00    0. |  0    27-JUN-91 00:00:00 14681. |   2
B|   0     00-JAN-00 00:00:00    0. |  0    27-JUN-91 00:00:00 19825. |   2
V|   1     23-JUL-91 01:37:53 13733. |  0    00-JAN-00 00:00:00    0. |   1
R|   1     23-JUL-91 01:31:39 15400. |  0    00-JAN-00 00:00:00    0. |   1
I|   1     23-JUL-91 20:27:46 20456. |  0    00-JAN-00 00:00:00    0. |   1
PERMISSION FLAGS:
     hadec = 1       filter   = 1   dome rot. = 1   dome_slit = 1
     focus = 1    mirror cov = 1    autofill  = 1    graphics = 1
  img dsp. = 0    flatfield  = 1   photometry = 1
       exp = 1        u exp  = 0        b exp = 0  sunset open = 0
TIME RECORDS:
Dewar flood 23-JUN-91 Last Warm 20-JUN-91 22:08:17 Cold 23-JUN-91 14:23:44
Dewar  pump 23-JUN-91 Fill       24-JUL-91 10:54:31 Fill Int.      13.00hrs
Database    14-JUL-91 10:42:29 Num. submits = 1   Dome flat Int. 24.00hrs
```

Fig. 7. The program INTERCEDE allows real-time control of the program UNATTEND through these variables and flags.

exercise the UNATTEND software when it is raining. The image display can be turned off if no one is at the observatory.

The information which UNATTEND uses to schedule the observations is contained in an ASCII "exposure" file for each star. This exposure file contains items such as stellar coordinates and offsets, ephemeris information, exposure times, and filters. These exposure files are assembled into a real-time database by the program BUILD-DB. This database resides in shared memory and can be accessed by UNATTEND to choose the next star. To change the parameters for a star or to add or delete a star from the program, the astronomer edits the appropriate exposure file, then runs BUILD-DB.

The photometry from each exposure is appended to a "photometry" file which is maintained for each program star. Figure 8 shows part of a photometry file. The header contains information appropriate to the whole exposure, and subsequent lines list information for each star which has been located and photometered in the image. All new photometry file entries are copied to the campus computer each dawn.

There are a number of additional programs which support UNATTEND but which are usually run in an attended mode. Examples include FOCUS, ATTEND, STATUS and APLOT. The FOCUS routine takes a series of offset stellar exposures with a range of focus settings on the same CCD frame, determines the best focus, and records that best focus and the telescope temperature to help build a telescope focus calibration curve. A program called ATTEND brings up a graphic image of the sky similar to that in Figure 5. Stars from any of several databases can be displayed, such as the Bright Star Catalog or the CCD-APT program star list. The observer can set the telescope to the object of interest by moving the cursor to the desired object. Exposures can be taken, encoder zero points updated, and flexure data obtained in this manner. A particularly interesting feature is the ability to overlay in real-time the expected star field from the HST Guide Star Catalog onto the CCD image as it is acquired at the telescope. ATTEND also incorporates image display and analysis functions. These functions can operate on the images which UNATTEND is acquiring and processing, in order to monitor

```
AQL V0794   APT#0137      Type NL           Filt=V     Exp=0240
20:17:33.9  -03:39:50    2000.0             MidHJD= 2448460.78116
20:17:33.9  -03:39:50    Set                MidUT   91-07-23  06:52:44
HA=  +0.82H  Alt= 45.6D  X=  1.40           Solar Alt= -28.9D
Lunar Alt= -22.3D   Phase= 0.40             Mag Limit 15.1  M
JD-JDlast= 4.894    Bias= 4144.8            Sky= 132.4 Sksig=   8.5 Skymag=-1.00
findX= 3.00   minpix=   6      diapX= 1.20  r1sky=15.0  r2sky=24.0
Xfwhm= 5.11   Yfwhm= 6.11      rstar= 3.7   #objs= 41   mzero=24.00
star  xcen  ycen    mag    merr   xfwhm yfwhm fndp  s-s_max  sky-bias  skymag
  1   772.9 721.9 12.937  0.004   4.92  5.15  170   1367.    134.7     24.63
  2   408.6 679.4 13.061  0.004   4.87  5.53  160   1131.    131.2     24.66
  3   435.0 270.4 13.580  0.006   4.72  5.46  129    741.    132.8     24.64
  4   463.1 503.1 13.920  0.007   4.72  5.22  109    537.    131.5     24.65
  5   730.6  90.3 14.008  0.008   4.89  5.48  101    504.    134.3     24.63
  6    77.0 183.7 14.104  0.008   4.88  5.71   96    443.    134.2     24.63
  7   385.0 423.1 14.157  0.009   4.70  5.31   93    413.    131.3     24.65
  8   393.6  54.7 14.194  0.009   4.76  5.87   93    408.    136.1     24.62
  9   670.8 216.6 14.371  0.011   4.22  5.14   75    364.    130.9     24.66
 10   109.5 388.8 14.489  0.012   4.66  5.40   77    308.    131.0     24.66
 11   307.7 760.1 14.587  0.011   4.85  5.55   79    295.    134.9     24.63
 12   723.1 605.1 14.505  0.014   5.17  5.48   67    336.    127.9     24.68
 13    51.6  80.1 14.722  0.013   5.07  6.06   85    230.    134.0     24.63
 14    99.4  82.8 14.897  0.014   5.20  5.85   66    192.    134.2     24.63
```

Fig. 8. A portion of the photometry data from a single image showing the header information and the entry for each star.

the automated image acquisition and reduction process. STATUS queries many different subsystems of the observatory for information on the state of various signals and mechanisms. Figure 9 shows the menu for STATUS and, as an example, the screen display that results from the selection of analog inputs from the menu. All these analog signals are recorded into an environmental log each 5 minutes by ENVIRON. This log can be accessed most conveniently by a program APLOT which plots the time history of any of these analog signals.

```
Enter #: 1=Dig. Inputs     2=Analog Inputs     3=Dates & Times
         4=Dome Pos.       5=Filter            6=Telescope
2
CHAN    SIGNAL               GAIN    VOLTS     VALUE   UNITS
 0      Focus                  8     0.010      0.01   Volts
 1      Telescope Temp.        2     2.971     24.12   Deg.C.
 2      CCD Temp.              4     1.274   -127.40   Deg.C.
 3      Relative Humidity      1     4.794     49.91   Percent
 4      Dewar Jacket Temp.     2     2.950     22.02   Deg.C.
 5      Precipitation          1     9.361      0.21   Percent
 6      Polaris                1     0.007      0.07   Percent
 7      Dome Temp.             2     2.963     23.29   Deg.C.
10      Dome Rack Temp.        2     3.007     27.71   Deg.C.
11      Cloud Cover            1     4.822     94.32   Percent
13      Precip. Plate Temp.    2     3.121     39.12   Deg.C.
14      Comp. Room Temp.       2     2.959     22.85   Deg.C.
15      Storage Dewar          1     7.300     89.43   Percent
```

Fig. 9. Screen display from the program STATUS showing the result of the selection of menu item 2.

ERROR HANDLING

Considerable attention has been given to the detection and handling of the inevitable errors and problems that occur in the astronomer's absence. This section discusses some of the techniques that we have used to address various error conditions.

The most serious errors involve telescope safety, and occur during the motion of various components such as the telescope, dome rotation, dome slit, liquid nitrogen autofill mechanism, mirror cover, and filters. We have several layers of error detection assigned to these motions. Nearly all of these motions are equipped with mechanical limit switches. Some of these hardware limits interupt the power to the appropriate mechanism, resulting in a hard stop representing the last line of defense. Other limits are expected to be reached and are used not only as stops but for confirmation that the motion has been completed.

Telescope limits for hour angle, declination, and altitude are provided. Hour angle and declination limits alone leave some low altitude areas of the sky unprotected, while altitude alone does not protect against under-the-pole targets that stress cables beyond their limits. Two layers of software checks on the telescope motions are provided. The lower level software which positions hour angle and declination makes sure that the requested position is within the mechanical limits, and the star selection software makes sure that the requested position is within the limits of the lower-level positioning software.

The action of each mechanical motion is confirmed before UNATTEND proceeds to the next function. A software timer surrounds each motion (or set of motions) with a time limit which somewhat exceeds the maximum expected time for the motion. If the timer expires before confirmation is obtained, an error is declared. In general two kinds of motion failure must be anticipated. Either the motion fails to start, or the motion started but failed to stop. For example, after requesting a new dome position, a check is made to make sure the dome is moving. Another check is made to ensure that the dome is stopped at the requested azimuth.

Some pairs of mechanical motions are interlocked in hardware and/or software to avoid undesirable conditions. The telescope will not move unless UNATTEND can confirm that both the liquid nitrogen probe and the mirror cover are retracted. Also, neither the mirror cover nor the liquid nitrogen probe will extend unless the telescope is in the home position.

When an error is detected the action taken by UNATTEND can take several forms. For non-fatal errors the situation is noted prominently in the mini log, an error count is incremented, and UNATTEND proceeds. If this error count accumulates beyond an unacceptable limit for a particular type of error, then either a safe-mode is declared, or UNATTEND is stopped and resubmitted. The number of resubmissions is noted by a variable in shared memory; if the number of resubmissions exceeds a preset limit, then safe-mode is declared.

Safe-mode performs the following functions. The telescope and dome are immobilized at their current position. The dome slit is closed and all the digital outputs are cleared. UNATTEND is stopped and not resubmitted. Finally, e-mail is sent to the astronomers advising of the safe-mode condition and the status of the systems.

Two uninterruptable power supplies (UPS) are provided. One powers the dome slit, the other powers the computer and related digital circuitry. UNATTEND queries both UPSs during each periodic check in both STANDBY and OBSERVE. If either UPS is found have switched to battery, then UNATTEND suspends its normal function and begins monitoring the UPS. If power has not returned within 10 minutes (the UPS will keep the system running for about 20 minutes) then safe-mode is declared. Since the battery of the dome-slit UPS will not have been discharged during the 10 minutes of backup power, we are assured of being able to close the dome slit in the event of a power interruption which is longer than the UPS can handle. If the line power returns within 10 minutes, UNATTEND resumes its suspended functions.

This amount of attention to error prevention, error detection, and error handling may appear somewhat excessive. Many of these error conditions have in fact never been encountered. However, many of the software functions such as the timers and the safe-mode routine are modules that are used repeatedly at different places in the software. The hardware error functions have admittedly been tedious to implement but are necessary to provide the protection that we think is appropriate for our considerable investment in this project. On several occasions, long forgotten safeguards have acted as planned to avoid downtime and/or damage.

ACKNOWLEDGEMENTS

This paper is concerned with the organization of the high level software, but this architecture would be useless without the numerous detailed hardware and software efforts of Brice Adams, Bill Kopp, Jay White, and Dave Vesper, whose contributions we gratefully acknowledge. This work is supported by the National Science Foundation.

REFERENCES

Genet, R. M., and Hayes, D. S. 1989, *Robotic Observatories* (Mesa: AutoScope Corp.).
Honeycutt, R. K., Adams, B., Grabhorn, R., Turner, G., White, J., and Vesper, D. 1989, in *Remote Access Automatic Telescopes*, ed. D. S. Hayes and R. M. Genet (Mesa: Fairborn Press), p. 105.
Honeycutt, R. K., Vesper, D. N., White, J. C., Turner, G. W., and Adams, B. R. 1990, in *CCDs in Astronomy II. New Methods and Applications of CCD Technology*, ed. A. G. D. Phillip, D. S. Hayes, and S. J. Adelman (Schenectady: L. Davis Press), p. 177.
Honeycutt, R. K. 1992, in *Proceedings of 1990 Pacific Rim Colloquium on New Frontiers in Binary Star Research*, ed. I.-S. Nha and K.-C. Leung (ASP Series), in press.

SOFTWARE FOR CCD PHOTOMETRY WITH THE PC

Ronald H. Kaitchuck
Unified Software Systems
P.O. Box 21294
Columbus, Ohio

ABSTRACT   With hardware prices declining, CCD photometry is becoming affordable to a larger number of people. This paper briefly describes a new inexpensive instrument , some of the characteristics of its CCD chip, and, at greater length, a software package designed for image and photometric reduction. Such developments may speed the evolution from photomultiplier tubes to CCDs as the detector of choice on APTs.

INTRODUCTION

Traditionally astronomical CCD photometry has been done with moderate to large size telescopes, using expensive cameras and reduced on large computers or workstations with software such as IRAF or DAOPHOT in addition to user-supplied programs. The hardware costs have limited CCD photometry to the well equipped professional observatories. But, now this situation is rapidly changing. Inexpensive CCD cameras are appearing on the market. This opens up the possibility of CCD photometry on smaller telescopes used by many universities, colleges and amateur astronomers.

There are many advantages of CCD photometry over conventional single-channel photomultiplier-based photometry. These include a broad spectral response (e.g., B, V, R, I) with a single detector, the ability to do precision differential photometry even through thin clouds (when the variable and comparison stars fall on a single CCD frame), and a multiplex advantage. The latter refers to the fact that many stars can be recorded on a single frame. This means that the variable and one or more comparison stars and sky background are recorded simultaneously, greatly reducing the actual time spent measuring the star field. This is especially useful for studies of star clusters. The disadvantages of CCD photometry include the loss of UV sensitivity and increased data reduction.

AN INEXPENSIVE SCIENTIFIC PACKAGE

Recently, Optec Incorporated and SpectraSource Incorporated have jointly produced a complete, yet inexpensive, CCD instrument for astronomical photometry. Optec refers to this camera as the SSP/Lynxx. The instrument consists of a Lynxx CCD camera mounted to an optical front-end assembly. Some of the features of this instrument are listed below.

> TC211 CCD (192x165 pixels)
> 12-bit A/D converter
> Electromechanical shutter
> Field of view eyepiece
> 5 - position filter slide (with optional stepper motor control)
> 2x focal reducer (optional)
> Control cards for the PC

As with most of the inexpensive CCD cameras now appearing on the market, the Lynxx camera uses the TC211 CCD chip manufactured by Texas Instruments-Japan. This chip was intended for medical applications, not astronomy. As a comparison the following table compares this chip with a Tektronix CCD chip currently in use at Kitt Peak National Observatory.

Characteristic	TC211	KPNO TEK
Type	Virtual Phase	3 phase
Pixel Size	13.75 x 16 μm	27 x 27 μm
Number of Pixels	192 x 165	512 x 512
Physical Size	2640 μm (0.10")	15,400 μm (0.5")
Read Noise	70-200 e^- (20 e^- DCS)	7.8 e^-
Full Well	150,000 e^-	470,000 e^-

While the TC211 does not match the capabilities of the current generation of the best research grade CCDs, the comparison is not nearly as unfavorable as it first appears. The most striking differences between these two chips is the small size of the TC211 and the read noise. The small size has both advantages and disadvantages. An advantage is that it reduces the problem of data storage. But a disadvantage is that it results in a rather small field of view. This is compounded by the fact that photometric work will require that the stellar seeing disk be spread over at least 2 - 3 pixels (i.e., a scale of at least 1 arcsecond per pixel). This is necessary due to the nonuniform pixel structure and dead space caused by the antiblooming gate. This means that, for the size telescopes that will probably be used with this instrument, the field of view will be 3 arcmin or less. As a result, for locations away from the galactic plane the field may not contain the variable and comparison stars simultaneously. But keep in mind that most variable stars are in or near the galactic plane. The read noise can be reduced by using a camera that uses double correlated sampling (DCS), which is available from SpectraSource. In many situations, the dominant noise source is the sky background or dark current (see below). When considering the TC211 it should be kept in mind that just a few years ago professional astronomers did very good science with RCA chips with read noise of about 70 e^- and fields of view of this same size.

In laboratory tests of the Lynxx camera we found a linear response function almost to the saturation limit of the A/D converter (4095 ADU). These tests were conducted by taking a series of exposures of a constant light source. Figures 1 and 2 show the mean counts per pixel (minus the bias) for a sample of 100 pixels near the center of the chip. Exposures ranged from under 0.5 to 20 seconds.

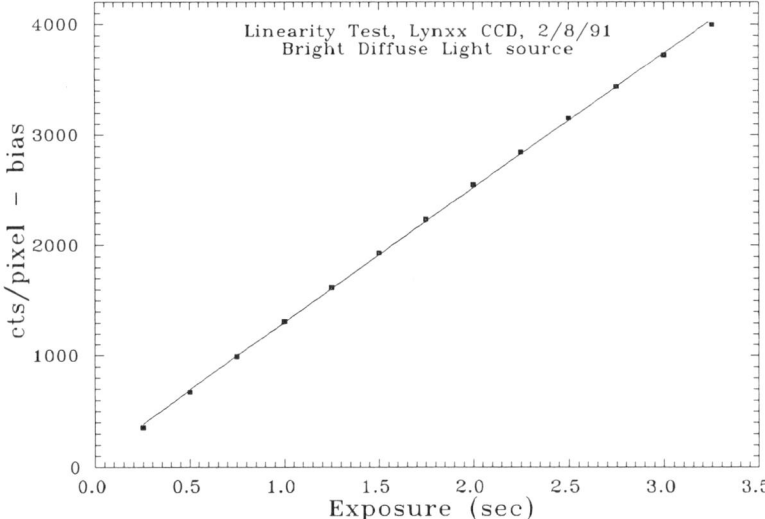

Fig. 1. Linearity test of the TC211 CCD chip with a bright constant light source.

For the two cameras tested, the read noise was found to be between 100 - 200 e^- per pixel. As stated earlier, this can be dramatically reduced with cameras that use DCS.

Tests of the dark current have also shown a linear response and with about 5 - 6 ADU/sec/pixel with the camera in an ambient temperature of about 20 C and the TEC cooling the chip to about -25 C. The results of this test are shown in Figure 3. Except for cold nights, the dark current will limit exposures to a few minutes duration. Because the dark current is appreciable, and the TEC is unregulated, good thermal data must be obtained while observing. This can be done either by obtaining frequent dark frames or using data from the covered columns of the chip to scale a good thermal frame obtained on some cloudy night.

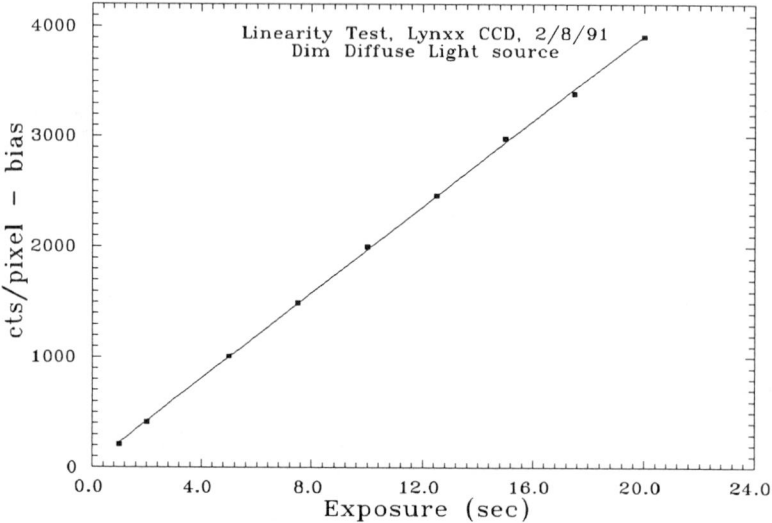

Fig. 2. Linearity test of the TC211 CCD chip with a dim constant light source.

Despite these limitations the TC211 is able to do some serious quantitative work, especially now that an inexpensive camera system is available.

COMPUTER HARDWARE

Since the SSP/Lynxx camera is controlled with an IBM-PC or compatible computer, this machine also becomes the natural choice for the image and photometric reductions. In addition to being inexpensive, the PC with a math co-processor is an efficient single-user platform for image reduction. In terms of speed, a 25 MHz 80386 machine is as fast or faster than a Micro Vax II computer in floating point operations. The image from the TC211 chip can be adequately displayed with a VGA display. The image storage requirements are less than 60 KB per image which means that extremely large hard disks are not required. Of course, there are disadvantages. For example, the 640 KB memory limit of DOS must be dealt with and the system is not multitasking.

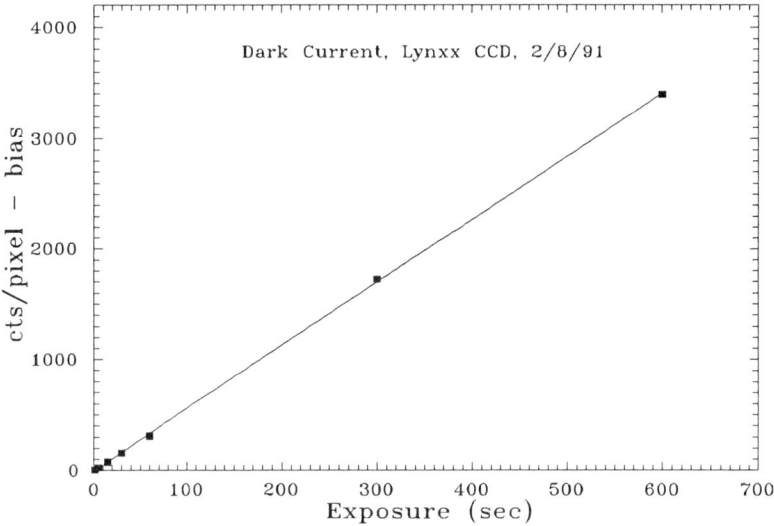

Fig. 3. Dark current test of the TC211 CCD chip.

SOFTWARE DESIGN

Because of the memory limitations of the PC, and the large size of any image reduction package, it is necessary to break the software into modules. A natural division is:

> Camera Control
> Image Reduction
> Photometric Reduction

For each of these modules it is possible to construct a "wish list" of desired features. While no two people would construct exactly the same list, many of the following features would be common to most lists.

A) Camera Control

Complete control of camera and filter positions to minimize manual operation.

Large dynamic range in display to best evaluate image quality.

Complete mouse controlled operation to eliminate fumbling with a keyboard in the dark.

Image files should have data headers to store information for later reduction.

Ability to make multiple exposures to reduce demands on user input.

Focus mode, where exposures are made and displayed rapidly.

Ability to coadd images to increase the S/N before storing.

Ability to display plots of the data along any row or column for scientific evaluation and to test for image saturation.

Ability to display stellar profiles in order to evaluate seeing and focus.

Option of performing a rough bias and flat field correction to evaluate image quality.

B) Image Reduction

Interactive, command driven structure, with keywords for added flexibility.

Prompts for missing command line information.

On-line help to jog the user's memory.

Scripts, so the user can create "commands" and automate the reduction.

Ability to spawn user programs.

At least three image buffers to allow image arithmetic (e.g. add image in buffer A to the image in buffer B and store result in buffer C).

Image buffers should be maintained on disk to free up memory for program and to allow for larger CCD chips in the future. There should be disk caching for speed.

Ability to view the image, any plots and the command line simultaneously.

Ability to rescale the image and view it in a gray-scale or false color mode.

Ability to add, subtract, multiply or divide an image by a constant or another image.

Commands to re-size an image.

Plots of the data along any row or column.

Extraction of instrumental, sky-subtracted, magnitudes via aperture photometry.

C) Photometric Reduction

Complete reduction of differential and "all-sky" photometry, including:

> The computation of instrumental magnitudes, airmass, and heliocentric Julian dates.
>
> Differential magnitudes, either instrumental or standardized, with or without extinction corrections.
>
> Both first-order and second-order extinction determination.
>
> Determination of the transformation coefficients to the standard system, and the conversion of instrumental magnitudes to the standard system.

THE SOFTWARE PRODUCT

The features described here are those of an image reduction package now available from Optec and SpectraSource Incorporated. This package was developed by professional astronomers at Unified Software Systems (Arne Henden, Ronald Kaitchuck, and Ryland Truax). In future releases we expect to add:

> Pull-down menus
> Super VGA support
> Crowded field photometry extraction
> Multiple file formats

A FITS conversion program can be ordered from Unified Software Systems for those who wish to use IRAF. Also, a version of this package is available (from USS) for users of cameras made by Photometrics Incorporated (e.g. the Star-1).

SUMMARY

There is now an inexpensive CCD instrument and software package available, producing a *complete* package to go from photons to reduced magnitudes and colors. This opens up the possibility of CCD photometry to many more individuals. For professional astronomers this package can be a supplemental research tool, or the primary one for those at smaller institutions. For the amateur astronomer precision differential photometry is now possible even with variable sky quality. Finally, a whole new range of student observing projects are now possible, many of which were just too difficult or time consuming to do before (e.g. determining the HR diagram for open star clusters).

It seems fairly certain that in the near future the CCD will become the detector of choice in APT systems. Such an APT-CCD system is already in operation in Indiana (see paper by R. Honeycutt, this symposium). While the hardware/software package discussed in this paper was not designed for APT operation, it does provide the basic tools for many more individuals to begin experimentation in this area.

AN AUTOMATED TELESCOPE SYSTEM FOR UNDERGRADUATE TEACHING

D. V. DELEO and R. L. MUTEL
Department of Physics and Astronomy, University of Iowa, Iowa City, Iowa 52242

ABSTRACT We describe the design and construction of a fully automated telescope for use by undergraduate astronomy students. The instrument's value as a teaching tool is discussed by a comparison with our past year's pilot undergraduate CCD imaging laboratory. A complete description of all components, with sources, is included.

INTRODUCTION

The University of Iowa Physics and Astronomy Department offers two introductory astronomy courses: a one-semester survey course which is directed toward the non-science major, and a more quantitative two-semester sequence intended for science majors. During the past academic year, we have been running a pilot program in the laboratory section of the two-semester course using a CCD imaging system. We have had considerable success with this system but, as previously discussed by Deleo (1991), there is much to be gained by further improvements.

The present system consists of two inexpensive CCD cameras mounted on existing telescopes in the department's rooftop observatory. With three different telescope configurations we are able to obtain 3.4', 8.1', and 33.8' fields of view. This allows for a wide variety of laboratory exercises ranging from solar and planetary studies to close contact binary system and open cluster studies. We also use the position encoder and guiding system SGT-MAX developed by Software Bisque of Golden, Colorado. This system consists of 8-bit shaft encoders and a database of 275,000 objects (including the complete SAO, NGC, and Messier catalogs, the solar system, plus several comets). Once the student has completed the alignment procedure, the software keeps track of the telescope position. The software displays a view of the sky (using superb graphics) with a crosshair indicating the current position of the telescope. The program may then be used to guide the student to any object in its database by watching the display as the telescope is moved.

Due to often severe weather conditions in Iowa City, it was necessary to locate the controlling PCs (one for the CCD cameras and one for SGT-MAX) two floors below the rooftop observatory. This requires that two students must work together at all times. One remains in the observatory to point the telescope while the second controls the CCD operations from downstairs. An intercom system was installed to maintain communication between the students. This setup, although a major improvement from the old-fashioned in-class laboratory exercises, caused some inconvenience for students. We also found that locating objects by manual slewing and searching occupied the majority of each laboratory's observing time, and caused confusion and frustration.

Based on these experiences, it was apparent that the ideal system for undergraduate labs would be a fully automated telescope (AT). During this past summer (1991) we began the design and construction of two robotic telescopes. The first (presently undergoing mount tests) is a prototype system which uses 8-inch Schmidt-Cassegrain (SC) optics, while the second (under construction) uses 14-inch SC optics with an attached fast ($f/3.5$) 4-inch refractor.

SITE LOCATION

The present location of the telescopes is on the roof of Van Allen Hall in downtown Iowa City. Light pollution and seeing conditions are often extremely poor at this location, but the logistics of getting students to a good site such as our Riverside Observatory have forced us to make do with an urban rooftop location.

Although the automated instruments will be tested on the rooftop, we plan to move both instruments to our Riverside Observatory (15 km from campus) by early 1992. We plan to run both the mount and CCD camera control program on a single multi-tasking PC which can be controlled with a high-speed (9600 baud) modem link to a control room on campus.

Precautionary measures are being taken to ensure that the instruments are not damaged by user negligence. Additional safety features, such as a weather monitoring system, will be added when the move to Riverside is made. Routine inspection and maintenance checks will be performed.

IMAGING SYSTEMS

Telescopes
It was decided early in the development stage that we should design and construct a small AT as a prototype of a larger system. For the prototype's optics we used one of the department's 8-inch Celestron SC telescopes. The compactness and light weight of the 8-inch SC optics made the overall size and torque requirements of the mount design straightforward and economical. The mount system for this prototype was designed and built in the Department during May–July 1991.

Our second instrument will utilize 14-inch Celestron SC optics which we also currently have in-house. We will piggy-back mount a 4-inch refractor for wide-field observations. We hope to complete this system in early October 1991.

CCD Cameras
The Department presently owns two CCD cameras which we have used extensively over the past year. They are both Lynxx PC cameras manufactured by SpectraSource Instruments of Westlake Village, California. This camera

has a 192 × 165 pixel imaging chip with a pixel size of 16 × 13.75 μm and a 12-bit dynamic range. A thermoelectric cooler cools the chip to between $-30°$ and $-40°$ C, which allows for a peak quantum efficiency greater than 50% at 650 nm and a dark current of less than 150 electrons s^{-1} pixel^{-1} with a charge-transfer efficiency of 99.998%. The readout noise is less than 80 electrons. The controlling software allows for false color viewing, full and subframe focussing modes, zooming, and axis flips. Flat field, bias, and thermal corrections may also be applied to the image. Intensity histograms may be viewed and contrast stretching performed on the image. Frame arithmetic, log rescaling, and photometric measurements may also be easily performed within the controlling CCD software.

One camera is equipped with an automated ten-position filter rack. This unit is fitted with $UBVRI$ filters as well as Hα, Hβ, [O III], and 520 nm narrowband filters plus a clear and offset filter. The second camera has two two-position filter holders which must be switched manually. The filters in the second unit are the Johnson B, V, and R plus a clear filter. This unit will be replaced with an automated filter rack as funding becomes available. Both of these units were supplied by Optec, Inc. of Lowell, Michigan.

For the 14-inch AT we have ordered a SpectraSource HPC-1 CCD system. This camera has a 1000 × 1018 pixel imaging chip with a pixel size of 12 × 12 μm and an 8/16-bit selectable dynamic range. A thermoelectric cooler cools the chip to $-45°$ C, providing a peak quantum efficiency of greater than 55% at 650 nm and a dark current of less than 5 electrons s^{-1} pixel^{-1} with a charge transfer efficiency of 99.998%. The readout noise is less than 10 electrons. The controlling software is similar to that of the Lynxx PC with some added features. There is a selectable 10/25/50 electrons ADU^{-1} gain and anti-bloom electronics. To this camera we will add an automated filter wheel (being built in-house) containing $UBVRI$ and clear filters. As additional funding becomes available, narrow-band filters will be added.

Field of View Specifications
The Lynxx-PC cameras have fields of view (FOV) of 33.8' and 8.1' with the 4-inch ($f/3.5$) and 8-inch ($f/11$ with 0.6× telecompressor) systems, respectively. The HPC-1 14-inch ($f/11$) system will have a 35' FOV with a 0.3× telecompressor (which is being designed and built by SpectraSource Instruments) or an 11' FOV without telecompression.

Software
In order to provide a user-friendly telescope control interface, we have decided to use a single windowing, menu-driven graphical user interface (GUI) for telescope and camera control. Since both the telescope and camera interface hardware is AT-bus specific, we were constrained to a PC hardware architecture. Since we require true multitasking, DOS-based windowing software (such as MS-DOS Windows or DESQview) was not adequate. Therefore, we chose a PC-based UNIX operating system with DOS emulation and Motif-GUI. Since the SpectraSource cameras graphics software is written in C, with video calls done via the BIOS, it should be relatively easy to rewrite the source code to be Unix compatible. This will allow both CCD cameras on our 14-inch/4-inch instrument to be run in separate windows simultaneously on a PC running under a Unix operating system. The mount control software is also written in

Unix compatible C and will run in a separate window. The motor controller card can slew and track from a preloaded sequence of commands so that real-time program interrupt servicing is not a serious problem.

In order to make best use of the CCD cameras' dynamic range, we required a Unix operating system which could support a 1024 × 728 SVGA screen with 256 colors. We are using ESIX (v. 4.0) with the Motif GUI Toolkit and VP/ix from Interactive Systems for DOS applications. This allows us to run our motion control software, which is a DOS application (discussed in a later section), on the same PC as our camera(s).

TELESCOPE MOUNT

Mount Design

Two critical concerns in designing the mounts were structural integrity and minimization of slippage and backlash in the drivetrain. By using 3/8-inch and 1/2-inch aluminum plates we were able to design a rigid fork and wedge assembly without encountering a serious weight constraint. Figure 1 shows the engineer's drawing of our 8-inch mount including optics, CCD camera, reduction components, and shaft encoders; the wedge adjustment assembly is not shown.

The first speed reduction component on both equatorial axes following the microstepping motor is a 12:1 timing belt system. This system is manufactured by Winfred M. Berg, Inc. of East Rockaway, NY. This timing belt system was chosen for its ability to provide a reasonable reduction in a very compact system with a high power transmission and low backlash, when compared to standard V-belt systems. The long life expectancy of the fiberglass-reinforced belts should allow for minimal maintenance requirements for this component.

Following the timing belt system is a 12:1 friction wheel assembly. A friction wheel speed reduction is virtually free of backlash and slippage. Placing it after the timing belt reduces any backlash or slippage from the belt by a factor of 1/12. The large friction wheels are made of T6061-T6 aluminum for ease of machining and to keep weight to a minimum. To ensure rigidity of the main friction wheel, which is 12 inches in diameter, three locator bearings were placed around the wheel running in a groove machined around the wheel's circumference. A cam assembly holds the small stainless steel friction wheel, one inch in diameter, so that the pressure at the interface may be easily adjusted. Additional speed reduction is provided by microstepping motor drivers as described below.

The microstepping motor drivers (by Oregon Micro Systems, available from CyberResearch) provide for a clockwise (CW) and counter-clockwise (CCW) limit switch which power down the appropriate motor and home sense switch which resets the index counts. We are using microswitches on the equatorial axes to avoid damage to the instrument should the tracking be inadvertently left on and to indicate when our home, or stowed, position is reached. All machined components are anodized to protect the mount from weathering.

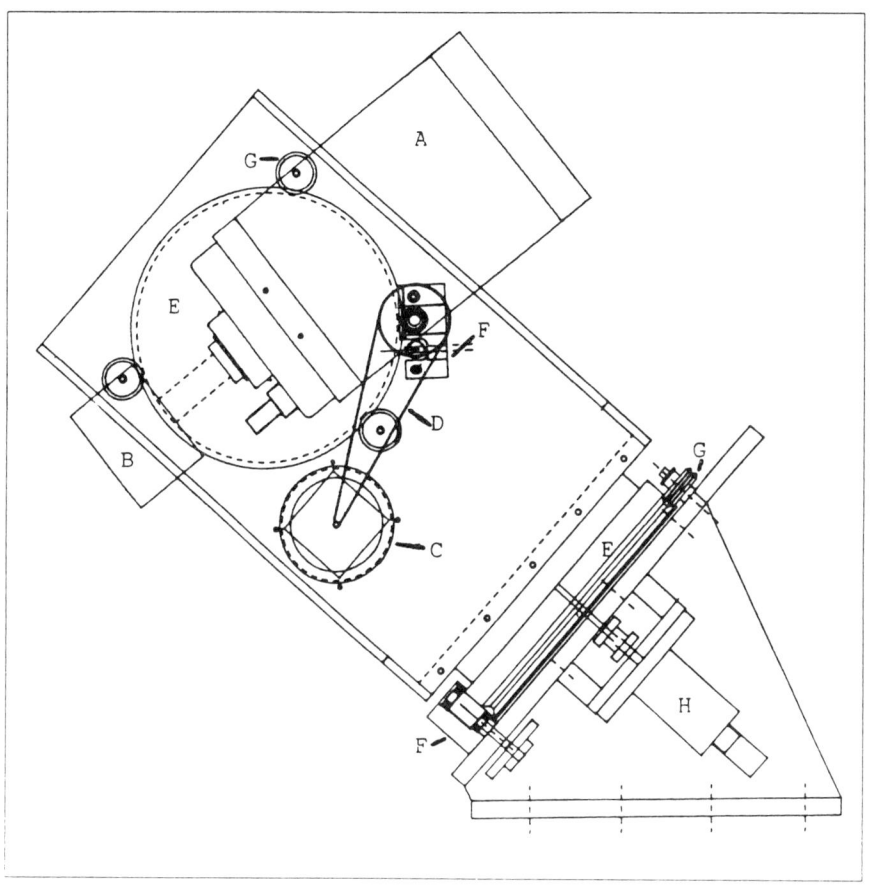

Fig. 1. Engineering drawing of 8-inch automated telescope. A- Celestron C8 SCT, B- SpectraSource Lynxx PC, C- Vexta stepper motor, D- 12:1 timing belt system, E- 12-inch friction wheel, F- 1-inch friction wheel and cam assembly, G- locator bearing, H- absolute position encoder. The wedge adjustment assembly is not shown in this diagram.

Motion Control Components
Motion control components are now readily available for PC applications. To simplify our purchasing we went to a distributor (CyberResearch) who carries various manufacturer's components from which a complete PC-based motion control system may be designed.

There are 3 motorized axes on the 8-inch instrument: hour angle, declination, and focussing. On the 14-inch/4-inch system we have 5 control axes: hour angle, declination, 2 focussing adjustments, and a filter wheel. A 4-axis intelligent stepping controller (CyberResearch ESH-344) was chosen for the first case and a 6-axis intelligent stepping controller (ESH-386) for the second. These controllers plug into any AT-bus expansion slot. In addition to the standard 8 control pins per axis, there are 16 user I/O pins available on these boards. The user may choose from among three velocity profiles: linear, cosine, and parabolic. The cards come complete with sample control programs in MS QuickBASIC, MS QuickPASCAL, and MS C. As discussed in greater detail in the following section, we have modified the C code extensively to meet our requirements.

For telescope focussing and our filter wheel, 40 volt, 2 amp, bi-polar chopper drivers (Intelligent Motion Systems) were selected. These allow for either full-step or half-step operation in either CW or CCW directions. The equatorial axes are driven by multi-resolution microstep motor drivers (Oregon Micro Systems). They allow for selectable microsteps of 10, 25, 50, or 125 per full step and operate in either CW or CCW directions. And, as discussed in the previous section, connections for CW and CCW limit switches and a home switch are provided.

On the hour angle axes we use NEMA-34 2-phase stepping motors manufactured by Vexta. The motors have a maximum torque rating of 250 oz-in. Each full step is 1.8° and thus with a maximum total reduction of 18,000:1 each microstep is 0.36″. For the declination and focus axes we use NEMA-23 2-phase stepping motors. The maximum rated torque of these motors is 50 oz-in. In half-step mode these motors are driven at 0.9° per half-step.

The 8-inch mount is equipped with 16-bit optical shaft encoders (BEI) with absolute positional accuracy of 50″ and PC interface. Since they are absolute encoders, once aligned they are accurate even after a power outage. Our 14-inch/4-inch instrument makes use of the encoder and sky display package of SGT-MAX discussed in the introduction.

Mount Control Software

Since the CCD software has been written in C, we decided to write all of our instrument software in C to maintain consistency. As we mentioned in the previous section, the intelligent stepping control cards come with a sample motion control program written in MicroSoft C. This code provides all of the routines needed to interactively move any instrument in any of the available axes.

The mount control program first asks the user whether he wishes to observe at the solar, lunar, or sidereal rate, followed by object selection either from a catalog or by manual entry of coordinates. The AT then immediately slews to the object and begins tracking. The program automatically precesses to current epoch and applies refraction and mount misalignment corrections from a precalculated pointing array. For lunar tracking routine, the program reads a data file which contains power series coefficients for position at the observing date from the *Astronomical Almanac 1991*. The solar tracking routine uses the twenty term Chebyshev series from the *Almanac for Computers 1991*.

We are now developing software to allow a sequence of observations with predetermined observing times and synchronized camera control. This program allows the user to enter a schedule of objects to be observed in one session. The objects are sorted by priority and hour angle. Priority is given to ensure that an object within a limited observing window is not missed. Sorting the objects by hour angle avoids wasting time by slewing back and forth on the celestial sphere. The telescope tracks each object on the observation list until the user pushes the appropriate function key which sends the command to slew to and track the next object in the list.

Telescope focussing is handled by pushing a function key to activate the focussing motor and then the "<" or ">" keys for moving the focus. When focussing is complete the "Q" key is pushed to deactivate the focus motor and return to normal operation. The focussing motor drives the focus knob through a 1:1 o-ring pulley system. When the end of the travelling range for the focus is reached, the o-ring slips. This avoids damaging the focus mechanism.

For our large format CCD camera's filter wheel assembly, a routine is accessed from a function key which allows the user to select a filter. A microswitch is located at the clear filter position which acts as the home position. The software keeps track of how far away from home the wheel has turned and thus knows what filter is currently in the optical path. When a request to switch filters is entered, the program determines which direction would require the smallest number of steps and then rotates to that position.

When the user has completed their night's activity, a single function key sends the telescope mount and filter wheel to its home position and exits the program. The user then goes to the telescope mount and moves the cover into place. When additional funds become available an automated cover will be added to the system and will be automatically moved into place upon startup and shutdown.

COMPUTER SYSTEM

As we discussed earlier, we have decided to run our instruments in a true multitasking environment so that all systems on each AT are controlled by a single PC. We chose a Unix operating system (ESIX) with the Motif GUI Toolkit and VP/ix, but several other PC-based Unix systems are also available. The computers are Hewlett Packard Vectra 20 MHz 80386 PCs with math coprocessors, but nearly any 386 PC with a math coprocessor would suffice. The VGA card and monitor supplied with the PC should support 1024 × 728 pixel graphics with 256 colors for adequate resolution of multiple windows and sufficient dynamic range of images.

SUPPLIERS LIST

CCD Camera
SpectraSource Instruments, 31324 Via Colinas, Suite 114, Westlake Village, CA 91362; (818) 707-2655.

Filter System
OPTEC, Inc., 199 Smith St., Lowell, MI 49331; (616) 897-9351.

Locator Bearings
Bishop Wisecarver Corp., 2104 Martin Way, Pittsburg, CA 94565; (510) 439-8272.

Motion Control Components
CyberResearch, Inc., 25 Business Park Drive, Branford, CT 06405; (203) 483-8815.

SGT-MAX Sky Display and Encoders
Jim's Mobile Inc., 1960 County Road 23, Evergreen, CO 80439; (303) 277-0304.

Timing Belt System
Winfred M. Berg, Inc., 509 Ocean Ave., E. Rockaway, NY 11518; (516) 596-1700.

Miscellaneous Software and Components
Contact the authors for additional information, drawings, motion control software, etc.

ACKNOWLEDGEMENTS

We would like to thank the undergraduate students who have made major contributions to our project: Peter Sauerbrei, Brian Rachford, and Terry Lo, who have developed all of our motion control software, and Robert Winsor, who machined, assembled, and aided in designing our mounts. Also, we are grateful to Everett Williams for the designing of these mounts. Finally, our thanks go to Adam Durst and William Rico, two advanced high-school students from New York who joined our team in a special summer program run by the Department of Science Education, and aided us in the automation of the CCD cameras, multiple object imaging runs, and converting our PC operating environment to a Unix operating system.

REFERENCES

Almanac for Computers 1991, U.S. Naval Observatory, Washington: U.S. Government Printing Office.
The Astronomical Almanac for the Year 1991, U.S. Naval Observatory, Washington: U.S. Government Printing Office.
Deleo, D. V. 1991, presented paper at the 21st Mid-America Regional Astrophysics Conference, University of Missouri, Kansas City.
Fundamental Astronomy, ed. H. Karttunen, P. Kroger, H. Oja, M. Poutanen, and K. J. Donner (New York: Springer-Verlag).
Smart, W. M. 1977, *Textbook on Spherical Astronomy*, 6th ed., revised by R. M. Green (Cambridge: Cambridge Univ. Press).

PROGRESS REPORT ON THE BERKELEY AUTOMATIC IMAGING TELESCOPE

MICHAEL W. RICHMOND, RICHARD R. TREFFERS, and ALEXEI V. FILIPPENKO
Department of Astronomy, University of California, Berkeley CA 94720

ABSTRACT The Berkeley Automatic Imaging Telescope (BAIT) team has made good progress during the past few months. Because our telescope mount and mirror set are still under construction, we have almost completely automated a backup telescope, and we are using it to test our software. The autoguider is working well, and we have taken our first test images, albeit with human operators present.

WHAT IS BAIT?

In 1989, one of us (AVF) started a project to build a telescope very similar to automatic photoelectric telescopes (APTs; Genet 1986; Genet and Hayes 1989), but equipped with better optics and a CCD camera instead of a photometer. With such an instrument, we will not only measure the brightness of isolated stars, but ones located in crowded fields (open and globular clusters). Stars considerably fainter than those observed with APTs will be accessible. Moreover, we will be able to study the morphology and brightness of extended objects such as comets. The goal is to broaden the very successful mode of automatic observations to new classes of objects.

Although specific details of our instrument may be found elsewhere (Richmond and Filippenko 1991), its main features are as follows:

(1) 0.8-m Ritchey-Chrétien optics placed in an AutoScope mount.
(2) A site at Lick Observatory, on Mt. Hamilton, near San Jose, California (altitude 1280 m).
(3) A 516×516 pixel Photometrics CCD, scale $\sim 0.6''$ per pixel.
(4) An autoguider, allowing long exposures.
(5) Fully automatic operation.

In many ways, the telescope will be a general-purpose instrument. Users may specify one of up to twenty different filters, and exposure times ranging from less than one second to more than an hour. We intend to commit the telescope to long-term studies of faint variable or ephemeral sources, such as novae and supernovae, quasars and active galactic nuclei, cataclysmic variables, asteroids, and other objects. Since a large number of objects can be observed on a clear night with an automatic telescope, we expect to provide data to a number of researchers, especially within the University of California.

STATUS AS OF JULY 1, 1991

Although several major components of our system are not ready at the present time, we have managed to find alternates so that we can test our control-system software. For example, we have encountered difficulties with our plan to renovate one of the domes at Lick Observatory; construction has not even started, and might not finish for 2–3 years. However, we have found a temporary location at Leuschner Observatory, about 20 km east of the U. C. Berkeley campus. While the sky brightness and seeing are worse than at Lick, Leuschner is much closer to campus and therefore more convenient to use while commuting to fix problems and work on the equipment. Of course, once our system is operational, it will require only occasional visits, and the more remote location will no longer be an inconvenience.

A minor setback to our schedule is the telescope mount itself. While our mount was constructed over six months ago, recent tests of an identical mount in Italy have shown that the secondary focus assembly will need to be modified slightly. Work on the electronics which control the mount is ongoing, but close to completion. We hope to have the full system delivered before the end of the year. Meanwhile, we have supplemented the control system of an existing 0.5-m Cassegrain telescope at Leuschner Observatory so that it can be driven to any part of the sky by a computer, using software which will eventually control our final telescope. Computers also control dome rotation, and they can open and close the dome slit. We have not yet added a weather station to the system, but soon we will do so.

We are still waiting for our mirror set, which is taking a very long time to complete. It is apparently much harder than originally thought to fabricate a lightweight mirror, similar to the ones in current APTs, having optics of imaging quality. We have specified that the mirrors must concentrate at least 80% of the light from a point source into a spot only $\sim 1''$ in diameter, which is about a factor of three smaller than that produced by the best mirrors in conventional APTs. On the bright side, the secondary mirror was finished a few weeks ago; thus, final polishing of the primary may at last begin. (Our optical shop, Star Instruments in Flagstaff, Arizona, decided to match the primary to the secondary's shape, rather than vice versa.) Fortunately, the 0.5-m telescope we are currently using is sufficiently large to be an adequate substitute for testing our instruments; in fact, operations should become considerably easier in the future, since the 0.8-m telescope will gather over twice as much light.

An annoying problem occurred in mid-June, when our main science instrument, a Photometrics PM512 CCD, did not seem to respond to low light levels (a most unfortunate trait in astronomical instruments). After returning the camera to the manufacturer, we learned that the chip itself, not the timing electronics, was at fault and needed to be replaced. We shall have to wait several months for the new chip, but we expect to be using the camera again by September of this year. Luckily, we have access to another CCD camera; despite being based on a somewhat noisier chip, the Thomson 7883, it will permit us to continue operations throughout the summer.

One of the brighter spots in our efforts has been the success of our autoguider. Although there are already several automatic telescopes which produce images, such as the Berkeley Automated Supernova Search Team (BASST) telescope (Smith et al. 1989; described at this symposium by Saul Perlmutter) and the Indiana CCD/APT (described here by Kent Honeycutt), they are limited to short exposures, lasting at most a few minutes, due to the inevitable tracking errors. Most of the scientific goals we hope to achieve involve the study of objects fainter than can be seen with such short exposures. Therefore, we have designed and built an instrument, the autoguider (see Appendix), which can acquire guide stars by itself and use them to keep the telescope centered precisely on its target. Tests have already shown that the autoguider can successfully find and guide on stars as faint as $m \approx 10$ mag. We plan to extend its ability to guide on stars down to $m \approx 12$ mag at our present site and with our current optics. Our overall goal is to reach $m \approx 15$ mag after we have moved to Lick Observatory and installed the new mirror. It should then be possible to find at least one guide star candidate for nearly any position in the sky (Richmond 1990).

Despite the above (and other) obstacles, we have managed to produce all the necessary instruments for testing the most crucial, yet least understood, component of our automatic telescope: the software that controls it. As various parts have come on-line, we have written simple programs for using them. At the present time, we can open the dome slit, point the telescope, take a picture with the CCD camera, and display the image — all from a remote computer terminal. Moreover, we have already written a set of programs, PC-Vista (Treffers and Richmond 1989), that allow us to reduce and analyze images in a way which can be automated more easily than some other types of reduction packages. What remains to be done is to combine these separate tasks, under the control of a single computer program, so that they are all executed in the correct order and at the appropriate times. It is this synthesis which we foresee to be the most difficult, yet most innovative and important, part of BAIT. As such, it deserves its own discussion.

CONTROL SOFTWARE

While it is not a simple task to control the motions of a telescope, or to command a photometer to take a reading and store the results away for future reference, computers have been successfully trained to do so for over a decade now (Honeycutt, Kephart, and Hendon 1978). However, figuring out *which* objects from a large target list ought to be observed on some particular night, and *when* they ought to be observed, is quite a different matter. Most of the current systems, for example, pick the stars they will observe on some given night well before it actually gets dark (see, for example, Chapter 9 of Genet and Hayes 1989). It is impossible for such a system to "change its mind" in the middle of the night. If, for example, it measures some variable star to have a much higher brightness than normal, indicating a dramatic flare, it does not go back and observe the star more intensively, through other filters, or make additional measurements later that night. In fact, it may be months before the data are sent back to the observer and the outburst is noticed.

In some cases, it is vital that the control system immediately recognize such deviations from the norm and take appropriate actions. However, we know of only one automatic telescope that is capable of making such judgments: the computer used by the BASST analyzes images as soon as they are taken, and if it finds a star near some galaxy which is not present in the reference image of that galaxy, it automatically tells the telescope to go back several times during the night and take additional pictures (Perlmutter *et al.* 1988; Smith *et al.* 1989). We plan to create a system with the same "intelligence," adding one more facility. Our control system will allow astronomers to submit requests to the telescope at *all* times, even in the middle of the night while it is busy observing. Therefore, if we learn of a new target that must be observed at once, we need not drive out to the observatory to change the current program, or even log in remotely to halt the control program and modify its target list. Instead, we may simply submit an observing request — at a higher priority than usual — in exactly the same way that we normally do, by electronically mailing the request file to the observatory computer. The request file could even be sent by another automatic telescope, without any human intervention. As soon as the request arrives at the site, the controlling program will incorporate that new target into the current target list and schedule it at an appropriate time, perhaps immediately.

We feel that this flexibility will become *very* important as the interaction between different telescopes becomes more common. For example, suppose the BASST telescope discovers a good supernova candidate in the middle of the night. Ideally, it could notify BAIT automatically, making it possible to obtain images through a wide variety of filters on the night of discovery, prior to spectroscopic confirmation of the object (which typically cannot be done until at least the following night). Similarly, imagine a network of three telescopes at widely separated sites, all dedicated (at least in part) to the study of some object which requires nightly measurements. Under current systems of telescope control, in order to guarantee that bad weather at one site does not prevent the crucial observation, all three telescopes would have to be ordered to monitor it; yet this would be not be the wisest use of time for oversubscribed telescopes if the weather were clear at all three, or even at only two, observatories. With a flexible scheduling system, by contrast, only one telescope (the easternmost, for example) might be responsible for the object, but if it had to close because of weather, *its computer* could call up one of the others and request an observation. In this way, all three telescopes could have different lists of targets, with no overlap, and yet still give users three times the chance of getting images of a few important objects.

Let us now describe the system we will use to request observations at BAIT. In essence, an astronomer first creates a *request file*, which contains some basic information about the observation he wishes to make. He sends the file to one of BAIT's computers by ordinary electronic mail (e-mail). BAIT will then check the request for possible errors (e.g., epoch of the coordinates omitted), and send an e-mail message back to the astronomer, telling him whether his request has been accepted or rejected. If the request contains no errors, it is immediately added to the list of possible targets for the appropriate night, which may include the current night. Whenever the object is observed successfully, BAIT will send an e-mail message giving the details of the observation (start time, airmass, exposure time, etc.) to the astronomer; it will also tell him exactly where he may find the image file (in FITS format) containing the data, so that he may retrieve it himself via FTP or some other sort of file-transfer program. Thus, the *observer* will be responsible for keeping track of his own data and archiving it properly. Moreover, the images will remain available only for a very short time after they have been made — perhaps four days, at most. We will keep a tape backup of all BAIT observations, but it will be inaccessible to the outside user, and we do not plan on reading the tape more often than roughly once per month. We hope that by forcing observers to keep up with the data flow, we may delegate much of the responsibility and record-keeping to them; in return, the observers may discover problems with our system as soon as they occur, and by bringing them to our attention immediately, prevent several weeks of data from being ruined.

As mentioned by several speakers at this symposium, astronomers have many different scientific goals in mind when observing celestial objects, and these goals can dictate the manner in which they observe. Someone studying long-term variability in red supergiant stars, for example, might wish to make one measurement per week for each of the stars in his program, while another, trying to measure the period of an RR Lyrae star, would need to have nine or ten observations per night, but only for a few weeks. In some cases, such as occultations by solar-system bodies, the observations *must* be made at a specific time on a specific night (or not at all). Clearly, it is not an easy task to schedule time on a telescope which may be working on a number of such different projects. Let us describe how we are planning to give scientists a great deal of flexibility when they request BAIT to look at their targets.

The request file that users send to our computer looks very much like a FITS header; see Figure 1.

```
OBSERVAT= 'Leuschner'
INSTRUME= 'B.A.I.T.'
OBSERVER= 'Michael Richmond'
MAILADDR= 'richmond@bkyast.berkeley.edu'
SENDMAIL= T
RA       = ' 12:31:36.9'
DEC      = '+02:56:28.0'
EPOCH    = 1950.0
EXPTIME  = 30.0
FILTERS  = 'R'
DAYSTART= '01/12/1991'
DAYEND   = '31/12/1991'
NUM-OBS  = 10
OBJECT   = 'NGC 4527 and SN 1991T '
END
```

Fig. 1. An example of a request file. Ten observations of NGC 4527 (containing SN 1991T) are to be made during the month of December, 1991.

It contains lines which consists of a *keyword*, followed in most cases by an equals sign ('=') and a *value*. Typically, a request file will have about eight or ten lines, which specify the target's right ascension and declination, an exposure time, and so forth. There are a number of different keywords astronomers may use to specify how BAIT ought to schedule their observations. The DAYSTART and DAYEND are common to all schemes; they contain the UT dates when observations of the target may start and end. (Note the convention, dictated by the Kitt Peak FITS header: the date is given as DD/MM/YYYY, not the usual MM/DD/YYYY.) If the two are identical, then the target is valid on only a single night. The NUM-OBS keyword declares on how many separate nights the telescope should observe the target. Thus, if a header contains

```
DAYSTART= '01/12/1991'
DAYEND   = '31/12/1991'
NUM-OBS  = 5
```

the telescope will try to make five observations of that object, equally spaced, between December 1 and December 31, 1991 UT, inclusive. In general, the observations are scheduled as close as possible to the meridian passage of the target, but the MAXAIRMA keyword may be specified to tell the computer how much (or how little) leeway it may use in deciding exactly when to make the observation.

In order to begin the observation at a specific time, the user may use the UT-START keyword:

```
DAYSTART= '01/12/1991'
DAYEND   = '01/12/1991'
UT-START= '10:30:00'
```

In this example, the observation will take place at 10:30 UT on December 1, 1991, or (if the weather is bad at that time) not at all. Alternatively, if some window in local sidereal time is desired, the LSTSTART and LSTEND keywords should be used as follows:

```
DAYSTART= '01/12/1991'
DAYEND   = '01/12/1991'
LSTSTART= '12:32:23'
LSTEND   = '13:32:23'
```

The usual mode of observation will be to move to an object only once on any night, although several exposures, through different filters for example, might then be made. However, by including the NUMPERNI keyword,

```
DAYSTART= '01/12/1991'
DAYEND   = '31/12/1991'
NUMPERNI= 10
NUM-OBS  = 5
```

a number of observations, at equally-spaced intervals throughout the window of accessibility during the night, may be requested. If used together with the NUMPERNI keyword, the value associated with NUM-OBS denotes the number of *nights* of multiple observations. Hence, a single request file with the information above would cause BAIT to produce *fifty* images! Obviously, we will need to install some kind of accounting scheme to prevent a few users from monopolizing the instrument.

As other managers of automatic telescopes have found (and as discussed by Robert Dukes at this symposium), not only must there be a system of priorities, so that some observations are given precedence over others, but the priorities may occasionally have to be "tweaked" in order to yield the best overall mix of observations. We will allow users to set the priority of their own observations, via a PRIORITY keyword, within some reasonable limits. Although we will try to devise a good scheme from the start, we will probably discover a satisfactory system for dealing with priorities only after several months of telescope use by many astronomers.

There are a few additional subtleties in our proposed request file format, but the above description should give the reader an idea of its flexibility. Unfortunately, such flexibility requires quite a lot of complicated programming, but to a large extent the users of BAIT will not be aware of it. By applying a small amount of thought to their observing requests, they will be able to get just the sort of data their research requires. In the future, we hope that astronomers will be able to create a single request file (or something similar) and mail it off to a number of different telescopes at once, yet receive data from the *single* observation that was made at *one* site.

RESULTS

Thus far we have had only a few weeks to test BAIT, but the first results are encouraging. Our primary concern has been the autoguider; without it, we would be restricted to the study of relatively bright ($m \lesssim 14$ mag) objects, and most of the targets of primary interest to us (supernovae and AGNs) would be invisible. Fortunately, early tests have shown that the autoguider is a success. Compare the images in Figures 2 and 3: the first shows a 300 second exposure of NGC 4527 and SN 1991T without any guiding, the second shows a similar exposure after turning on the autoguider. (SN 1991T is the brightest star in the field, close to the visible edge of NGC 4527. Note that north is *left* and east is *up* in these images. See Filippenko *et al.* 1992.) Clearly, our hardware appears to be capable of doing its job; we hope that our software, too, will prove equal to the challenge.

ACKNOWLEDGEMENTS

None of the work described in this paper would have been possible without the National Science Foundation's Presidential Young Investigator Award to AVF (NSF grant AST-8957063). Our research is also supported by NSF Cooperative Agreement AST-8809616 (Center for Particle Astrophysics, U. C. Berkeley), NSF grant AST-9115174, CalSpace grant CS-96-91, and the University of California at Berkeley. AutoScope Corporation, IBM, Sun Microsystems, and Photometrics Ltd. provided generous donations. We have had helpful conversations with numerous people, especially Russ Genet and Jack Borde. MWR thanks Joseph Shields for many insightful discussions, without which he would have given up long ago.

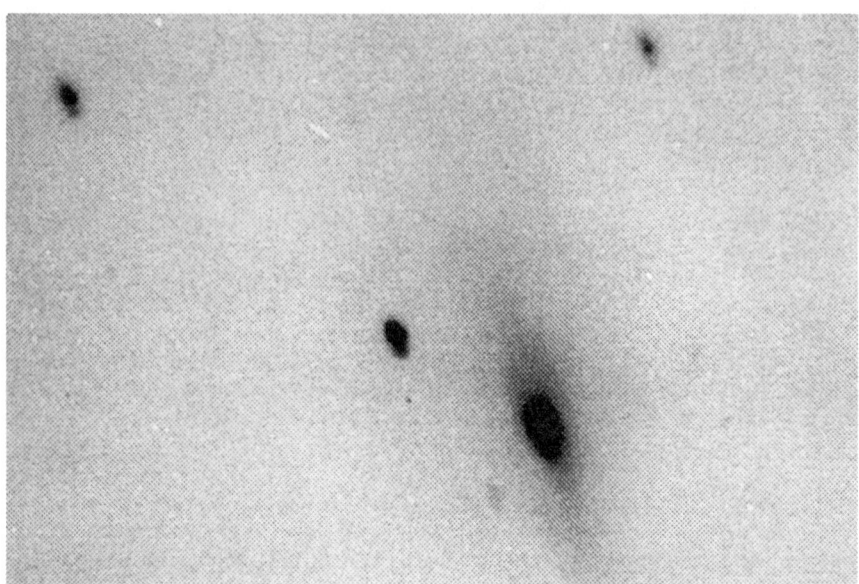

Fig. 2. A 300 second exposure of NGC 4527 (plus SN 1991T) without guiding.

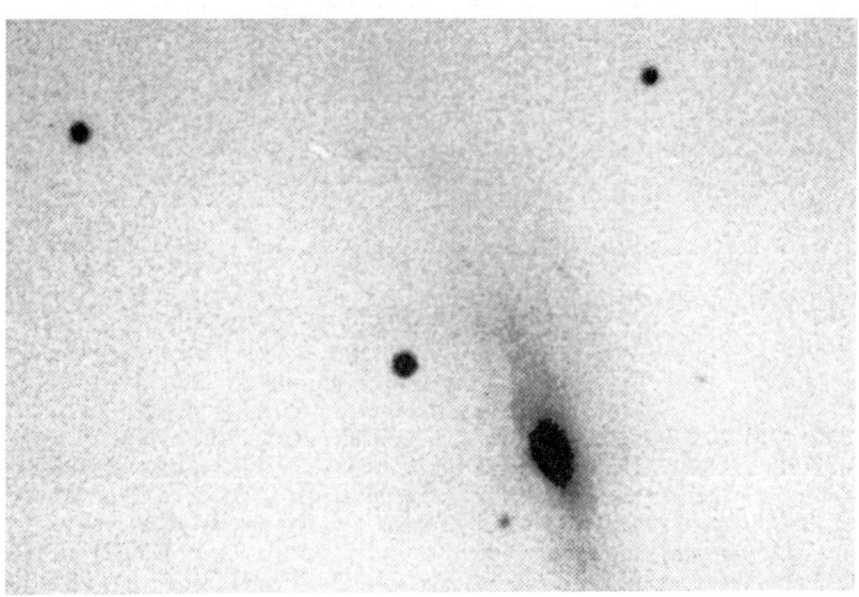

Fig. 3. A 300 second exposure of NGC 4527 (plus SN 1991T) with autoguiding.

APPENDIX: THE AUTOGUIDER

Although the article in this volume by Treffers, Richmond, and Filippenko (1992) presents details concerning the construction of our autoguider, here we briefly explain the method by which it finds guide stars and employs them to keep the telescope tracking properly.

For guiding we use a region of the sky centered upon, but not including, the current target. Therefore, with our present 0.5-m Cassegrain telescope, all the guide star images suffer from off-axis aberrations. (This will not be the case with the final 0.8-m AutoScope.) The most serious aberration is coma; near the edges of the guiding area, stars resemble tiny crescents more than circular disks. Fortunately, we have been able to track well even on very asymmetric images. As long as the shape is constant, it suffices for guiding.

When a target is selected for observation, the computer looks for possible guide stars by searching the Hubble Space Telescope Guide Star Catalog (GSC) (Lasker et al. 1990; Russell et al. 1990; Jenkner et al. 1990) for any nearby stars which fall within our guider's field of view. Since the GSC is reasonably complete down to $m \approx 14$ mag (Jenkner et al. 1990), the computer almost always finds at least one candidate. However, we have noticed that the magnitudes listed in the GSC, which are measured from photographic plates designed to yield a "V" bandpass centered around 5700 Å (Lasker et al. 1990), sometimes appear not to correspond well to the guider's output; that is, some stars listed at, say, $m = 11$ mag, appear to be brighter than others listed at $m = 10$ mag. In retrospect, it is not surprising that the guider's CCD camera, which is most sensitive around 6500 Å, should yield photometry significantly different from that in the GSC, but it did confuse us for a while.

Having selected a guide star candidate, the autoguider takes a ten second exposure centered on the candidate's position. If it cannot find a star in the resulting image, it picks another guide star candidate and tries again. If one or more stars are found, however, it selects the brightest and moves itself (on a movable x-y stage) so that the star falls within a small ($\sim 14'' \times 14''$) box near the CCD's center. It then begins taking shorter exposures of only one or two seconds, each time finding a sky value and subtracting it from all pixels, and calculating the centroid of the light in the box. We use a simple algorithm,

$$c_r = \sum_c \sum_r I(r,c)(r - r_0)$$
$$c_c = \sum_c \sum_r I(r,c)(c - c_0),$$

where $I(r,c)$ is the number of sky-subtracted counts in the pixel at row r and column c, to determine the centroid of the light distribution in both the row and column dimensions. The autoguider then sends the differences between the center of the box and the centroid of the light, for both rows and columns, to our main computer, which translates those differences into offsets in right ascension and declination. The main computer moves the telescope by an amount which will bring the light back to the center of the box, or possibly some fraction of the full amount. (We are concerned that moving the full amount might lead to some kind of growing oscillations, although tests so far don't show any.) After the telescope has moved to the new position, the guider takes another picture, and the process begins again. One full cycle takes about two seconds if a one-second guide exposure is used.

We need to conduct many more tests before we can draw any definite conclusions, but the first results show that we can guide well on a star which the GSC lists as $m = 10$ mag. We will need to use much fainter stars when pointing to fields far from the Galactic equator, but we believe that, even with the small telescope and poor seeing at Leuschner, we will be able to push the faint limit down to $m \approx 12$ mag without losing any image quality.

REFERENCES

Filippenko, A. V. 1992, *Ap. J. (Letters)*, **384**, L15.
Genet, R. M. 1986, in *The Study of Variable Stars Using Small Telescopes*, ed. J. R. Percy (Cambridge: Cambridge Univ. Press), p. 235.
Genet, R. M., and Hayes, D. S. 1989, *Robotic Observatories* (Mesa: AutoScope Corporation).
Honeycutt, R. K., Kephart, J. E., and Hendon, A. A. 1978, *Sky and Telescope*, **56**, 495.
Jenkner, H., Lasker, B. M., Sturch, C. R., McLean, B. J., Shara, M. M., and Russell, J. L. 1990, *A.J.*, **99**, 2082.
Lasker, B. M., Sturch, C. R., McLean, B. J., Russell, J. L., Jenkner, H., and Shara, M. M. 1990, *A.J.*, **99**, 2019.
Perlmutter, S., Crawford, F. S., Muller, R. A., Pennypacker, C. R., Sasseen, T. P., Smith, C. K., Treffers, R., and Williams, R. 1988, in *Instrumentation for Ground-Based Optical Astronomy*, ed. L. B. Robinson (New York: Springer-Verlag), p. 674.
Richmond, M. W. 1990, *Proceedings of the Electronics Oriented Astronomy Seminar 3*, ed. J. Sanford (OCA Publications), p. 68.
Richmond, M. W., and Filippenko, A. V. 1991, in *Advances in Robotic Telescopes*, ed. M. A. Seeds and J. L. Richard (Mesa: Fairborn Press), p. 81.
Russell, J. L., Lasker, B. M., McLean, B. J., Sturch, C. R., and Jenkner, H. 1990, *A.J.*, **99**, 2059.
Smith, C. K., Crawford, F., Muller, R., Pennypacker, C., Perlmutter, S., Sasseen, T., Williams, R., and Treffers, R. 1989, in *Automatic Small Telescopes*, ed. D. S. Hayes and R. M. Genet (Mesa: Fairborn Press), p. 47.
Treffers, R. R., and Richmond, M. W. 1989, *Pub. A.S.P.*, **101**, 725.
Treffers, R. R., Richmond, M. W., and Filippenko, A. V. 1992, in *Robotic Telescopes in the 1990s*, ed. A. V. Filippenko (San Francisco: Astron. Soc. Pacific), p. 115.

TECHNICAL DESCRIPTION OF THE BERKELEY AUTOMATIC IMAGING TELESCOPE

RICHARD R. TREFFERS, MICHAEL W. RICHMOND, and ALEXEI V. FILIPPENKO
Department of Astronomy, University of California, Berkeley, CA 94720

ABSTRACT The control system and computer configuration of the Berkeley Automatic Imaging Telescope (BAIT) are discussed. The system uses three IBM PCs to control the telescope, the CCD camera, and the autoguider, while a workstation running UNIX is the master control computer. The design of the camera back and autoguider is also discussed.

GOALS AND PROBLEMS

BAIT is described in detail by Richmond and Filippenko (1991). A brief summary of the project, as well as a progress report, can be found elsewhere in this volume (Richmond, Treffers, and Filippenko 1992). The scientific goals of BAIT include the photometric monitoring of supernovae, AGNs, comets, and other variable or ephemeral objects, as discussed by Filippenko and Richmond (1991). To accomplish these goals, long exposures will be needed, especially when narrow-band filters are used. In addition, since much of the data will be processed automatically, very accurate telescope pointing is desirable.

There are many different problems facing an automatic imaging telescope. First of all, a large number of devices (e.g., telescope motors, CCD cameras, filter wheels) must be controlled simultaneously. Observations must be scheduled taking into account the vagaries of weather, seeing conditions, and the constantly changing astronomical environment (e.g., supernova discoveries). CCDs generate large amounts of data, and in many cases it is beneficial to process some of it on-line to reduce the storage problems. Furthermore, there are issues of safety, such as preventing the dome from opening in a rain shower.

Here we present a technical description of BAIT. We also discuss some of the problems we have encountered, along with our solutions. Note that ultimately, BAIT will consist of a 0.8-m telescope on an AutoScope mount, equipped with a Photometrics CCD, and located at Lick Observatory. However, since the modifications to the site have not been funded, and the mirror and camera are not yet finished, we have proceeded with existing equipment at nearby Leuschner Observatory (owned and operated by the University of California at Berkeley). In this paper we concentrate on the present system, not on our future plans.

THE LEUSCHNER BAIT

We are using a network of computers to control the observatory. Most of the devices are controlled by commercially available cards that are designed to plug into an IBM PC bus. Communications are done over a local Ethernet connection with TCP/IP protocol. All critical real-time control is done by specialized cards. All safety is done in hardware.

0.5-m Telescope

The telescope is a classical Cassegrain with a focal ratio of 11.5, built by Tinsley Laboratories in 1952 and moved to the current location in 1968. The mount is a classical fork with a 512-tooth right ascension drive. The declination drive uses a ball screw as a tangent arm drive.

We have recently modernized the drive and added the features needed for unattended operation. Where possible, we have used commercially available cards and as few ancillary electronics as possible. The telescope drive uses D.C. printed circuit motors which are controlled using a Galil encoder feedback circuit. In order to make the telescope track, we built an "electronic differential" which adds a constant pulse rate from a counter card (AMD 9513) to the quadrature waveforms from the encoders. This circuit fools the motor controller into moving the telescope at a rate that is difficult to set otherwise. All A.C. power switching is done with solid state (hockey puck) relays driven by TTL logic.

The telescope has limit switches in hour angle, declination, and altitude. In addition, there is an emergency limit switch which completely terminates the telescope power if the telescope altitude goes below 5°.

The dome slit is opened using a motor powered by a 12-volt battery charged by a solar cell. To avoid complicated slip rings, we transmit the dome opening signal via two loop antennas wrapped around the dome. When the dome shutter is closed, a signal is transmitted down, indicating that the dome is stowed. In the event of a power failure, the dome will close automatically. A signal from the weather station on the nearby telescope dome of the Berkeley Automated Supernova Search Team (Smith *et al.* 1989) is used to determine whether it is safe to open our dome. It combines the inputs from a rain sensor, and signals whether the wind speed or humidity exceed preset limits.

We have not yet developed a good way of removing the mirror cover, but work is in progress. Currently, we simply store the telescope at a low altitude and hope that bird feathers don't float in.

CCD Camera

The CCD chip is a Thomson TH7883 which has 576×384 square pixels of size 23 μm. The BAIT camera (Fig. 1), which was built in-house, is cooled to $-100°$ C by a cold finger dipped into a thermos bottle containing liquid nitrogen. The camera electronics incorporate a 16 bit analog-to-digital converter and integrating correlated double sampling. An IBM AT controls the camera through a single high speed serial coax link using Manchester encoding and commercial integrated circuits.

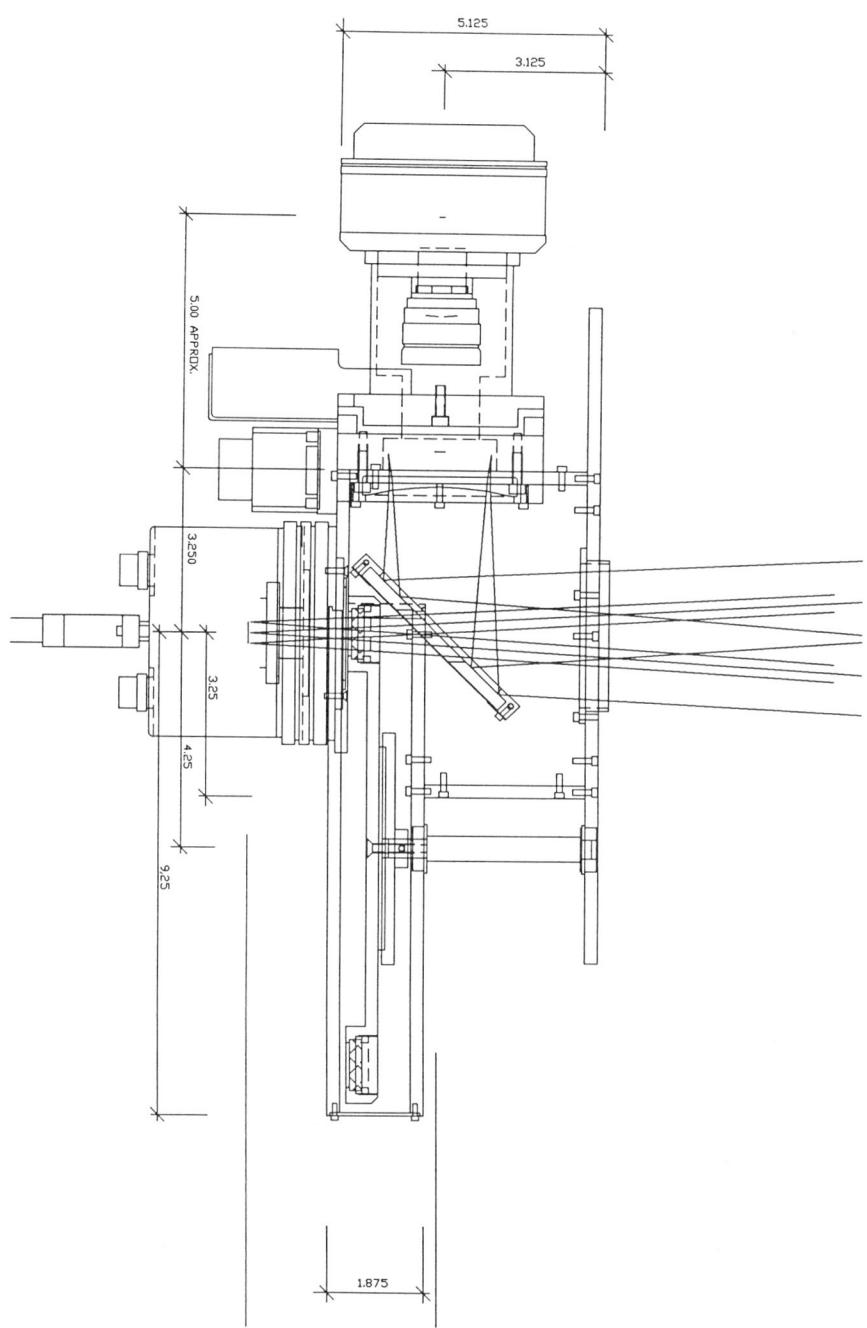

Fig. 1. The BAIT camera, with autoguider at the top and science imaging CCD to the left. The main pickoff mirror has a hole in its center.

BAIT CAMERA DESIGN

The BAIT camera was designed to fit in the space provided by the AutoScope telescope mount, to have a filter wheel with 20 positions, and to have a self-contained camera for acquisition and autoguiding. Also, we require no vignetting of the science beam.

As shown in Figure 1, the autoguider picks off a field from a mirror with a hole in it. A plano-concave field flattener is used to correct for the highly curved field of the Ritchey-Chrétien without refocusing. Ray tracing calculations have shown that without this lens, the images would be 5" in diameter. This field flattener is *not* ideal for the 0.5-m Cassegrain telescope, having moderate field curvature and serious coma at the edges.

Our autoguider employs a SpectraSource "Lynxx Plus" thermoelectrically cooled camera. A 25 mm C-mount lens is used to refocus and compress the scale by a factor of ~ 2, making one pixel subtend roughly 1". The autoguider is mounted on a 2×2 inch square translation stage so that it can be moved around the accessible field automatically. Stepper motors with Hall effect home position sensors are used to move the stage. The autoguider can acquire stars from an area of almost 0.2 square degrees. The operation of the autoguider is described by Richmond, Treffers, and Filippenko (1992).

The twenty-position filter wheel holds filters having a diameter of 1 inch. Unfortunately, the filters have to be placed quite close to the focal plane, due to the large number of filters required and the small space available. The filter wheel can be removed by one screw without dismantling the camera, so that filters can be changed or cleaned without affecting the rest of the instrument. The wheel is belt driven by a stepper motor. A small magnet in the wheel triggers a Hall effect sensor to determine the "home" position.

ORCHESTRATION

The main control computer has to do many things simultaneously. It must (1) schedule observations, (2) analyze data, (3) accept remote logins to change schedules or inspect telescope operation, and (4) log data and error reports. Since UNIX is explicitly designed for multi-tasking, we have chosen it as the operating system for the master computer.

In operation, each PC is running its own "server" program that listens to the Ethernet for messages. These messages are ASCII strings in which the first word is the function and subsequent words are keywords or values specific to that function. When the PC program receives a command, it goes away and does the requested task; it cannot listen for other requests during this time. When finished, it returns with either the keyword "done" and some additional information, or the word "ERROR" and a summary of what went wrong. Timeouts have been built into each function to report an error in the event of some failure. A sample page from the manual documenting some of these functions is shown in Figure 2. We treat this protocol as purely for our own convenience in debugging and logging. The protocol that remote observers will eventually use is described by Richmond, Treffers, and Filippenko (1992).

On the UNIX machine a single "server" program communicates with all these remote machines. This process itself listens to a "UNIX socket" and

TSERVER (local)   MISC. REFERENCE MANUAL PAGES   TSERVER (local)

The 'nbytes=' keyword initiates the image transmission. See the CCD command for the format. If no exposure time was specified, then the previous image will be transmitted.

The 'dark' keyword keeps the shutter closed, however, waits for the specified time before proceeding.

To display a 12 bit image on a 8 bit screen, usually the data are shifted right by 4 places. However, to increase the contrast, you can specify a smaller number. Note: the shift value is stored as a program static and will remain until it is either changed or the guider server program is restarted (at which time it goes back to 4).

help

In addition if the keyword 'help' is listed as an argument to any of the commands, e.g. 'point help', the argument list is echoed back.

mirror [open] [close]

Controls the mirror cover on the telescope. Returns with 'done mirror xx' where 'xx' is either 'open' or 'close' depending on the status after the command is issued. If 'open' or 'close' is requested and the status disagrees, an error message is issued.

move_sec [mils=%f] [abs]

Moves the telescope secondary by +/- yy.y mils (one thousandths of one inch). This command reports back the current absolute position of the secondary. If the 'abs' flag is specified, the secondary will move to the *absolute* position specified by the 'mils=', as opposed to the relative move.

observatory [open] [close]

Performs all the operations needed to get the observatory ready for taking data. It opens the mirror, turns on the drive power, and opens the slit. If this command is used without arguments it returns the keywords, 'open', 'closed', or 'mixed'; the latter keyword means that some of the states are closed while others are in the open state.

offset [ra=%f] [dec=%f] [cos]

Offsets the telescope from its present position where the values are in decimal degrees. The dome will not be adjusted and the current epoch will be used. The maximum offset is 5.0 degrees. If the 'cos' keyword is specified the motion is corrected (increased) by 1./cos(declination).

point ra=hh:mm:ss.s dec=dd:mm:ss epoch=yyyy [home]

Positions the telescope to the RA and Dec specified. The epoch to which the coordinates are referred must be specified. The telescope will respond with 'done' when pointed to within its tolerance. If the telescope drive is not turned on an error message will be issued.

The **home** keyword sends the telescope to the 'home' position and resets the local telescope encoders. This switch should only be used if there has been a pointing problem.

position_filter [num=%d] [init] [home=%d]

This command controls the filter wheel. The command always reports back 'done position_filter present_position=n' where n is the slot number between 0 and 19. If the 'num=xx' argument is specified in the command line, the filter is moved to position xx. The command waits for the filter to rotate, before returning. In the event of a power failure or after rebooting, the filter wheel returns to position 0. The 'init' keyword sends the filter back to the magnetic 'home' position and returns the string 'done filter hit=ff' where ff is the number of pulses off the expected value for the home. The 'home' value is used for the initial positioning and tells the number of pulses between the magnetic home position and the center of the filter. Hopefully this value does not have to be adjusted often.

power [on] [off]

Controls the power to the telescope drive. It always reports back 'done power xx' where 'xx' is either 'on' or 'off'. If the power was unable to do as requested an error message will be issued as appropriate. In the event of a reboot, or power failure in the telescope, the power is turned off.

Fig. 2. A sample page from the manual that documents the functions of the main control computer.

forwards the messages to the appropriate remote machine. We have written a simple "client" program to forward these messages so they can be transmitted from the command line.

Procedures

We have adopted the UNIX philosophy of writing small programs that can be invoked from the command line. For image processing we have ported programs originally written for the IBM PC as PC-Vista (Treffers and Richmond 1989) to UNIX. We use X-windows to display the images; although the 1-bit display is marginal, with dithering we can at least find out whether the system is functioning. We then combine the various disciplines of telescope control, star finding, and image processing into procedures written in C-shell. A sample procedure to find the best bright star is shown in Figure 3.

```
Jul  2 16:00 1991   acquire Page 1


#!/bin/csh
#Purpose:       to find brightest star in Guider field and center it
#Arguments:     EXPTIME
#
if($?ECHO) set echo=1
set fname=acquire                         # scratch file for image
if ($#argv == 0) then
        set EXPTIME=10
else
        set EXPTIME=$1
endif

guimage time=$EXPTIME outfile=$fname      # take a picture

sky $fname pix=5              #determine the sky value for the finder

#find all the brightest stars in the field
# BUT ignore the edges of the frame!
box 11 sr=5 sc=5 nr=155 nc=182
stars $fname sky=sky minsig=3 maxfwhm=10 box=11 > acquire.temp

if (-z acquire.temp) then     # bomb if no stars found
        echo no stars found
        exit (1)
endif

#isolate only the brightest
awk '{                        \
        if($4>s){             \
                s=$4          \
                row=$2        \
                col=$3        \
        }                     \
}                             \
END {print row,col,s}         \
'  acquire.temp > best
#
set loc=`cat best`
set off=`findoffset row=$loc[1] col=$loc[2] guider`
tx offset ra=$off[1] dec=$off[2] cos  | grep "done"
if( $status) exit 1
sleep 1                                   #let vibrations damp out
```

Fig. 3. Procedure for finding the brightest guide star not near an edge.

TECHNICAL DESCRIPTION OF THE BERKELEY AIT

```
#automatic telescope configuration file for the B.A.I.T
#observatory parameters
OBSERVAT='Leuschner'
OBSLAT   =     37.91833          / leuschner latitude in decimal degrees
OBSLONGH=       8.1437972        / leuschner longitude in hours
OBSELEV =    304.                / leuschner elevation (m)
#
INSTRUME='B.A.I.T.'
#
#telescope parameters
TELESCOP='Twenty inch'
TELAREA =  1778.0                / telescope collecting area (square centimeters)
TELSCALE=    33.6                / scale in arc seconds per mm
TELLIMN =    75.0                / North declination limit (decimal degrees )
TELLIMS =   -28.0                / South dec lim
TELLIME =  -105.0                / East hour angle limit
TELLIMW =   105.0                / West hour angle limit
TELLIMA =    10.0                / altitude limit
#
#ccd camera parameters
#DETECTOR='Photometrics PM512'
#CCDPIXSZ= '25.0 25.0'           / pixel size along columns and rows (microns)
#CCDSEC  = '[1:512,1:512]'       / number of columns and rows
DETECTOR='Lynxx PC Plus CCD'
CCDPIX  = '13.75 16.0'           / pixel size along columns and rows (microns)
CCDSEC  = '[1:192,1:165]'        / number of columns and rows
CCDREADN=  104.0                 / CCD read noise, in electrons
CCDGAIN =   54.0                 / CCD gain, electrons per ADU
CCDWELL = 150000.0               / CCD pixel full well (electrons)
CCDROT  =  180.0                 / CCD rotation angle (degrees)
CCDHAND =   -1.0                 / +1 if (x,y)::(ra,dec) same handedness
#                this is a pure guess for the Lynxx dark current
CCDDARKC=   10.0                 / CCD dark current, electrons per sec per pixel
#
#define shutter parameters
SHUTMIN =   .015                 / minimum shutter open time in seconds
#
#filter parameters
#the number is the filter bolt number which starts at 0
#define the names are CASE sensitive so that 'u' can be distiguished from 'U'
#filter format is 'name    eff.lambda(Angstroms)   equiv.width(Ang)'
NFILTERS=4                       / number of assigned filters
FILTER00='U    3650    700'  / numbers a guess, not measured
FILTER01='B    4400   1000'  / numbers a guess, not measured
FILTER02='V    5500    900'  / numbers a guess, not measured
FILTER03='open 6500   4200'  / PM512 QE response, from the Photometrics specs
#
#guider parameters
GURANGE ='50.8 50.8'         / guider size in mm
GUBLOCK =' 20.0 20.0 8.0 '   / vignette size and blur in mm
GUMAG   =1.0                 / magnification
GUROT   = 180.0              / guider rotation angle (degrees)
GUHAND  = -1.0               / guider handedness
GUPIX   = '13.75 16.0'       / guider pixel size column and row (microns)
GUSEC   = '[1:192,1:165]'    / number of columns and rows
GUCEN   = '25.4 25.4'        / x,y location of center of main field
# limits on input request file parameters
MAXPRIOR=  1                     / most important priority (for most peop
MINPRIOR= 99                     / least important priority
MAXOBS  = 1000                   / max number of observations in a single
# the next line MUST be at the end of the file
END                              / must be last line
```

Fig. 4. Configuration file which stores parameters associated with the telescope and system.

Configuration File

We store many of the parameters associated with the telescope and the system in a configuration file (Fig. 4). Programs that need to know the telescope latitude, for example, will access this file to read it. We find this technique safer than the more traditional ".h" files which have to be recompiled to take effect. Where possible, we have used NOAO FITS keywords to describe the values.

ACKNOWLEDGEMENTS

This project could not have begun without the NSF Presidential Young Investigator Award (grant AST–8957063) to AVF, and without a generous donation from AutoScope Corporation. We have also received equipment donations from IBM, Photometrics Ltd., and Sun Microsystems, as well as financial support from NSF Cooperative Agreement AST–8809616 (Center for Particle Astrophysics, U. C. Berkeley), NSF grant AST–9115174, CalSpace grant CS–96–91, and the University of California at Berkeley. We have had helpful conversations with numerous people, especially Russ Genet and Jack Borde. Many of our developments have utilized experience and knowledge gained by the Berkeley Automated Supernova Search Team, which is led by Richard Muller and Carl Pennypacker. In particular, Robbi Smits has contributed greatly to the modernization of the 0.5-m telescope. We thank Jerome Hudson for his insights into telescope control philosophy, and Suzanne Jacoby for providing the NOAO keywords. The excellent machining of the camera by Lockwood Associates is appreciated.

REFERENCES

Filippenko, A. V., and Richmond, M. R. 1991, in *Advances in Robotic Telescopes*, ed. M. A. Seeds and J. L. Richard (Mesa: Fairborn Press), p. 89.

Richmond, M. R., and Filippenko, A. V. 1991, in *Advances in Robotic Telescopes*, ed. M. A. Seeds and J. L. Richard (Mesa: Fairborn Press), p. 81.

Richmond, M. W., Treffers, R. R., and Filippenko, A. V. 1992, in *Robotic Telescopes in the 1990s*, ed. A. V. Filippenko (San Francisco: Astron. Soc. Pacific), p. 105.

Smith, C. K., Crawford, F., Muller, R., Pennypacker, C., Perlmutter, S., Sasseen, T., Williams, R., and Treffers, R. 1989, in *Automatic Small Telescopes*, ed. D. S. Hayes and R. M. Genet (Mesa: Fairborn Press), p. 47.

Treffers, R. R., Richmond, M. R. 1989, *Pub. A.S.P.*, **101**, 725.

THE EXPLOSIVE TRANSIENT CAMERA - - AN AUTOMATIC, WIDE-FIELD SKY MONITOR FOR SHORT-TIMESCALE OPTICAL TRANSIENTS

ROLAND K. VANDERSPEK
Massachusetts Institute of Technology, Department of Physics and Center for Space Research, 77 Massachusetts Ave, Rm. 37-527, Cambridge, MA 02139-4307

GEORGE R. RICKER
Massachusetts Institute of Technology, Department of Physics and Center for Space Research, 77 Massachusetts Ave, Rm. 37-535, Cambridge, MA 02139-4307

JOHN P. DOTY
Massachusetts Institute of Technology, Department of Physics and Center for Space Research, 77 Massachusetts Ave, Rm. 37-541, Cambridge, MA 02139-4307

ABSTRACT The Explosive Transient Camera (ETC) is a wide-field sky monitor designed to detect short-timescale (1-10s) celestial optical flashes. It consists of two arrays of wide-field CCD cameras monitoring ~0.4 steradian of the night sky for optical transients with risetimes of ~1-10s and peak magnitudes $m_V < ~10$. The ETC was designed to be completely automated in order to make year-round observations with minimal human intervention. A small, powerful 68000-based computer controls all aspects of observations, including roof motion, CCD readouts, and weather sensing: under software control, the ETC is able to perform all the functions of a human observer automatically.

INTRODUCTION

The Explosive Transient Camera (ETC) is a wide-field astronomical instrument able to detect optical transients of moderate brightness ($m_V < ~10$) and short risetime ($\tau \sim 1\text{-}10$ seconds). It was developed as a means of detecting transient optical radiation from gamma-ray bursts (Ricker, *et al.* 1984); however, as the ETC is sensitive to *all* sources of short-timescale optical transients, it will conduct a systematic survey of the night sky for optical transients with risetimes of the order of one second. Because such a survey has not yet been performed, the possibility of serendipitous discoveries of new classes of astronomical objects by the ETC is very promising.

SCIENTIFIC GOALS OF THE ETC

The ETC program was conceived after the discovery by Schaefer (1981) of transient optical radiation on an archival photographic plate of a small gamma-ray burst (GRB) error region. Subsequent discoveries of archival transient images (Schaefer, *et al.* 1984; Moskalenko *et al.* 1989) support the hypothesis that transient optical radiation is emitted from the within GRB error boxes; however, no optical transient has yet been detected from a GRB *during* outburst.

The primary goal of the ETC is the detection of an optical counterpart to a gamma-ray burst. Since the operational phase of the Gamma Ray Observatory began in May, 1991, the ETC has been operating contemporaneously with the Burst and Transient Source Experiment (BATSE) on GRO. Should BATSE detect a GRB in a field which the ETC is observing, the ETC will be able to confirm the existence or non-existence of an optical counterpart to that GRB. If a celestial optical transient is seen at the time a GRB is detected, it will provide conclusive evidence of the existence of optical counterparts to GRBs. If a GRB occurs and no optical counterpart is seen, then stringent limits on the ratio of γ-ray to optical flux in GRBs ($F_\gamma/F_{opt} > \sim 10^5$) and the risetime of any optical radiation of accompanying GRBs may be set.

In addition to monitoring the night sky for optical counterparts to gamma-ray bursters, the ETC will be performing the first systematic survey of the night sky for transients of m < ~10 and risetime $\tau_{OT} \sim$ 1 - 10 seconds. Previous experiments to measure the rate of celestial optical transients have yielded no detected transients in 0.3 sr-hr (Schaefer, Vanderspek, Bradt, and Ricker 1984) or 3.0 sr-hr (Vanderspek 1985) of observations, to a limiting magnitude of m ~ 10. The ETC is expected to monitor ~1-2 sr-hrs of sky per clear night, resulting in ~200 sr-hrs of sky monitoring per year: these observations will result in stringent new limits on the rate of celestial optical transients.

During the course of standard observations, the ETC stores up to four exposures from each CCD during a night. As the field-of-view of the entire instrument is ~0.4 steradians, this means that ~1-1.5 steradians of the night sky will be imaged to a limiting magnitude of m ~ 11 each night. Over the course of a year, the entire declination band spanned by the ETC (roughly 0° to 45°) will be imaged multiple times, creating a vast archive of image data (~2-4 Gbytes): each part of this declination band will be imaged an average of ~60 times. These data will be analyzed at MIT for optical variability in field stars over timescales of hours to months.

ETC INSTRUMENTATION

The ETC consists of sixteen wide-field CCD cameras, each capable of detecting optical transients to a limiting magnitude of $m_V \sim$ 10. The cameras operate in pairs: two cameras monitor the same part of the sky, so that sources of false optical transients in one camera, such as cosmic rays, are recognized and rejected because of the absence of a confirming report in a second camera.

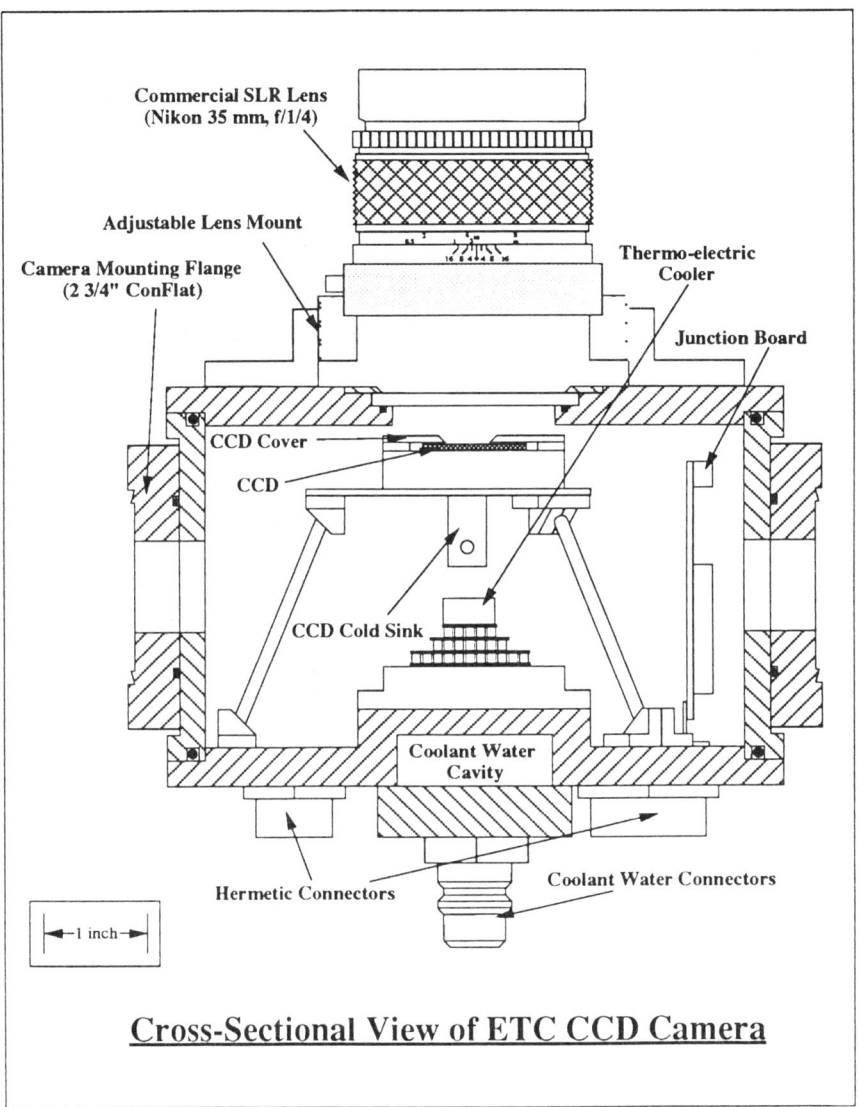

Figure 1: A cross-sectional view of an ETC CCD camera. The CCD is in the focal plane of a commercial 35mm format lens. The CCD is cooled by two thermoelectric coolers (the tinned-copper braids coupling the CCD heat sink to the coolers have been omitted for clarity). The heat generated by the thermoelectric cooler is removed by water passing through a cavity in the back plate of the camera. The camera is mounted to a common vacuum manifold via the 2 3/4" vacuum flange. Signals to and from the CCD pass through hermetic connectors in the camera back plate.

Each ETC camera consists of a cooled CCD in the focal plane of a commercial 35mm-format SLR lens (see Figure 1). The field-of-view of each CCD camera is 14° x 11°, with each pixel subtending 2.2 arc-minutes: the total field-of-view of the ETC is over 1200 square degrees. The cameras are mounted in two sets of eight on two telescope drives in a manner that fixes the declination of each camera and the relative right ascension of each set of eight cameras (see Figure 2); however, the two telescope drives are independent, and can steer each set of eight fields-of-view to any hour angle. During observations, the fields-of-view are oriented in a manner which minimizes the zenith distance of the cameras: a typical arrangement of fields-of-view is shown in Figure 3. The ETC cameras are housed in a dedicated building on the summit ridge of Kitt Peak. The building is covered with a fully-retractable roof, which allows the ETC instrumentation access to the full sky above 30° altitude (see Figure 4 for a schematic representation).

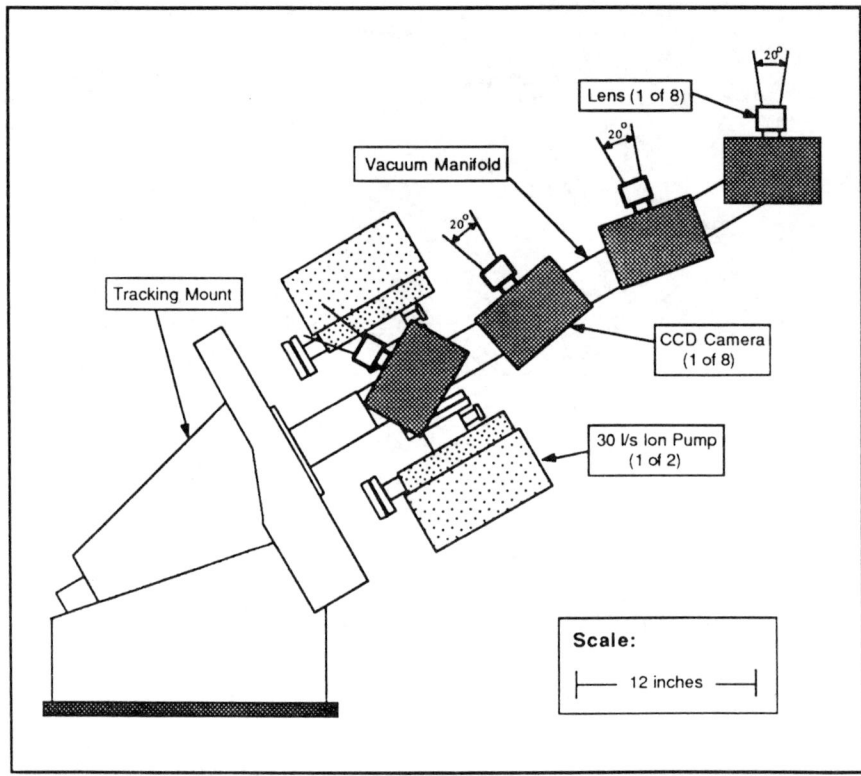

Figure 2: A side view of one eight-camera ETC module. The cameras are mounted to a common vacuum manifold: two ion pumps mounted on the manifold maintain vacuum in the eight cameras. The manifold is the extension of the polar axis of the sidereal drive; thus, when the drive is moved, the hour angles of the eight cameras are changed. The angle of each camera about its mounting point, and thus its declination, is fixed.

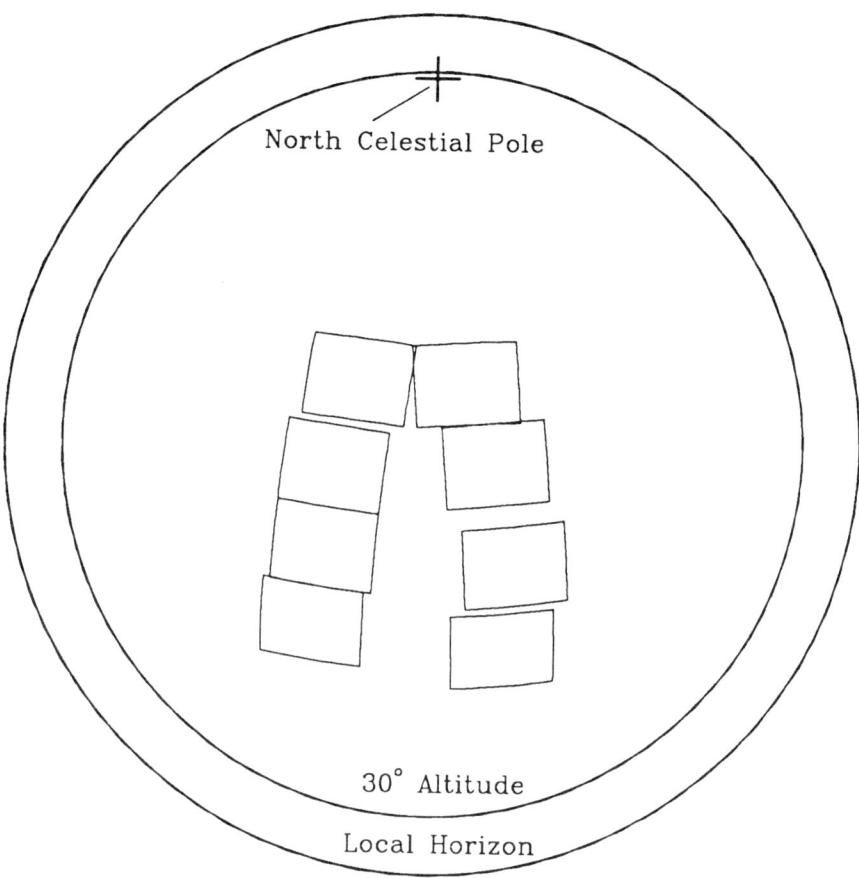

Figure 3: The present layout of the ETC CCD camera fields-of-view on the night sky. The view shown is a zenith projection from Kitt Peak: the outer circle represents the horizon, and the inner circle an altitude of 30° (2 airmasses). The declinations of the centers of the cameras are roughly 4°, 16°, 28°, and 40°. The field-of-view of each CCD camera measures 14°x11°.

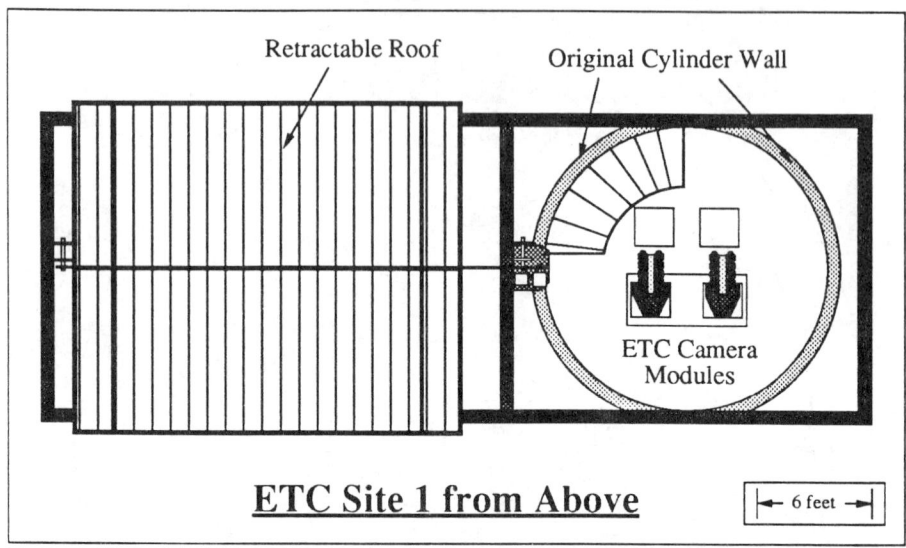

Figure 4: A schematic view from above of the ETC site on Kitt Peak. In this view, the roof is open, showing the two sets of eight ETC CCD cameras and their associated electronics.

An overview of the ETC computer system is shown in Figure 5. The primary control computer is the Overseer Computer: software running in the Overseer Computer determines the flow of ETC observations. The CCDs are read out and the CCD signal is amplified and digitized by the front-end electronics, while the CCD image data are analyzed by the Trigger Processors. The ETC instrumentation, including the sidereal drives, weather system (including precipitation, windspeed, and lightning sensors), and roof control are accessed through the peripheral hardware controller. The Overseer Computer communicates with these and other peripheral computers over RS-232 serial lines.

OPERATION OF THE ETC

The ETC has been designed to be completely automatic, in order to reduce the trouble and cost of constant human attention to the instrument. Because the ETC is "intelligent", no human observer is needed on site in order to conduct observations. The ETC computers have complete control of all of the ETC instrumentation, and so are able to open and close the roof, slew the telescope mounts, and sense weather conditions. The ETC software determines when and how to operate, and makes the decisions regarding operating conditions that an observer on site would.

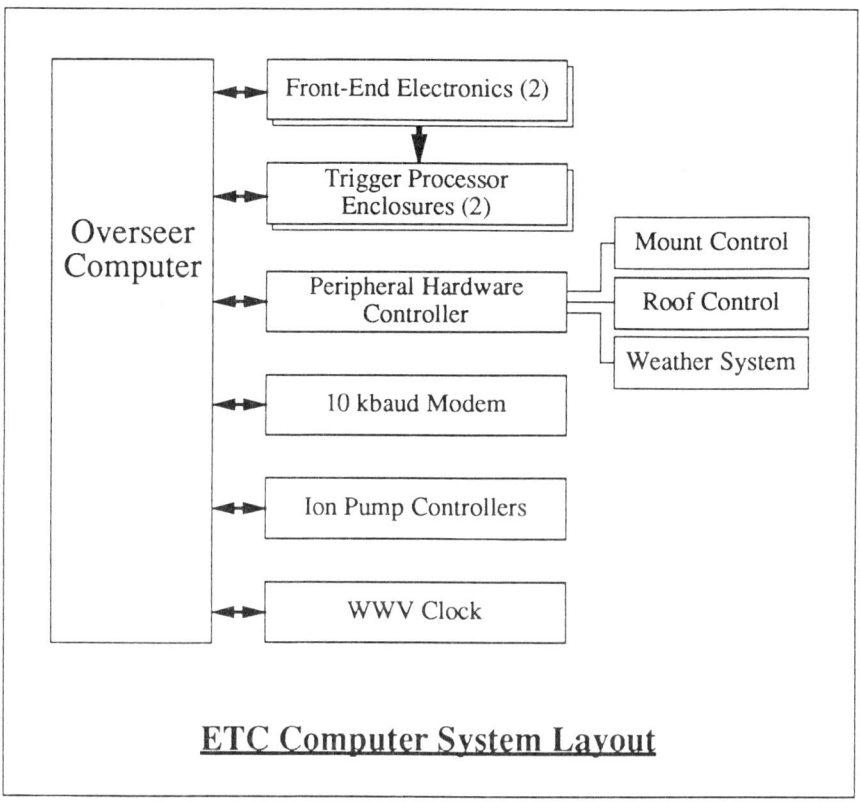

Figure 5: A block diagram of the ETC computer and peripherals system. The Overseer Computer communicates with the other ETC computers over RS-232 serial lines. On command from the Overseer Computer, the Front-End Electronics read out the CCDs: the CCD image data are then analyzed in real time by the Trigger Processors. The Overseer Computer controls the actions of the sidereal drives and retractable roof through the Peripheral Hardware Controller.

A flow chart describing the decision-making processes in the Overseer Computer at the beginning of a night is shown in Figure 6. During the day, the ETC is idle, and the Overseer Computer waits for astronomical twilight. At astronomical twilight, the Overseer Computer checks the weather: if the weather is bad (precipitation has been detected or the measured windspeed is too high), the Overseer Computer pauses for a short period (~15 minutes) and checks again. If the weather conditions are acceptable, the Overseer Computer moves the cameras to their starting hour angles, opens the roof, and checks the night sky for clouds. If the sky is cloudy, the roof is closed, the cameras are brought to their idle positions, and the Overseer Computer pauses, as above; if the skies are clear, observations begin.

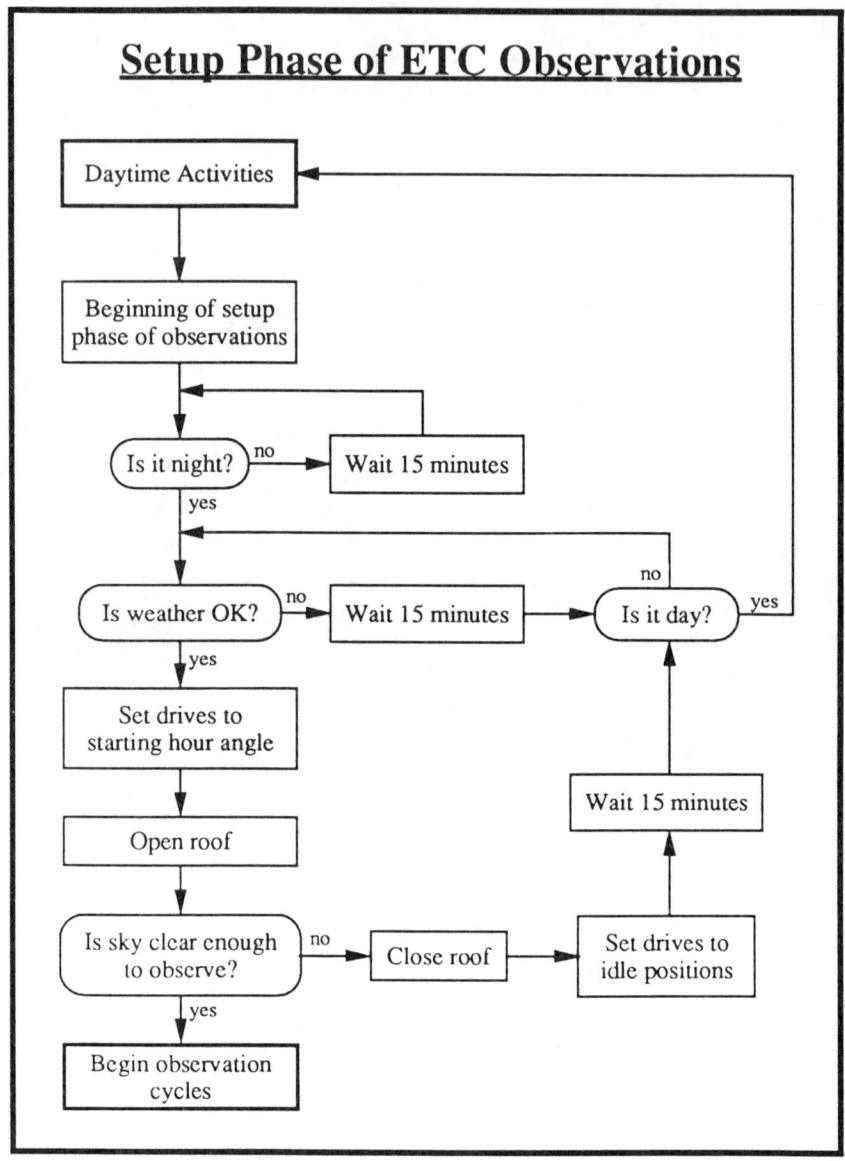

Figure 6: A flow chart of the decisions made by the Overseer Computer at the beginning of a night. The Overseer Computer is programmed to act as a human observer on site would: thus, it first checks the weather conditions, then the sky conditions, before beginning operation for the night.

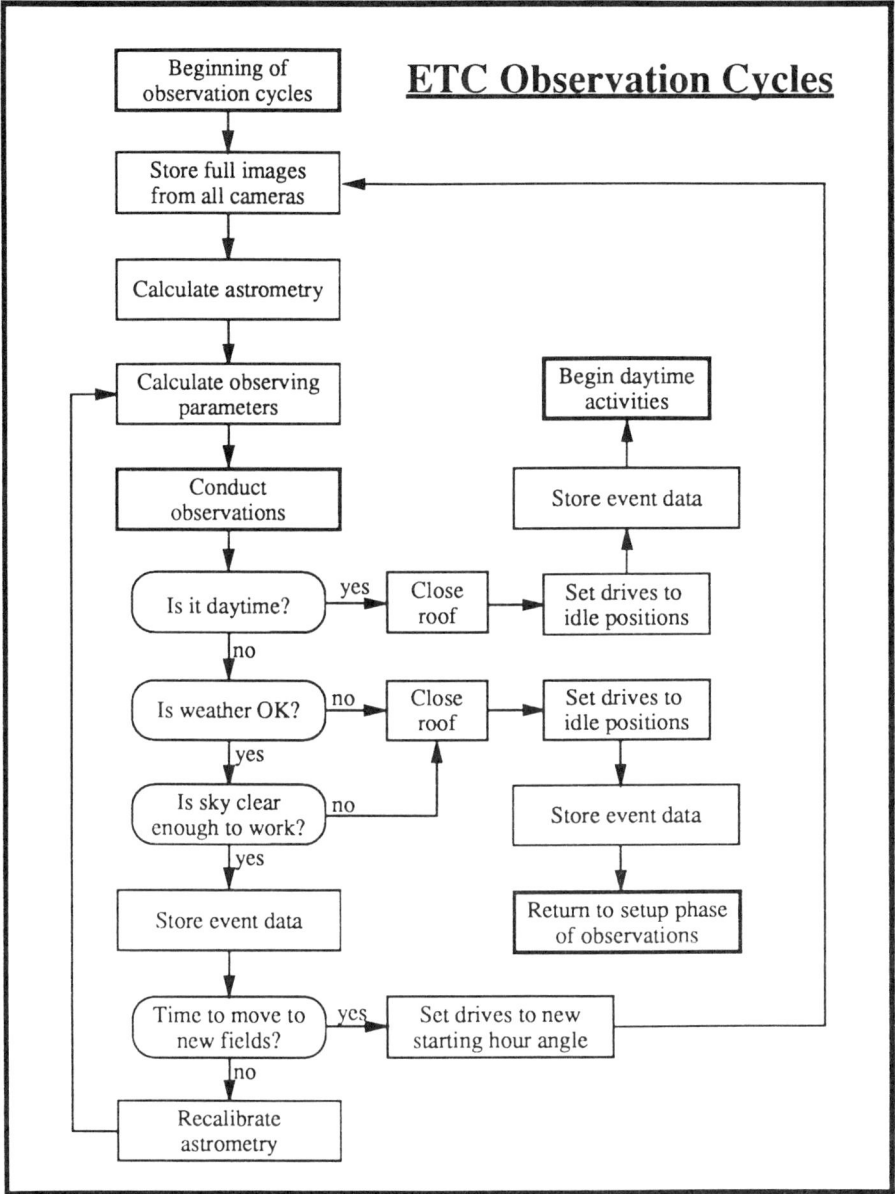

Figure 7: A flow chart of an ETC observations cycle. Observations begin with the storage of full images and the calibration of observing parameters. The phase labelled "conduct observations" generally lasts ~30 minutes and is interrupted in order to check the observing conditions, store event data, refresh the observing parameters, or, if appropriate, to move the cameras to a new hour angle. After daybreak, the Overseer Computer performs daytime tasks and then waits for night to fall.

ETC observations are broken up into a series of "observation cycles", each of which lasts roughly one to two hours. An observation cycle consists of a period of instrument calibration followed by a period of observations. The observations are a continuous search for optical transients, in which consecutive, contiguous exposures of the night sky are analyzed for optical transients. The observation periods are interrupted on a regular basis (every 30 minutes or so) to store collected data, recalibrate the camera pointing, and, if necessary, to slew the cameras to monitor a new set of fields-of-view and begin a new observation cycle. The observing conditions, such as sky clarity and weather conditions, are checked regularly during the observation cycles to protect the instrumentation from inclement weather. ETC observations end at astronomical twilight: at this time, the roof is closed, the cameras are brought to their idle positions, and data are archived. Once the ETC has been shut down for the night, the data collected during the night are prepared for transfer to MIT. A flow chart describing an observation cycle is shown in Figure 7.

The method used by the ETC to detect optical transients is straightforward: compare consecutive CCD images in a search for 'new' stars or 'suddenly brighter' stars. When observing, the ETC's CCD cameras take continuous, contiguous, precisely-timed exposures of the night sky simultaneously. The image data from each CCD camera is analyzed by a dedicated Trigger Processor, which compares consecutive images for sudden, point-like brightenings. Any detected brightenings are reported to the Overseer Computer, which tabulates and correlates flash reports from all cameras. If two cameras monitoring the same field-of-view detect a brightening at the same celestial coordinates, the brightening is considered a *bona fide* celestial optical transient, and data from the event is stored. If no events are detected, no action is taken; in either case, the reduction of CCD image data stays on its precisely-timed schedule. Data taken by the ETC is stored on hard disk and later transferred to MIT over the Internet.

The data collected consist of 9x9 pixel subarrays centered on the location of the flash event, as well as 9x9 pixel subarrays centered the locations of known SAO stars in the frame, for photometric and astrometric calibration. Because the image frame exposed before the flash event is in Trigger Processor memory at the time of the detection, image subarrays from the flash location *before* the event can be collected. Image subarray data is taken from both detecting cameras for the ten exposures following the event detection.

The coordinates of any celestial optical transients detected by the ETC are reported immediately to the Rapidly Moving Telescope (RMT - - Teegarden *et al* 1984), located near the ETC on the summit of Kitt Peak. The RMT consists of an 18 cm telescope pointing down at an elliptical, two-axis gimballed mirror; the mirror allows the field-of-view of the telescope to be slewed to any part of the night sky within one second with arcsecond precision. On the receipt of flash coordinates, the RMT will slew to the location of the optical transient and begin taking short ($\tau_{exp} = 1.5$ seconds) CCD images of the transient. The CCD system in the RMT will allow it to monitor the lightcurve of any optical transient to $m_V \sim 15$.

PRELIMINARY RESULTS

Initial analysis of ETC data taken in January and February, 1991, shows no unusual optical transients; *i.e.* no transients conclusively associated with celestial sources. Many optical transients were detected in that time, but these were associated primarily with artificial Earth satellites. These events, although part of the "noise" of the observations, serve as proof that the ETC is working correctly, and that it is sensitive to optical transients of m ~ 10.

The archive of full ETC images presently contains roughly 300 images, covering the declination band of 0° to 45° and right ascensions from ~8^h to ~16^h. Work is underway at MIT to develop software which will analyze these images for long-timescale variables.

FUTURE WORK

In the coming years, we will expand the capabilities of the ETC by 1) deploying the cameras at the extremes of a large baseline, in order to reduce the rate of false triggers by artificial Earth satellites, and 2) adding more cameras to increase sky coverage. We will also continue our efforts to analyze the performance of the ETC, to be able to make changes to its *modus operandi* to improve its efficiency and sensitivity.

Establishing a second ETC site will dramatically reduced the rate of false triggers by orbiting satellites and debris, as shown in Figure 8. A second ETC site has always been planned for the ETC (Ricker, *et al.*1984); however, the large rate of triggers by satellites has lent particular urgency to this move, since the effective observing time of the instrument as a whole is significantly affected at present. During observations made in the winter of 1990-91, an average of 50 events from satellites were detected each night; observations made in May 1991 show that this rate is significantly higher (~10^3 events/night) when the ETC cameras are not pointing into Earth shadow. At present, we are considering two options for the second ETC site. The southwest ridge of Kitt Peak would provide a baseline of 1.4 km which would permit the ETC to recognize all satellites at altitudes below 1500 km. As a large fraction of the population of orbiting satellites is found at altitudes less than 1500 km (Su and Kessler 1985), this separation would reduce the rate of events caused by orbiting satellites by a factor of ~10. However, as the population of orbiting satellites (and orbiting debris) is increasing and expanding at an alarming rate (Kessler and Cour-Palais 1978), we may find that a baseline of ~100 km is required: siting camera on another mountain (*e.g.* Mt. Hopkins or Mt. Lemmon) might be necessary.

We plan to increase the field-of-view of the ETC during 1991-2 by two new extra-wide-field CCD cameras on the existing telescope drives. These cameras are based on a hybrid 840 x 840 CCD developed for use at MIT's Lincoln Laboratories (Burke, Mountain, Daniels, and Harrison 1987). The Lincoln Labs CCD has a detector area *nine* times the area of one of the present-generation ETC CCDs: this means that a single new-generation camera will have a larger field-of-view than an entire bank of eight of the present ETC cameras! The field-of-view of this new CCD camera will be 36° x 36°: with one new-generation camera in place, the field-of-view of the ETC will be more than doubled. A possible arrangement of the completed ETC fields-of-view is shown in Figure 9.

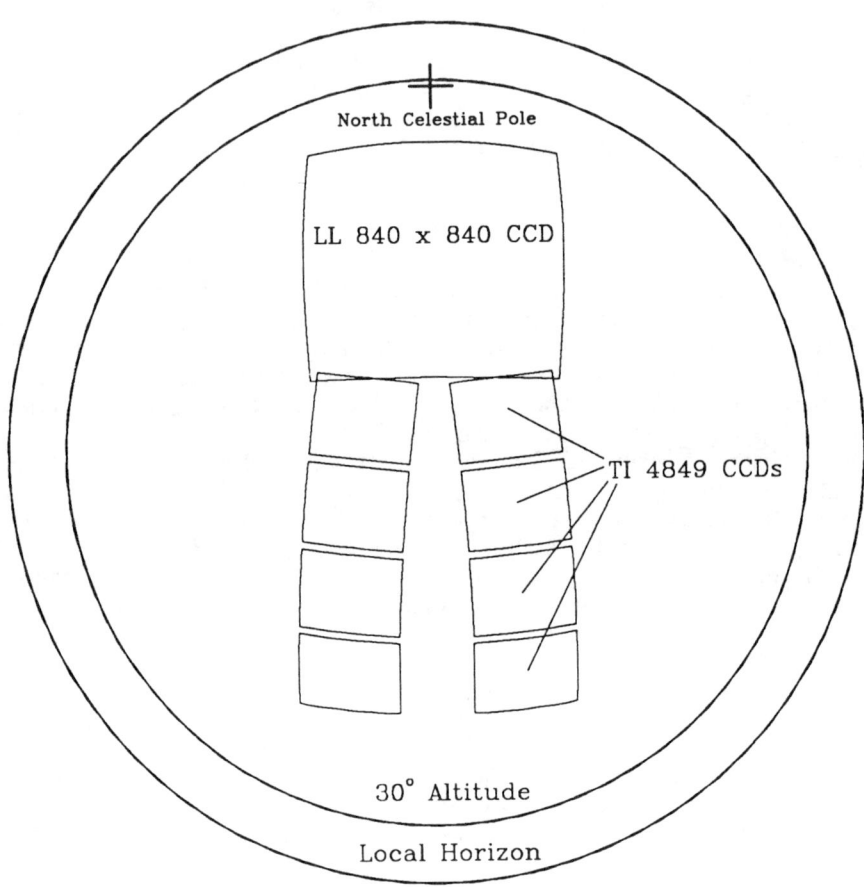

Figure 8: The probable arrangement of the fields-of-view of the ETC with the installation of the new-generation CCD camera. As in Figure 2, the view is a zenith projection from Kitt Peak, and the outer and inner circles represent the local horizon and an altitude of 30°, respectively. The new-generation camera will utilize a CCD measuring 840 x 840 pixels, while the existing cameras are equipped with devices measuring 390x292 pixels. The smaller fields-of-view measure 14° x 11°, while the large field-of-view measures 36° on a side.

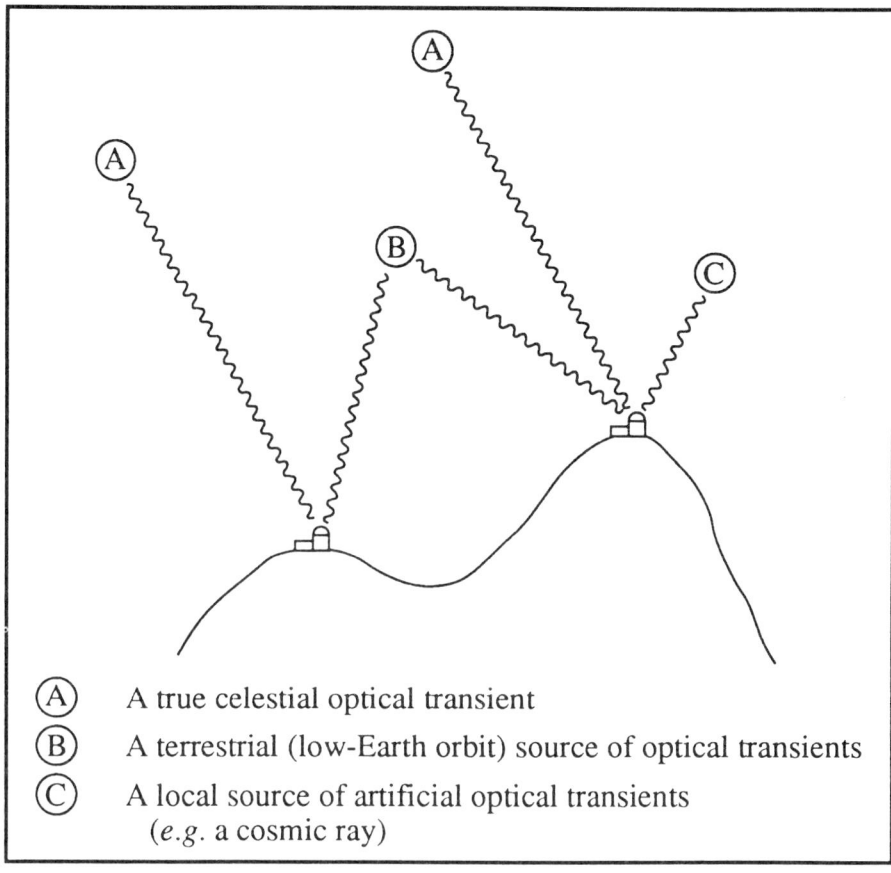

(A) A true celestial optical transient
(B) A terrestrial (low-Earth orbit) source of optical transients
(C) A local source of artificial optical transients
   (*e.g.* a cosmic ray)

Figure 9: A schematic representation of the parallax and time-coincidence methods of rejecting false optical flashes. Event A, a true celestial optical transient, is detected at both ETC sites with no measureable parallax. Event B, a terrestrial source of optical transients (such as an artificial Earth satellite), is detected at both ETC sites, but is recognized by the difference in the apparent celestial coordinates at the two sites. Event C is local to one site (a cosmic ray), and is rejected because of the absence of a confirming event at the other site.

ACKNOWLEDGEMENTS

The ETC project has benefitted from the advice and assistance of many people over the years. At MIT, Gerard Luppino, Patrick Mock, Steven Rosenthal, Peter Tappan, John Vallerga, Dan Zachary and David Breslau have all made significant contributions to the ETC. The downtown and mountain staff at Kitt Peak have also been instrumental in the progress the ETC has made: specifically, we would like to thank Sidney Wolff, Bob Barnes, and Don Versluis downtown, and Kurt Cramer, John Scott, and Hal Halbedel on the mountain for their support of the ETC program.

This project is supported by NASA grant NSG-7339.

REFERENCES

Allen, C.W. 1976, *Astrophysical Quantities*, The Athlone Press.

Burke, B.E., Mountain, R.W., Daniels, P.J., and Harrison, D.C. 1987, *Optical Engineering* **26**, 890.

Kessler, D.J., and Cour-Palais, B.G. 1978, J. Geo. Res. **83**, 2637.

Moskalenko, E.I., *et al.* 1989, Astron. Astrophys. **223**, 141.

Ricker, G.R., Doty, J.P., Vallerga, J.V., and Vanderspek, R.K. 1984, in *Instrumentation for Astronomy V*, SPIE Vol. 445, p. 370.

Su, S.Y., and Kessler, D.J. 1985, in *Adv. Space. Res.* Vol. 5, No. 2, p. 25.

Schaefer, B.E., *et al.* 1984, Ap. J. **286**, L1.

Schaefer, B.E., Vanderspek, R.K., Bradt, H.V., and Ricker, G.R. 1984, Ap.J. **283**, 887.

Teegarden, B.J, Cline, T.L., and von Rosenvinge, T.T. 1983, B.A.A.S. **14**, 885.

Vanderspek, R.K. 1985, Ph.D. Thesis, Massachusetts Institute of Technology.

THE RAPIDLY MOVING TELESCOPE: A PROGRESS REPORT

S.D. Barthelmy, T.L. Cline, B.J. Teegarden, T.T. von Rosenvinge

NASA Goddard Space Flight Center
Code 661
Greenbelt, MD 20771

ABSTRACT A ground-based optical telescope system has been constructed with the capability to locate fast optical transients that may be associated with Gamma Ray Bursts (GRB). The instrument has been integrated and operated during a shakedown period at GSFC, Maryland. Results of 35 hours of "stare mode" data are presented. The telescope has the proven capability to slew to any point on the night sky within 1.0 second, track that position with better than one arcsecond stability, and image a 9x12 arcminute field of view with one arcsecond angular resolution with 1.5 second time resolution. The telescope-CCD camera system has a sensitivity of 13th magnitude for transients and 14th mag for field stars. In the 35 hours of operation many single frame transients of instrumental and optical origin have been observed; no two-sequential frame astrophysical transients have been identified. The combined rate of instrumental transients (predominantly sea-level muons) is 7.2/hr and of optical transients (satellite glints, airplane strobe lights and meteors) is 5.1/hr. The RMT will operate in conjunction with the MIT Explosive Transient Camera survey instrument at Kitt Peak National Observatory, Tucson. The RMT is now being installed at Kitt Peak. Full operation will begin this summer.

1) INTRODUCTION

Gamma ray bursts were first observed twenty years ago with orbiting instruments designed to detect man-made nuclear explosions. Since then, all observations of these celestial gamma ray events have been with spacecraft or balloons above the atmosphere, but to date the astronomical source objects have not been identified. We have been developing a ground-based alternative approach based on the discovery (Schaefer, 1981) of an archived optical transient from a gamma ray burst source direction. Our new technique, based on studying optical transients, will make possible burst source location determinations that are much more precise than can be made with detectors of gamma radiation and more timely than can be made with spacecraft systems. Using optical astronomy to study gamma ray burst sources has several other advantages. In particular, ground instruments are cheaper than satellites and have longer duration than balloon flights. The duty cycle is unfortunately lower, but since we need only identify one or a few sources, rather than a survey, this is not a severe disadvantage.

1A) Scientific Justification

Gamma ray bursts (GRB) are transient cosmic events lasting from a fraction of one second to several tens of seconds in time. Their detected intensities are similar to those of solar gamma ray flares, yet their sources are at cosmic distances. In spite of this implication of a truly immense absolute intensity, their source objects have not been identified. The distribution of their source directions is consistent with isotropy, with no galactic disk clustering or other directional clue. The events occur at random times; until recently their detection rate was from ten to one hundred unrelated events per year, depending on the instrument sensitivity. (See survey and review articles, e.g., in texts edited by Liang and Petrosian, 1986; Woosley, 1984; Lingenfelter et al., 1982). Gamma Ray Observatory (GRO), with its more sensitive BATSE instrument,in its initial operation has seen about one GRB per day (G.J. Fishman,private communication). But GRO will not be able to locate the bursts accurately enough to make source object identifications possible.

The discovery of an archived optical transient (recorded in 1928) from the exact direction of a known classical gamma ray burst (detected in 1978; Cline et al., 1980) was made ten years ago (Schaefer, 1981). Several more such identifications (Schaefer et al., 1984; Moskalenko, et al. 1989) and other, looser or less direct, inferences (Pedersen et al., 1984; Hartmann et al., 1989) have also been made. However, it has not yet been possible to simultaneously observe an optical transient and a gamma ray burst. Nevertheless, if the same cosmic source that emits gamma ray bursts also produces optical bursts that are bright enough to be observed, whether the two transient types are simultaneous or not, then that source object can be identified optically. The great advantage of optical measurements over observations made with gamma ray instrumentation alone, of course, is the improvement expected in directional accuracy: less than 1 arcsec, compared with tens of seconds to minutes of arc with long-baseline burst timing networks.

It may never be possible to anticipate either the time or the arrival direction of a gamma ray burst, in order to pre-orient any sort of collimated, sensitive telescope. However, an optical device like the RMT can be reoriented and stabilized during the occurrence of the transient itself, assuming that the optical transients have durations similar to many of the gamma ray bursts (less than one to several tens of seconds). Thus, improvements in gain and resolving power over an isotropicly sensitive survey system may be possible with a high resolution, small field-of-view telescope . We therefore feel that our concept of a rapidly moving telescope, that can make precise observations of celestial optical transients, is an entirely new technique that may represent a breakthrough in the instrumentation of astronomy.

The Rapidly Moving Telescope (RMT) will measure many GRB parameters that are vital to both theorists and other observers. The RMT will position the burst to an accuracy that is 7, 4, and 3 orders-of-magnitude better in area than GRO, ETC, and the very best interplanetary network positions, respectively. This unprecedentedly high accuracy will allow for very deep searches for a quiescent counterpart at all wavelengths. One such study is the ratio of the gamma ray to optical brightness, a value which is very constraining to the theory of both optical flashes (Melia et al.1986; Melia 1988; Woosley 1984) as well as gamma ray bursts (Schaefer and Ricker 1983). Another is the comparison of the gamma ray and optical light curves. The relative smearing and time delays will identify the location of the flash as coming from a companion star, an accretion disk, or near the neutron star (cf. Lewin and Joss 1981).

1B) Approach

The known archived optical transients, thought to be associated with gamma ray bursts, shows them to be possibly so bright as to be visible to the human eye, assuming a one- to several-second duration. This result provides us with the assurance that the instrumentation we are developing will be more than adequate. Of course, the distribution of total event luminosities and durations, and the event intensity time history per event are several of the many unknown parameters of the optical transients. The rate of occurrence and/or rate of association with gamma ray bursts and the rate of possible repetitions are other unknowns. Nevertheless, assuming that optical transients of the sort found to be associated with bursts are not anomalous or rare, then we can be confident of the success of our telescope design.

The Rapidly Moving Telescope (RMT), relies on a sky monitor that will adequately identify celestial optical transients from the background noise. This device must measure the approximate source direction of the candidate transient and transmit this information to the RMT quickly enough so that the RMT can then be pointed towards that source and stabilized within a total time delay less than the duration of the optical burst. The RMT can acquire a target quickly because we move only a flat mirror for pointing. The sky monitor is called the Explosive Transient Camera (ETC) and is being built by MIT (Ricker et al., 1983). The ETC and the RMT are housed at Kitt Peak, each in its own dedicated structure.

The primary operating mode of the RMT is to observe optical transients detected by the ETC. To predict the number of GRB optical flashes that will be detected, we need to know the gamma ray to optical brightness ratio and the number of bursts as a function of gamma ray brightness. Both of these quantities are poorly determined, but for the best estimates the RMT/ETC system should detect roughly 20 bursts per year (Schaefer 1985a). These bursts should be bright enough in the gamma ray region to be detected simultaneously by the GRO satellite. It is this simultaneity that will guarantee the GRB nature of the observed flashes.

During the time between ETC transient coordinate detections, the RMT will operate in "stare" mode, where it watches a previous GRB error box waiting for a repeat burst. There are currently eleven known GRB error regions that are both visible from Kitt Peak and small enough to fit in the RMT's field of view. These positions are well spread out over the sky, so that the stare mode operation can follow a schedule where at least one burster is observable at any time. The recurrence time scale for optical flashes from gamma ray bursters is roughly one year, based on the four archival optical transients (Schaefer and Cline 1985). The sensitivity of the RMT for detecting a burst is several magnitudes better than the ETC, so any relatively faint bursts from the watched source will be spotted. Since the dynamic range of the luminosity function of the flashes is unknown, faint flashes might be quite common. The connection between the observed flash and the gamma ray burst phenomenon will be provided by the position inside the error box, and the elimination of local and instrumental events by detailed analysis of the image shape as well as its appearance in two successive frames.

2) **INSTRUMENT DESCRIPTION**

2A) *Overview*

The RMT is shown schematically in figure 1 and the primary characteristics are listed in Table 1. It consists of an 18 cm diameter fixed telescope looking down at an Az-El axis gimballed flat mirror. During observations the weather cover over the mirror mount and doors under the telescope enclosure open. The mirror mount servo electronics and data analysis computers are located in a building near the tripod. Rotation of the mirror ±90 degrees in azimuth and ±45 degrees in elevation allows viewing of all the sky except the 20% that is blocked by the tripod legs and the telescope enclosure. When the ETC detects an optical transient, it sends the corresponding coordinates (RA & Dec) to the Mirror Mount Servo computer. These coordinates will be accurate to ±2 arcminutes, which is sufficient to place the target within the 9x12 arcmin RMT field of view (FOV). The Servo computer then does the coordinate transformation to Az-El, calculates the slew profiles for each axis for the maneuver, does the slew, and then tracks the target at the sidereal rate. Simultaneous with the slew maneuvers, the Analysis computer starts the CCD camera control computer to start a sequence of approximately 100 1.5 second integration exposures.

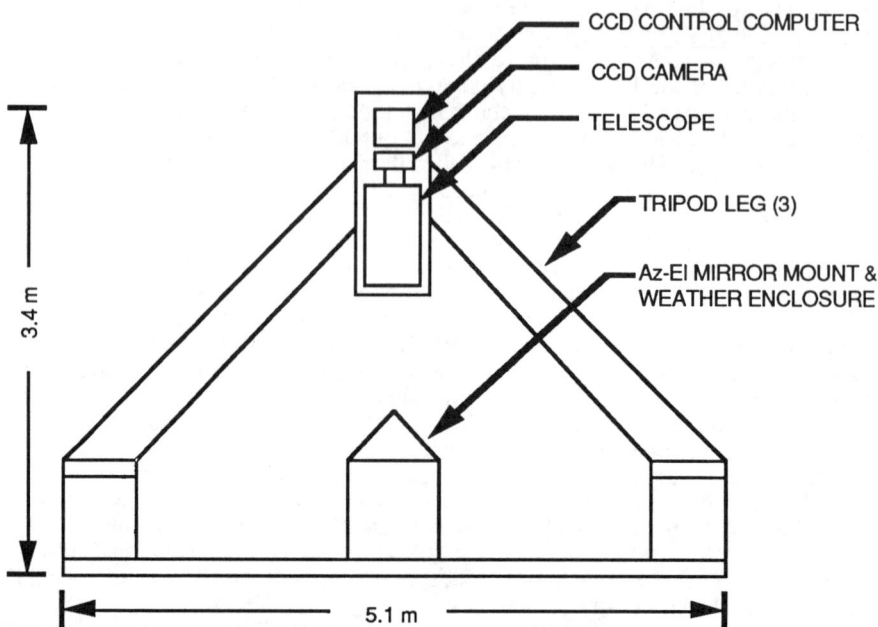

Fig. 1. The RMT instrument consists of a fixed telescope and CCD camera looking down at a Az-El gimballed flat mirror.

TABLE 1 - RMT CHARACTERISTICS

CCD
- Type: Texas Instr., 584x394, in frame store mode
- Pixel Size: 22x22 microns; 1.8x1.8 arcsec
- Integration Time: 1.5 sec (1.2 minimum)
- Readout Noise: 35 e$^-$ rms per pixel @ 100K pix/sec
- Operating Temperature: -40 to -50^0 C
- Quantum Efficiency: 0.75 @ 6000 Å
- Sensitivity: 14th mag

Telescope
- Type: f/14.3, 18 cm dia, Maksutov-Cassegrain
- Throughput: 0.8 (averaged over 3500-7000 Å)
- FOV (w CCD): 9x12 arcmin

Mirror
- Dimensions: 19 x 25 cm, quasi-elliptical outline
- Surface: λ/8, flat, Aluminized, SiO coated

Mirror Mount
- Max Acceleration: 1000 deg/sec*sec
- Max Velocity: 260 deg/sec
- Max Slew time: 1.0 sec
- Tracking Accuracy: 0.3 arcsec rms (typical); 1.0 (windy)
- Positional Accuracy: 2.5 arcsec over all 360 deg, 3σ

2B) CCD Camera & Control Computer, & Telescope

The CCD chip is a Texas Instruments model 4849, 584x394 pixel array, which we use in "frame store" mode. The array is divided into upper and lower halves with the lower half masked off from the telescope beam path. After the appropriate exposure integration time, the contents of the upper half are rapidly shifted into the lower half (3 msec) for the much slower readout and digitization (1.2 sec). This frame store method allows for simultaneous exposure (upper half) and readout (lower half); achieving more than a 99.7% observing duty cycle. The pixel intensity signal passes through a digitally switched four gain-level amplifier and is then digitized by a 12 bit ADC. The digital pixel data are sent to the image analysis computer over a high speed serial link. To reduce the thermal dark current, the chip is cooled thermoelectrically to a typical operating temperature of -40 to -50 °C. The CCD chip is contained in a vacuum housing so that contaminants do not condense on the open face of the cold chip and so that heat paths are minimized. The CCD is mounted at the focal plane of a ruggedized Questar 18 cm diameter, focal length 254 cm, focal ratio 14.3, Maksutov-Cassegrain telescope. The resultant pixel size is 1.8 arcsec. The total system throughput quantum efficiency is 60%.

2C) Two-axis Mirror Mount & Hybrid Servo Electronics

The mirror mount is a two axis azimuth-elevation gimbal system. Each axis has a brushless direct drive motor and a differential optical shaft encoder (resolution of 0.07 arcsec). The servo electronics to control the mount is a hybrid design of analog and digital electronics and software algorithms. There is separate circuitry for the azimuth and elevation axes, but the software algorithms are implemented on a single control computer (80386 pc machine). The PC machine receives transient coordinate packets from the MIT ETC instrument, converts them to azimuth and elevation, calculates and executes the slew maneuvers to that transient location. This hybrid system monitors and corrects each axis 400 times a second. Absolute calibration of the encoders over their entire 360 degree range is ±2.5 arcsec (3 sigma). The servo system design has closed position and velocity loops and an open acceleration loop. The dynamic range of the velocity loop is quite large (10^6), ranging from 250 deg/sec during a slew maneuver to a few arcsec/sec during tracking. The mirror is a flat, quasi-elliptical outline, light-weighted design. It has a $\lambda/8$ flatness, aluminized surface with a SiO coating.

2D) Data Collection & Image Analysis Computer

Figure 2 shows the flow of data from the CCD Control computer through the High Speed Serial Link to the Data Analysis computer. CCD exposure commands are sent from the Analysis computer to the CCD computer over an RS-232 link. After readout digitization (12 bits), the pixel information is transmitted over the serial link and DMA-ed into a frame buffer on the Analysis computer. Here it is threshold compared to a previously stored background frame and the above-threshold pixel information is stored on disk along with periodic full (no thresholding) reference frames and housekeeping information. The high speed serial link is capable of transferring 2 MBytes/sec, but because of readout noise limitations of the CCD we run it at 200 KBytes/sec. The analysis computer is a 68000-based multibus machine running AT&T UNIX V.2 with all the real-time and post-processing software written in the C language. The full data set (image plus housekeeping,~50MB/night) is sent back to GSFC via internet. All the post-processing analysis is done on a SUN SparcStation (see section 2G).

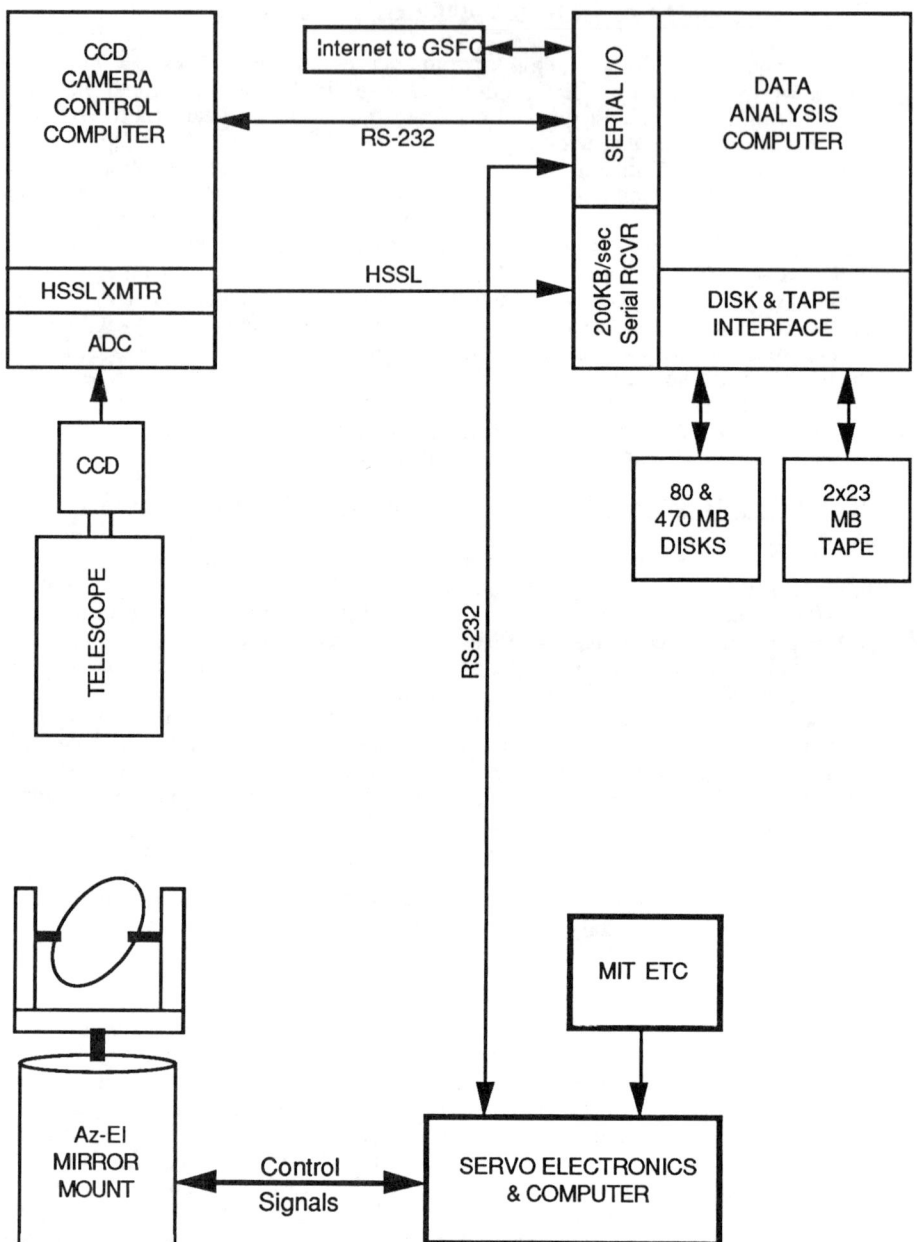

Fig. 2. The data analysis computer sends "start exposure" commands to the CCD control computer every 1.5 sec and the pixel intensity image data comes back across the High Speed Serial Link (HSSL). The MIT-ETC sends transient coordinate packets to the mirror mount servo computer, and it in turn directs the slewing of the mirror mount to the target and sends the coordinates and some other housekeeping information to the analysis computer for storage.

2E) Housekeeping & Control Subsystem

The RMT system is automated as much as possible, requiring the minimum of human intervention and servicing. To that goal we have implemented a system of sensing the environmental parameters and the state of the instrument, and a control system to operate the instrument. The housekeeping part monitors the weather (wind speed and direction, precipitation, dew point and condensation, and lightning activity). Once good weather and night time have been determined, the control system initiates a sequence of powering some of the electronics and computer subsystems and opening the mirror mount protective housing and the telescope doors. The status of the various subsystems is then checked for state of health, and then some preliminary observing is done to adjust the telescope focus and determine the cloud cover. If all checks pass then an observing program is begun (stare mode target selection and action upon any ETC transient coordinates). Periodically throughout the course of the night's observing, the weather, cloud cover, focus, and instrument state of health are rechecked. Appropriate action is taken for any parameter out of tolerance. This may include self-adjustment or the termination of the observing session. Daily reports will be sent back to the GSFC via computer phone link including the status of the instrument, the observing program, and a synopsis of the observing data.

2F) System Operation Time Line For Transient Mode Operation

One of the most critical RMT parameters is the maximum source acquisition time. From figure 3, the worst case acquisition time is 11.25 sec. This corresponds to the maximum delay for ETC detection (10.2 sec max) and maximum slew maneuver for the RMT (1.05 sec max). The minimum acquisition time is 0.25 sec with an average of 5.75 sec.

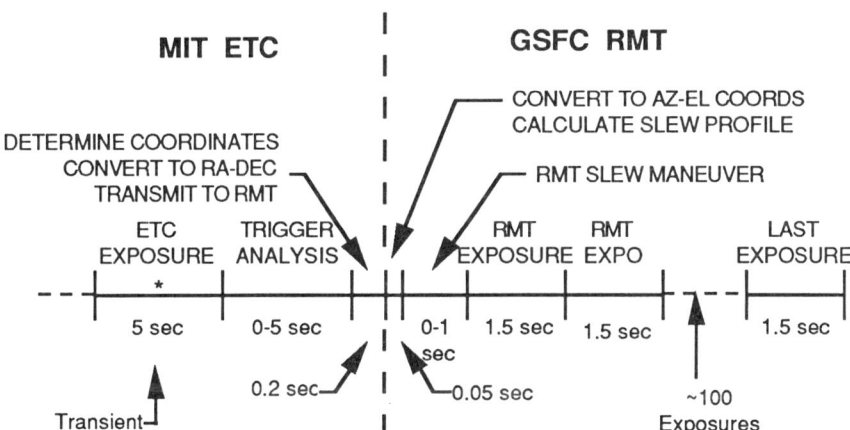

Fig. 3. A transient (*) event sequence starts with the ETC discovering a transient (a brightening of a pixel from one exposure to the next; 0-5 sec). The ETC then sends (0.2 sec) these coordinates to the RMT mirror mount servo computer where they are transformed into Az-El (0.05 sec), then the mount slews both axes simultaneously (0.0-1.0 sec), and a sequence of about 100 exposures are made of the transient position.

The ETC delay has a part due to the transient's relation to its exposure integration sync sequence(0-5sec) and a part from the detection analysis(0-5sec). This latter part is also variable because the delay time is related to exact position of the transient in the sequence of the processed ETC pixel stream. Once the RMT has received the transient coordinates from the ETC, it will switch from the preplanned "stare" mode operation to "transient" mode. This involves the transformation of the ETC RA-Dec coordinates into Az-El coordinates, calculating the slew profiles to slew from the current position to the transient position, executing the slew maneuver(0.05-1.05sec), and then tracking the transient target. A sequence of ~100 1.5 sec exposures will then be made.

2G) Data Analysis (Real-time & Post-processing)

As described in section 2A, there are two modes of operation for the RMT instrument. The data analysis procedure described below will be for the "stare" mode, which is the predominant mode based on the amount of data generated. It is also the mode for which we currently have any data (operating without the ETC at GSFC). The "transient" mode data analysis will be sketched in the last paragraph.

After the full frame image data is transferred from the CCD Control computer to the Analysis computer, there are three stages in the data analysis. Currently the first stage happens in real-time and the latter two happen off-line. The first step is the threshold "sifting" of each 1.5 second integration frame. About once every three hours a background frame (1.5 sec integration with the shutter closed) is taken and this is used to produce a threshold frame. Then for the sequence of stare mode frames each pixel is compared to its mate in the threshold frame. If it is of greater intensity, then its address and intensity are stored (along with some housekeeping information). At this point nothing is known about adjacent pixels, that is, how they are related spatially in true image structure. Every 10 minutes a full reference frame (12 sec integration) is stored so that background and fainter field stars can be recorded.

The next two steps are post-processing operations. The thresholded pixel data for each successive sifted frame are reconstructed into a full frame. Because of the thresholding in the first stage, this reconstructed image frame is ~99.6% empty. Now the pixels can be analyzed for the nearest neighbor relationship. All adjacent pixels that are part of a larger structure are, together, designated as "objects". These are typically field stars or in the case of a true optical transient some newly appeared object within the field of view. Those pixels which have no nearest neighbor pixels in the reconstructed image, called "single" pixels, are in general the result of CCD high sigma readout noise fluctuations.

The final stage is then to compare these objects in successive frames to see if new ones appear and when they disappear. In stare mode cosmic ray events and most optical transient events are rejected by requiring that a transient last for a minimum of two successive frames. True optical transients are further studied for their shape, light curve, position, motion, and other properties to determine if they are of astrophysical origin.

Currently, it takes less than half real-time to complete stages two and three. However, stage three is not finalized as the large number of "spurious" transients, due to just sub-threshold field stars that pop up above threshold, is unacceptable. More sophisticated algorithms are needed to identify and reject these events. While it is fully intended that the RMT run as a fully computer-automated, hands-off system, the final analysis of the set of "transients" resulting from the automated computer analysis will be done by a human in the loop.

As for the analysis of the ETC transient mode data, it will depend on the class(es) of events the ETC triggers on and their rates. The sequence of ~100 transient exposures will be thresholded just like the stare mode frames, because of data storage speed and size limitations. However, methods of circumventing this problem are being investigated so that full-frame, unthresholded data with its increased sensitivity can be obtained.

3) RMT INSTRUMENT PERFORMANCE

During the spring and summer of 1989, the Rapidly Moving Telescope was operated at Goddard Space Flight Center - in a remote site where optical nuisances are minimized - in the stare mode, since there was no ETC trigger available. This operation served as a shakedown period to fine tune the instrument operational procedures and data analysis procedure. The results of the instrument performance and the transient results are discussed below.

3A) CCD Magnitude Sensitivity & Angular Resolution

During the integration and testing phase of operation at the GSFC, we have attained a sensitivity of 13th magnitude for field stars and transients in the 1.5 second exposures in stare mode. The limit is 14th magnitude for the 12 sec full reference frame exposures. The observed stellar image FWHM has been 4 to 10 arcsec. This large number has two contributions. The first is that the Questar telescope has a temperature dependent focus. A 10°C change in the ambient air temperature (which is common during the course of a night's observations) will increase the FWHM by several arcsec. We are in the process of implementing a mechanism that will allow the analysis computer to automatically focus the telescope. The second contribution is the local seeing at suburban sea level Maryland. At Kitt Peak the average seeing is about one arcsec. The auto-focussing and better seeing at Kitt Peak should yield another magnitude improvement in sensitivity.

3B) Mirror Mount Position & Tracking Accuracy

The mirror mount, as explained in section 2C, is a high performance system. The 10^6 dynamic range in speed and position require tight mechanical and electronic tolerances. Figure 4a shows the position error (difference from the servo commanded position profile) from an actual 40 degree slew maneuver. Figure 4b is the position error at the end of the slew transitioning into normal sidereal tracking. The mount gets behind during the first half of the maneuver (a fraction of a degree) and then over-corrects in the second half. By the end of slew (0.56 sec) the mount is within a few arcsec of the commanded position. This position error during tracking was under conditions of 15 mph winds. Note the wind gust glitch at ~2.2 sec in the tracking mode, and that the mount fully recovered within 150 msec. The wind upper limit is 20 mph, as the servo stability of the mount quickly degrades for higher wind speeds. The tracking error has nominal 0.3 arcsec rms under no wind conditions and 1.0 arcsec rms (2.0 arcsec p-p) at 20 mph. The wind speed conditions at Kitt Peak are such that a 20 mph limit will allow us to observe on 80% of the nights. During normal stare-mode operation, exposures are rejected if the tracking error exceeds a threshold of 3 arcsec for more than 1% of the exposure interval.

The mount has two other contributions of positional error. The shaft encoder errors are less than 2.5 arcsec over the entire range with a value less than 1 arcsec over any 5 deg span (3 sigma). Because the mount also deviates mechanically from an ideal Az-El mount, a functional form describing the deviations is fit and used after the initial RA-Dec to Az-El coordinate transformation to make for the slew maneuver and during tracking. The functional form is shown in Eqns 1. Table 2 lists the physical origin of the twelve terms of the error correction equation. Included in this mechanical alignment correction is a correction for the atmospheric refraction. The magnitude of these corrections is ~10 arcmin for Az and ~1 arcmin for the El axis.

After applying the corrections the residual absolute pointing error is ~20 arcsec. This will be reduced by at least a factor of ten when the final alignment sightings are made at the Kitt Peak installation. This will make the pointing error equal to the pixel angular resolution. Any transient detections will always have field stars within the FOV.

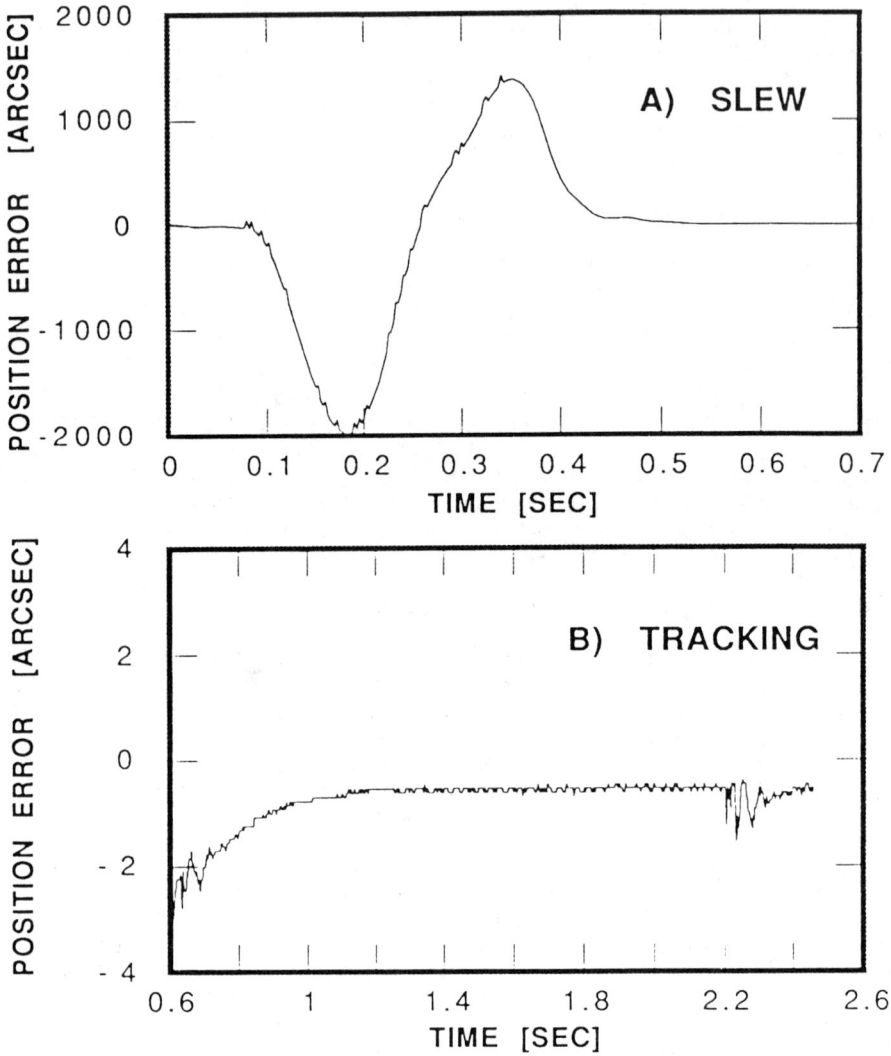

Fig. 4. This is the position error for a 40 degree azimuth slew maneuver (an elevation maneuver is similar). Panel A is the error during the actual slew maneuver (0.0 to 0.56 sec) and for a short time thereafter. The mount gets behind during the first half and then over-corrects during the second half, but approaches zero error at the actual end of the maneuver. Panel B shows the tracking error for the 2 seconds after the above slew. Within 1 second from the start of the slew the mount is stabilized at ~0.5 arcsec tracking error (due to static wind loading ,15 mph). The glitch at 2.2 sec is due to a wind loading transient. Within 150 msec the mount has fully recovered.

El_Corr = E1 + E2*COS(Az) + E3*SIN(Az) +
 E4*Atm_Pressure*COT(El) + 5*SIGN(El)   (1a)

Az_Corr = A1*SEC(El) + A2*TAN(El) + A3 + A4*COS(Az)*TAN(El) +
 A5*SIN(Az)*TAN(El) + A6*COS(Az) + A7*SIN(Az)   (1b)

TABLE 2 - ERROR CORRECTION TERM ORIGIN

PARAMETER	EXPLANATION
E1	El axis offset
E2	Az axis tilt in the East-West plane
E3	Az axis tilt in the North-South plane
E4	Atmospheric refraction
E5	Hemisphere sign dependent offset
A1	Traverse
A2	Non-orthogonal Az and El axes
A3	Az axis offset
A4	Az axis tilt in the East-West plane
A5	Az axis tilt in the North-South plane
A6	Telescope optical axis tilt in the East-West plane
A7	Telescope optical axis tilt in the North-South plane

3C) Observed Transient Rates

We distinguish two major types of "transient" events. Those which last only for a single exposure frame (present in the n-th frame but not in the n-1 or the n+1 frames) and those which last for at least two sequential frames. From the analysis of the 35 hours of stare mode data on a target near Polaris from the GSFC operation, only single frame transients have been found -- no multi-frame transients have been observed. Transients can be further broken down into two subtypes: instrumental and optical. The instrumental events are characterized by "image" sizes less than the true instrumental resolution (including the atmospheric seeing); typically a single pixel. Optical events are those with a proper image size (FWHM = 4-10 arcsec, see section 3A) and originate somewhere above several thousand feet altitude (airplane lights, meteor trails, satellite glints, and true astrophysical origins). The next four sections discuss the observed single frame transients. The first three (cosmic rays, CCD readout noise, and electronic noise bursts) are the instrumental type and the final section discusses the optical type.

3C1) Cosmic Rays

Sea level muons and electrons resulting from high energy cosmic ray interactions high in the earth's atmosphere produce the largest fraction of transient events in the CCD. The rate is, on average, one in 320 frames (7.2/hr). They are characterized by image shape profiles with large aspect ratios and always in the horizontal position angle (oriented along a row). They are further characterized by a single pixel with a large intensity (10 to 100 sigma above background) at the left-most pixel and a sharp fall off in intensity for the pixels trailing off to the right. This sharp distribution in pixel intensities is the result of two processes within the CCD chip structure. Typically, the ionization of the cosmic ray is confined within the volume of a single pixel and this results in the large intensity single pixel (at the left end of the image). The low intensity pixels to the right are the result of imperfect charge transfer efficiency (CTE = 0.9997) during the readout process. Occasionally there will be two or three pixels with a moderate intensity instead of the single large intensity pixel. This is caused by the cosmic ray's ionization energy spanning across a pixel's charge collection volume due to the non-zero stopping length of the delta rays, and because the cosmic ray penetrates below the depletion layer in the CCD and the ionization diffuses across to other pixels. Because this image signature is easily recognizable, we will be able to reject those chance coincidence two-frame transient events where one of the frames was a cosmic ray transient and the other was areal optical transient frame.

3C2) CCD Readout Noise Fluctuations

The threshold value used in the sifting of the stare mode exposure sequence is adjusted to balance the trade-off of minimizing the number of single-pixel upward-fluctuating pixels recorded versus the reduced sensitivity. As the threshold value is increased the number of single pixels decreases in an approximate gaussian distribution, but then the intensity required for a true optical transient is also increased. Typically, the threshold is set slightly greater than 3 sigma above the average background value. This yields of order 50 pixels per frame. These pixels are easily identified in the post-processing analysis because they are single pixels (no nearest neighbors -- not in an "object") and their intensities are just slightly above the mean threshold value. Because the number of these noise pixels varies in time due to CCD temperature drifts and night sky intensities, the threshold frame is recalculated several times during the night to rebalance these two effects.

3C3) Electronic Noise Bursts

This last contribution to the instrumental category of transients is of unknown origin. It manifests itself as a "burst" of single pixels during the thresholding process. Because it is a burst of noise, these single pixels are confined to a band of rows during the readout process, and there can be further clustering within a single row as well. The rate of occurrence is about once every 10,000 frames. At this point the source of these events is not known. It could very well be a GSFC "local" phenomenon. Since the rate is very low and the characterization is easily recognizable, no major effort has been made to identify the source other than standard electronic interference filtering and rejection techniques.

3C4) Optical Transients (yet to be identified)

This last category contains transient events which we believe to be optical and to originate at least more than 10,000 feet from the telescope. This would include airplane lights, meteors, satellite glints, and astrophysical sources. To date we have not identified any transient event which we believe to be astrophysical in origin. Based on the 35 hours of stare mode operation at GSFC, the rate of these optical transients is one in 500 frames (5.1/hr). Figure 5 shows one event of this category. This is a black & white (non-gray scale) display of a "reconstructed" stare mode frame image (see section 2G for a description of the reconstruction process in the data analysis). Only those pixels above threshold appear in the display (black). In the frame there are three field stars (circular objects), many single pixels, and the transient object (upper right corner and non-circular diagonal image). The array of pixel intensities for the transient object is shown in the panel above the CCD image frame. The single row trailing off to the right is an example of the imperfect CTE (see section 3C1) and is the result of the very intense pixel at the upper left edge of the image (intensity equal to 1435). Using typical numbers for the apparent angular speeds of satellites and airplanes, we believe this transient to be caused by a satellite, although an airplane scenario cannot be ruled out.

The transient in figure 5 can easily be identified as non-astrophysical because of its elongated shape. However, 90% of the transients in the "optical" category are circular in shape and have FWHM equal to the field star image profiles. Of these transients about half occur in pairs with angular separations ranging from 10 to 400 arcsec. The recurrence time for airplane strobes is much too long to cause these pairings. The distribution of intensities of these optical transients are most abundant at the 13th magnitude sensitivity limit, falling off in abundance with none seen greater than 8th magnitude. As with the above transient, these are believed to be satellite glints from satellites with small products of apparent angular motion and glint duration time. A much weaker argument based on the apparent magnitude of a verified recorded airplane strobe light flash also favors the satellite glint origin. Their stellar-like appearance mandates the two sequential frame requirement for identifying true astrophysical source transients.

0	0	0	0	195	111	0	0	0	0	0	0	0	0
0	0	0	153	1435	133	93	74	78	75	56	49	44	0
0	0	0	93	428	49	41	0	0	0	0	0	0	39
0	0	0	186	51	0	0	0	0	0	0	0	0	0
0	0	0	151	0	0	0	0	0	0	0	0	0	0
0	0	0	331	58	0	0	0	0	0	0	0	0	0
0	0	356	157	50	0	0	0	0	0	0	0	0	0
0	0	277	59	0	0	0	0	0	0	0	0	0	0
0	354	150	45	42	0	0	0	0	0	0	0	0	0
0	200	43	0	0	0	0	0	0	0	0	0	0	0
172	118	0	0	0	0	0	0	0	0	0	0	0	0
289	84	51	0	0	0	0	0	42	0	0	0	0	0
74	0	0	0	0	0	0	0	0	0	0	0	0	0

```
RMT

MAJ_CY       10
FR         1926

89Mar26 23:59:10
XCLK    52:11.522

THRESH      335
N_PIX       884
MEAN      350.1
SIGMA     110.4
AzTL       1.76
ElTL       1.76
B_LOSS        0
T_WAIT       75

GAIN          0
OFFSET       60
HT_CURNT      0
COOL1
COOL2
AMBIENT
Tccd        -43
Tset        -40
INT        1500
ROW     1   288
COL     1   394
```

Fig. 5. This is an example of the type of optical origin. This is a display of a "reconstructed" exposure (frame 1926 in the sequence taken in stare mode operation at GSFC on 26 MAR 89). In this frame there are three field stars (circular objects), many single pixels from readout noise, and the transient (upper right corner, non-circular). Inserted in the top panel is a listing of the pixel intensities for the sub-array of pixels for the transient. Only those pixels above threshold (38) are stored and appear in this display. This transient lasted for only one frame. It is most likely a satellite glint.

4) RMT FUTURE PROGRAM

The instrument has been disassembled and shipped to Kitt Peak. During several trips the tripod, telescope, and mirror mount were assembled and installed at our newly constructed facility. The facility consists of a building for housing the servo and analysis computers, other associated electronics, laboratory work space, the RMT mounting piers, and the ETC instrument enclosure. The electronics and computer integrations are complete. The automation, initial optical alignment and shakedown period will start shortly. The MIT-ETC instrument will be connected for joint operation this summer of 1991. Quarterly visits are planned for preventive maintenance and calibration checks.

REFERENCES

Cline, T. L., et al., 1981, Ap. J. (Letters), 246, L133-140
Hartmann, D, et al,, 1989, Ap. J., 336, 889
Lewin, W. H. G. and Joss, P. C. 1981, Space Sci. Rev., 28, 3.
Liang, E. P. and Petrosian, V. 1986, Gamma-Ray Bursts (New York: Am. Inst. Physics).
Lingenfelter, R. E., Hudson, H. S., and Worrall, D. M. 1982, Gamma Ray Transients and Related Astrophysical Phenomena (New York: Am. Inst. Physics).
Melia, F., Rappaport, S., and Joss, P. C. 1986, Ap. J. (Letters), 305, L51.
Moskalenko, E. I. et al. 1989, Astron. Ap., in press.
Pederson, et al., 1984, Nature, 312, 46
Ricker, G.R., et al.,1984 in High Energy Transients in Astrophysics, ed. S.E. Woosley (New York: Amer. Inst. Physics)
Schaefer, B. E. 1981, Nature, 294, 722.
Schaefer, B. E. 1985a, Astron. J., 90, 1363.
Schaefer, B. E. 1989, Ap. J., 337, 927.
Schaefer, B. E. 1989b, in Reference Encyclopedia of Astronomy and Astrophysics, ed. S. P. Maran (New York: Van Nostrand).
Schaefer, B. E. and Cline, T. L. 1985, Ap. J., 289, 490.
Schaefer, B. E. and Ricker, G. R. 1983, Nature, 301, 43.
Schaefer, B. E. et al. 1984, Ap. J. (Letters), 286, L1.
Schaefer, B. E. et al. 1987, Ap. J., 320, 398.
Warner, B. 1988, High Speed Astronomical Photometry (Cambridge Univ.).
Woosley, S. E. 1984, High Energy Transients in Astrophysics, (New York: Am. Inst. Phys.).

PART III.

FUTURE USES OF

ROBOTIC TELESCOPES

THE USE OF ROBOTIC TELESCOPES FOR DETECTING PLANETARY SYSTEMS

WILLIAM J. BORUCKI
NASA Ames Research Center, Mail Stop 245-3, Moffett Field, CA 94035

RUSSELL M. GENET
Fairborn Observatory, 3435 E. Edgewood Ave., Mesa, AZ 85204

ABSTRACT   Increases in the precision photometric measurements and the development of fully automatic photometric telescopes (APTs) now make possible the detection of Jupiter-size objects around other stars. The two most promising approaches are to: a) Search for brightness variations of binary stars that have their orbital plane nearly in our line of sight, and b) Measure the rotation periods of stars that are known to have dark companions and have very accurate values for the limb velocity, Vsini. The recent development of APTs makes it possible to routinely observe over one hundred stars each night without the expense of human operators. Differential photometry with these telescopes, based on rapid comparison among a group of stars, provide a precision of 2 parts in 1000. Transits of solar type stars by planets the size of Jupiter or Saturn produce brightness variations of approximately 10 parts per 1000. Consequently transits by planet-sized objects or by brown dwarfs should be readily recognized and be distinguishable from that of spots on the surface of the star. The spots can also be recognized by the short period of their repetition and by their characteristic duration as they cross the face of the star. The observation of spots on stellar surfaces allows an estimate to be made of the rotation period of each monitored star. A three year observation program can be expected to detect the transit of large planets around solar-like stars and to determine the masses of the dark companions discovered by radial velocity techniques. If Jupiter-sized dark companions in inner orbits are common and if the monitoring is continuous, then about 18 discoveries are expected.

INTRODUCTION

Because the formation of a planetary system is expected to be a consequence of the star formation process, it is expected that most stars are accompanied by planets. However, direct evidence that this is the case has been elusive so far. Determining if other stars are accompanied by planets is of direct human interest as well as scientific value. From a human viewpoint, it would be somewhat disconcerting to rationalists if ours were the only planetary system. In the words of the Solar System Exploration Committee (1986); "This search, even in the unlikely event that it provides evidence of our own solar system's uniqueness in the universe, will have a profound impact on both scientific theories and philosophical ideas about our place in the universe. If planets are common, then our theories are good, and life (and intelligent life) may be common in the universe. If planets are rare, our theories are in serious trouble, and humankind must face the possibility that we are alone in the universe."

From a scientific viewpoint, if we knew what proportion of stars were home to planetary systems, what types of stars tended to be accompanied by planets, and what their sizes and orbital characteristics were, we would be able to more fully understand how stellar and planetary systems form and evolve. In fact, one of the principal conclusions of the 1983 Workshop on The Detection and Study of Other Planetary Systems (Black, editor, 1983) was that "Evidence concerning the frequency of occurrence and detailed structure of other planetary systems would profoundly affect efforts to understand the origin of the solar system, and would be of great value in understanding the problem of star formation."

OVERVIEW OF APPROACHES TO PLANETARY DETECTION

Black, 1980, and Tarter et al., 1985, present summaries of the various techniques that have been considered for detecting other solar systems. Thus we present only a brief synopsis of the various methods. Each of the techniques to find other solar systems makes severe demands on technology, has serious drawbacks, and is best suited for a particular aspect of the search.

Because the central star is so much brighter than any of its planets, and because of light scattering by the Earth's atmosphere and by the telescope optics, the method of direct imaging (Smith and Terrile, 1984, Terrile, 1988) requires a large spaceborn telescope even to find Jupiter-sized planets (Tarter et al., 1985). Nevertheless, direct imaging of large planets might be done by future generations of large telescopes that employ new methods to reduce scattered light from their optical surfaces and block out the light from the primary star. Although the planets would not be resolved, and would be point-like specks, it is possible that spectroscopic techniques could be used to determine the atmospheric composition of these planets. If oxygen- or water- bearing atmospheres were found, the discovery would imply the possibility of life on those planets. Unfortunately this method is by far the most difficult and expensive method, and could only provide data on a few, very close by stars. However it shows a great deal of promise for investigating dust disks around stars which might be the remnant of the formation of planetary systems.

The astrometric method uses long-term measurements of a star's position to search for cyclic variations that indicate the presence of orbiting planets or dark stellar companions. This method has been used for many years by ground based observatories to determine the movement of stars, but the accuracy is limited to a few milliarc seconds by the presence of the earth's atmosphere. An engineering study has shown that it should be possible to increase the instrument precision by a factor of 100 or more and thus reach a precision of 0.01 milliarc second. With this precision it should be possible to find Jupiter-size planets around hundreds of nearby stars (Gatewood, 1987, and Reasenberg et.al., 1988). Future generations of a space-based astrometric telescopes might be based on an interferometric approach which should increase the resolution and allow less massive planets to be detected.

Radial velocity techniques (McMillan, et al. 1986) measure the reflex velocities of stars due to orbiting companions. These techniques have already detected four or five stars that show the presence of a dark companion that might be a planet rather than a brown dwarf (Campbell et al., 1988, Latham et al., 1989). Although the radial velocity technique has already found several stars with a dark companion, the technique cannot determine the mass of the companion unless the inclination angle "i", of the orbital plane from the observers line of sight can be determined. In principle, this angle can be determined from a quantity called "Vsini" and the rotation period of the star (Borucki and Summers, 1984, and Campbell, 1985). (Vsini is the product of the velocity V, of the limb of the star in the observer's line of sight and the sine of the inclination angle.) With a determination of the rotation period and Vsini, the inclination angle can be determined from :

$$i = \arcsin\left(\frac{t \text{ Vsini}}{\pi d}\right) \qquad (1)$$

where t is the stellar rotation period and d is the diameter of the star. The stellar diameter can be estimated from a knowledge of the spectral and luminosity class of the star. The rotation period can be estimated from the measured photometric variation due to inhomogeneities from starspots or plages if a high enough photometric precision can be attained.

Of course the photometric determination of the stellar rotation period is not sufficient; Vsini must also be measured by an appropriate high resolution spectrometer. Note that Soderblom (1985) has warned that the tabulated values of Vsini are sometimes too uncertain to obtain useful inclination angles in this way. Only high resolution measurements of Vsini that represent the actual motion of the stellar limb, rather than other phenomena, are useful.

Rosenblatt (1971), Borucki and Summers (1984), and Borucki et al. (1985) discuss high-precision photometric methods that have the potential of detecting planets by searching for the variation of stellar brightness and color that occurs when a planet transits its star. The magnitude of the brightness reduction is proportional to the ratio of the planet's area to that of the star. When observing stars the size of the Sun, the decrease in light amounts to 1% for giant planets such as Jupiter and Saturn, 0.1 % for planets like Uranus and Neptune, and 0.01 % for Earth-sized planets. A characteristic color change with an amplitude approximately 0.1 of that for the brightness reduction would also accompany the transit and could be used to verify that the source of the brightness variation was a planetary transit rather than some other phenomenon. This color change is caused by the variation of limb darkening with wavelength. Another way to verify that the transit was caused by a planet is to look for a repetition. Because the photometric technique strongly favors the detection of planets with short period orbits, such a wait should be short, typically one year or less. Further, from measurements of the transit time, a prediction of the next transit can made. The transit times range from 8 hours for planets with an orbital radius of 0.4 AU (Mercury's orbit) to 30 hours when the orbital radius is 5.2 AU (Jupiter's orbit). The transit time is proportional to the ratio of the stellar diameter to the orbital period of the planet. The transit time is only slightly affected by the size of the planet.

The photometric method only works for those objects whose orbital plane is near the line of sight between the Earth and the central star. Because the probability of observing a transit during a single observation is small, the search program must be designed to continuously monitor many stars so that a significant discovery rate can be attained. With the advent of techniques to make photometric precision of 2 millimagnitudes and better and the development of automatic photometric telescopes (APT) that acquire and make over 600 photometric measurements per night, systems now exist that can collect large amounts of high precision data at low cost. We will show that the use of this approach to search for major planets and brown dwarfs is now practical.

As suggested earlier, there is another way in which photometric observations can contribute to a determination of the frequency of planetary systems in our galaxy. The second method is to determine the rotation periods of stars already known to have dark companions from radial velocity observations. Here the objective is to estimate the mass of the dark companion by determining the inclination angle from measurements of Vsini and t. (See Equation 1.) At present, the radial velocity technique detects the presence of dark companions, but cannot determine whether these companions are dark stellar companions or massive planets. The distinction between the two possibilities is based on mass. Those companions larger than 0.1 solar masses have enough mass to initiate hydrogen burning and are clearly stars and those objects with the mass of Jupiter (i.e., 0.001 solar masses) or less, are clearly planets. Dark companions with intermediate masses, often called "brown dwarfs", can be considered to be either a failed star or a supermassive planet. If the mass of the dark companion can be established from the stellar rotation period and Vsini, then, for the first time, we should be able to determine whether these already observed companions are planets, brown dwarfs, or stars. Of course the photometric determination of the stellar rotation period is not sufficient; Vsini must also be measured by an appropriate high resolution spectrometer.

The photometric method only works for those objects whose orbital plane is near the line of sight between the Earth and the central star. Because the probability of observing a transit during a single observation is small, the search program must be designed to continuously monitor many stars so that a significant discovery rate can be attained. With the advent of techniques to make photometric precision of 2 millimagnitudes and better and the development of automatic photometric telescopes (APT) that acquire and make over 600 photometric measurements per night, systems now exist that can collect large amounts of high precision data at low cost. We will show that the use of this approach to search for major planets and brown dwarfs is now practical.

HIGH PRECISION PHOTOMETRY

Although space-based radiometric measurements of the Sun already attain a relative precision of 1 part in 100,000 (Willson et al., 1981), current stellar photometry, which must observe much dimmer objects through the terrestrial atmosphere, obtain a far lower precision. Even at good observing sites, common techniques often attain a precision of no better than 1 part in 100 (Young et al., 1991). However with careful attention to detail, the frequent observation of extinction stars, the use of filters that are closely spaced in wavelength, stable filters and detectors, careful choices of comparison and check stars, and special attention to the best data reduction techniques (Young 1974, 1988), photometerists (Waelkens and Rufener, 1985, Radick et al., 1987, Lockwood and Skiff, 1988) attain a season-to-season precision of 500 (i.e., 2 parts in 1000 or 2 millimagnitudes). Such precision is adequate for the determination of the rotation rates of many stars and to detect transits by Jupiter-sized objects.

DETECTION PROBABILITIES

Let p be the probability that the orbital plane is properly oriented so that transits can be observed, d be the stellar diameter and D the orbital diameter. In Borucki and Summers (1984) it was shown that:

$$p = d/D \qquad (2)$$

Equation (2) shows that p is largest for close-in planets of large stars. Table 1 gives p for planets in our solar system.

TABLE 1 Probability of a favorable alignment as a function of orbital radii for planetary systems with an arbitrary orientation relative to our line of sight

Orbital Radius (AU)	Planet	Probability
0.4	Mercury	0.012
0.7	Venus	0.0064
1.0	Earth	0.0046
1.5	Mars	0.0030
2.8	asteroid belt	0.0017
5.2	Jupiter	0.00089
9.5	Saturn	0.00049
19.2	Uranus	0.00024
30.1	Neptune	0.00015
	Sum =	0.03

It is clear from Table 1 that if stars are chosen with randomly oriented planetary orbital planes, then the probability of observing a planetary transit is only about 3% and that the largest probabilities are associated with the innermost planets. If other planetary systems are like our own in having only small planets in short-period orbits, then we must sum only those probabilities for planets that are sufficiently large that they can be detected by the proposed observing system. Although space based systems should be able to detect Earth-sized planets, ground-based observations will be affected by the noise introduced by observing through the Earth's atmosphere which will make it difficult to detect planets smaller than Uranus and Neptune (Borucki et al. 1984). If we assume that large planets do not form in inner orbits, then either many stars must be monitored or we must find a method of choosing stars most likely to have their planetary orbital planes near our line of sight. Note that the probabilities tabulated above assume that the observations are conducted for one orbital period. For the inner planets, this is a fairly short period, but for the outer planets, this period is so long that the expected yield during a several year search would be insignificant. However, if other planetary systems have large planets or brown dwarfs in inner (i.e., 1.5 AU or less) orbits, then monitoring 100 solar type stars each night should produce a discovery rate of approximately 2 discoveries per year. Thus a search conducted for a period of five years should provide a useful estimate of the frequency of Jupiter-size companions in inner orbits.

Although it is possible to use automated photometric telescopes (APTs) to observe many stars each night, a more promising approach it to choose stars that have their planetary orbital planes near our line of sight so as to increase the likelihood of detections. In particular, if our observations are confined to binary stars that have their orbital planes within "i" degrees of our line-of-sight, then the probability p, of a detection at any orbital radius is increased by factor of $(1/\sin i)$. This value is derived from a consideration of the fraction of the celestial sphere that a transit can be observed if the orbital plane is randomly distributed over all angles, compared to the fraction of the sphere that is delimited by $\pm i$ degrees. Thus;

$$p = d/(D \sin i), \qquad (3)$$

where d is the diameter of the star, D is the diameter of the planetary orbit, and i is the inclination of planetary orbits relative to the orbital plane of the stars. Note that p can never exceed 1 for any orbit. If we observe only binaries with i less than or equal to 3 degrees and assume that most of the planets will be within 3 degrees of the invariable plane (as is the case for our solar system), then the probability of observing a transit is increased by a factor of 19 over what it would be if stars with arbitrarily oriented orbital planes were observed. Table 2 shows the probability per star of observing a transit of a planet at various orbital radii from its primary. For such observations, the probability of observing a planet within the orbit of Mars increases to 50% per star observed. (These calculations assume that all stars have planetary systems.) However, most of the probability is associated with the innermost planets. Again, space based observations look quite promising in that they should be able to detect planets as small as Earth and Venus and should be able to quickly determine if such planets are common. Ground based observations will be successful only if larger planets form within a few AU of the central stars. If planets large

TABLE 2 Probability of a favorable alignment as a function of orbital radii for observations within 3 degrees of the orbital plane

Orbital Radius (AU)	Planet	Probability
0.4	Mercury	0.226
0.7	Venus	0.122
1.0	Earth	0.088
1.5	Mars	0.058
2.8	asteroid belt	0.0314
5.2	Jupiter	0.0188
	Sum =	0.544

enough to be detected can form only near the orbit of Mars, then the probability of a transit observable from a ground based system is only about 5% and if Jupiter-sized planets are only found at 5 AU from the central stars, then the probabilty of a detection is only 2%. Note that Schneider and Chevreton (1990) have shown that the probability of a transit occurring in a binary system is not substantially increased from that for a single star. Only the number of transits per planetary orbit is increased. A difficulty with searching for planets around binary stars is that it is not known whether planets can form near such stars. Numerical calculations by Pendleton-Schisler and Black (1983) suggest that stable planetary orbits are possible if the diameters of the planetary orbits are less than 1/5 or greater than 5 times the distance between the stars. Whether Jupiter-mass companions actually form in binary systems can only be determined from observations. There are over 40 known solar-type binaries brighter than twelfth magnitude with orbital planes within 3 degrees of our line of sight. Consequently a photometric search for major planets with orbital radii less than 1.5 AU should be able to determine whether or not binary stars have major planets in inner orbits. To detect the small variations in stellar brightness that are characteristic of rotating stars and planetary transits, not only must the precision of the measurements be maintained, but many observations are necessary as well.

ROBOTIC TELESCOPES

The photometric detection of planets requires that one or more telescopes be dedicated to observations for a period of several years. If the telescopes had to be operated manually, the cost of the many telescope operators would make photometric planetary detection rather expensive. However, completely automatic photoelectric telescopes are now well developed and are in routine operation for a number of research programs.

For example, the Automatic Photoelectric Telescope (APT) Service at the Fairborn Observatory currently has seven robotic telescopes are housed under one large roll-off roof in a totally automated observatory. Four of these telescopes have apertures of 0.75 m and they were designed specifically for automatic photometry. They slew rapidly, find each star and center it with a CCD camera, and check that each star has the correct magnitude. Because the telescopes are designed to do nothing but automatic photometry, they readily make over 600 ten second observations per night, even during the short nights of summer. (During a winter night, its is not unusual to obtain over 1100 such observations.) Of course, the actual number of stars that can be monitored depends on the desired precision and the frequency that each star must be reobserved. Integration times of approximately 40 seconds are required to get scintillation noise down to 1 part per thousand and integration times for faint stars are larger than for bright stars. Stars must be observed in at least two colors to correct for color dependent extinction. Comparison and check stars must also be observed.

It has been suggested that various robotic telescopes around the world could be formed into a Global Network of Automatic Telescopes (GNAT). A global network can overcome the limitations of the day/night cycle, poor weather, equipment failures, etc., and thus allow continuous coverage or sampling of stars for extended periods of time. Continuous (sampled) coverage of stars suspected of having detectable planets would allow more powerful conclusions to be drawn about the absence of planets than would be possible with intermittent coverage via a single telescope, where a transit might have occurred during the daytime, during bad weather, or during a time when the system was not operational. Furthermore, although an observation of only the beginning or the end of a transit is sufficient to detect the presence of a companion, it would be very helpful to observe the entire transit so that the orbital period could be estimated and the time of the next transit predicted. Because transits periods are expected to range from several hours to a day, it is unlikely that the entire transit will be observed from a single telescope site.

Within a few years there should be a sufficient number of automatic telescopes in place to make possible a limited amount of nearly continuous observations. However, as the bulk of the observing time on these systems is likely to be devoted to local (i.e., non-networked) observing projects, the number of objects that could be continuously observed will be quite small. To keep a larger number of objects under nearly continuous observation will require a global network that is primarily dedicated to such observations.

EXPECTED RESULTS FROM A GROUND BASED PHOTOMETRIC SEARCH FOR DARK COMPANIONS

We have identified two situations that ground based telescopes can exploit to give some insight into the frequency of planetary systems around other stars. Observations of binary stars that have orbital planes within a few degrees of our line of sight appear very promising. Stars which are known to have dark companions from previous radial-velocity investigations should also be monitored. For the first situation, we seek to observe transits. For the second situation, our goal is to determine the rotation periods and to use these periods with spectroscopic measurements to deduce the masses of the dark companions.

So that the signal from a transit will be large enough to readily be detected, the ratio of the diameters of the primary to the companion must not be much larger than a factor of ten. Since the companions that we are searching for (whether the companion is a brown dwarf or a major planet) will be about the size of Jupiter, primaries larger than twice that of the Sun are not acceptable targets. Hence observations of giant stars or spectral classes earlier than A5 are not likely to be successful. The advantage of observing spectral classes somewhat earlier than that of the Sun is that they show very little variability and thus the transit of relatively small companions should be detectable. The advantage of the later spectral types is that the primaries will be smaller and thus the transit signals from Jupiter-sized objects will be larger. For dM5 stars, the primary is only one tenth the area of the Sun. Consequently even objects as small as Uranus and Neptune will provide large signals. Furthermore, if the formation of the giant planets is dependent on the stellar heat flux at the protoplanet's position, then such planets will form much closer to a dM5 star than Jupiter did to the Sun. For a giant planet orbiting a dM5 star at a distance for which the planet experiences the same stellar flux as that experienced by Jupiter around our Sun, the orbital period will be less than one year. Thus not only is the absolute probability of a transit occurring enhanced because d/D in Eq. 2 is increased, but the probability per year of observation is also increased by an even larger factor. We estimate that probability of observing a transit of a giant planet around a dM5 star is of order 0.5% per year if each star has a planetary system. This probability can be increased by an additional factor of 19, if binary stars with their orbital plane within a few degrees of our line of sight, can be found.

To get the photometric precision necessary to detect a transit, differential photometry should be used. Therefore a comparison star (and, occasionally, a check star) must also be observed for each target star. Both comparison and check stars must be observed so that any variability in these stars can be recognized and removed. To minimize corrections, these stars are chosen to match the brightness, color, and location of the target star as closely as possible. Because many of these stars are expected to show brightness variations due to the presence of spots or plages, it should be possible to determine the rotation rates for many of these stars. This data should add to our information on the absolute rotation rates of many stars. A list of some of the binaries that would make good targets is given in Table 3. Because of the wide spacing between the visual binaries, separate planetary systems could exist about both stars. In the case of the short period eclipsing binaries, planetary orbits are expected to be stable (Pendleton-Schisler and Black, 1983) at five times the distance of separation of the two stars. For stellar orbital periods of 5 days and a combined stellar mass of twice solar, this criterion implies planets with orbital radii greater than 0.37 AU will be stable; i.e., all the planets in our solar system, including Mercury, would be in stable orbits if our Sun were replaced with a double star with a period of five days or less.

There are approximately 5 groups in North America conducting radial velocity searches with sufficient precision to detect brown dwarfs or Jupiter-mass companions (Campbell, previously of Dom. Astrophys. Obs.; W. Cochrane, MacDonald Obs.; R. McMillan, Univ. Ariz.; G. Marcy, Lick Obs.; D. Latham, Harvard-Smithsonian Center for Astrophysics). Other groups outside the US are also making such measurements. Three groups have found some evidence of companions with masses of the order of ten Jupiter-masses (Latham et al., 1989, Campbell et al., 1988, Marcy and Benitz, 1989). If we assume that more stars will be discovered as each group improves its precision and as the observation period of their measurements increases, then it is clear that a substantial, and increasing, number of stars should be observed photometrically to determine the rotation rates. At the present time there is a effort to greatly expand these radial velocity searches because of their promising results. If we assume that each group will discover an average of four stars with dark companions, then observations of approximately 20 stars should be conducted for approximately a year to get trustworthy estimates of their rotation periods. If new telescopes are put into operation for these searches, then the rotation rates for a 100 or more stars will be needed. For each such star, a comparison star, and a check star must also be measured to obtain the necessary photometric precision. However, these stars do not need to be monitored as frequently as the stars that are being monitored for the occurrence of a transit.

TABLE 3   Candiate binary stars

Star	Binary type	Inclination Angle(deg)	Period
RY Gem	Eclipsing	87.2	9.3 d
Aql v889	Eclipsing		11.1 d
RS Ceph	Eclipsing	85.4	12.4 d
RW Per	Eclipsing	86.4	13.2 d
EY Ori	Eclipsing	90	16.8 d
Delta Cass	Eclipsing		759 d
Alpha Com Ber	Visual Binary (VB)	90	26 yr
Peg Sigma 536	VB	90.0	26.5 yr
Lac H925	VB	90.0	30 yr
Tau 1080	VB	92.6	40 yr
Ori a847	VB	90.0	48 yr
HD199766	VB	92.8	101 yr
Dra 2384	VB	85.5	137 yr
Peg 37	VB	89.0	140 yr
37 Peg	VB	88 < i < 92	150 yr
HD1624	VB	88.5	163 yr
And Sigma21	VB	87.0	450 yr

Because APTs are designed to do nothing but automatic photometry, they readily make hundreds of ten-second observations per night. Nevertheless, there are real limits to the number of stars that can be monitored during a single night with a single channel photometer. To reduce the scintillation noise to 1 millimag for an airmass of 1.5, the integration period must be approximately 40 seconds long. A comparison and a check star and the sky must also be observed. Some observations must be with a B and V filters to determine the color correction to the extinction coefficient. However, it is not necessary to measure the check star as frequently as the comparison star and it's not necessary to make measurements in two colors at every observation. Because of the changing sky conditions, it is important to keep individual integrations to no longer than 10 seconds so that all the stars in the group and the sky can be observed within a 60 second interval. Consequently, the time required to observe the entire group of stars and sky and to measure the detector dark current and to slew between the stars within the group and to slew to the next group of stars will average approximately 5 minutes. If 15 groups are monitored for transits throughout the night, then the observations can be repeated each 1 1/4 hours. If another quarter hour each period is added for observing stars that are being monitored to determine their rotation rates, then the repetition interval will be 1 1/2 hours. This interval should be short enough to allow the time history of the transits to be determined.

The expected number of detections is calculated for orbital radii for planets in the orbits of the solar system planets. It is not known what orbits in other planetary systems will be occupied, nor what the size distribution of the planets will be. Hence the expected number of detections must be considered speculative. Only the actual observations of other planetary systems can determine whether our solar system is representative or unusual. Nevertheless, this table allows us to estimate the expected number of detections for various assumptions about the radii of planetary orbits. The calculations assume that all stars have a planetary system with planets large enough to be detected and that only solar-type stars are used for the comparison and check stars. The combined mass of the binary stars is assumed to be two solar masses. Table 4 assumes that a set of 15 binary stars with orbital planes within 3 degrees of our line of sight and an additional 30 single stars used as comparison stars are monitored on a nearly continuous basis for a period of four months and that a new set of 15 binary stars and reference stars is chosen each four months. Thus, over a one year period, a total of 45 binary stars and 90 reference stars could be monitored. An additional 18 stars (and their reference and check stars) can also be observed once each night for four months to determine their rotation rates, but these observations are likely to be too sparse to observe the within-night brightness variation needed to demonstrate that a transit has occurred.

Table 4 implies that about 18 companions should be detected if binary stars have large planets or brown dwarfs within 0.5 AU of center of mass. Even if none are found, a useful estimate of the frequency of such companions can be made. Even out to orbital radii of 1.5 AU, useful statements can be made about the frequency of companions. Although continuing to observe these same stars would not increase the probability of detection for companions with orbital periods of four months or less, continued observation would increase the probabilities of companions with longer period orbits. A three year period could be expected to triple the number of observations of companions that have orbital periods of a year or more.

TABLE 4   Expected number of dark companions observed in one year of monitoring as a function of orbital radius

Orbital Radius (AU)	Single Stars	Binary Stars
0.4 (Mercury)	1.071	10.2
0.7 (Venus)	0.194	4.2
1.0 (Earth)	0.140	1.9
1.5 (Mars)	0.048	0.65
2.8 (Asteroids)	0.010	0.14
5.2 (Jupiter)	0.002	0.03
Sum =	1.5	17

It should be noted that Table 4 does not account for a variety of important conditions. Necessary corrections to Table 4 include: 1) visual binaries can have a planetary system around each of the two stars and therefore they have a detection probability twice as large as assumed in Table 4, 2) unless several telescopes widely spaced in longitude are used, the monitoring will miss some of the transits because of bad weather and because short transits that occur during daylight would be missed, and 3) telescopes will be out of service for repairs for a portion of the time. Therefore, unless a global network is employed, the usefulness of the results could be seriously compromised.

Lockwood and Skiff (1988) found rotation rates for 25% of the solar-like stars that they observed because of the high precision (1 part in 500) that they were able to attain.

Because the same precision has been demonstrated with APTs, the determination of the rotation rates for about 35 stars is expected. For those stars with dark companions and with the appropriate measurements of Vsini, the masses of the dark companions should be determined.

SUMMARY

Increases in the precision of photometric measurements and the development of fully automatic photometric telescopes now make possible the detection of Jupiter-size objects around other stars. The two most promising approaches are to: a) Monitor binary stars that have their orbital plane nearly in our line of sight, and b) Measure the rotation periods of stars that are known to have dark companions and have very accurate values for Vsini. The recent development of APTs makes it possible to routinely observe approximately one hundred stars each night without the expense of human operators. Differential photometry with these telescopes, based on rapid comparison among a group of stars, provides a precision of 2 parts in 1000. Transits of solar type stars by planets or brown dwarfs the size of Jupiter or Saturn will produce brightness variations of approximately 10 parts per 1000. Transits of smaller planets orbiting spectral class M stars should also be detectable by ground based telescopes. The observation of spots on stellar surfaces allows an estimate to be made of the rotation period of each monitored star. A one to three year observation program can be expected to detect the transit of large planets or brown dwarfs, if such objects are common, around solar-like stars and to determine the masses of the dark companions discovered by radial velocity techniques. If the monitoring is continuous, then a useful estimate of the fraction of stars with dark companions in inner orbits should be obtained.

REFERENCES

Black, D.C. (1980). *Space Sci. Revs.* 25, p. 35.

Black, D.C., editor (1983). Science Workshop Report to the National Aeronautics and Space Administration. NASA Ames Research Center, Moffett Field, CA 94035.

Borucki, W.J., and A.L. Summers (1984). *Icarus* 58, p.121.

Borucki, W.J., J.D. Scargle, and H.S. Hudson (1985). *Ap. J.* 291, p. 852.

Campbell, B. and R. F. Garrison (1985). *Pub. Astron. Soc. Pac.*

97, p. 180.

Campbell, B., G.A. Walker, S. Yang (1988). *Astrophys. J.* 331, p. 902.

Gatewood, G.D. (1987). *Astron. J.* 94, p. 213.

Latham, D.W., T. Mazeh, R.P. Stefanik, M. Mayor, & G. Burk (1989). *Nature* 339, p. 38.

Lockwood, G.W. and B.A. Skiff (1988). Final Report AFGL-TR-88-0221, Air Force Geophysics Lab.

Marcy, G. W., and K. J. Benitz (1989). *Ap. J.* 344, p. 441.

McMillan, R.S., P.H. Smith, J.E. Frecker, W.J. Merline, M.L. Perry (1986). *S.P.I.E.* 627, Instrumentation in Astronomy VI, p. 2.

Pendleton-Schisler, Y., and D.C. Black (1983). *Astron. J.* 88, p. 1415.

Radick, R.R., D.T. Thompson, G.W. Lockwood, D.K. Duncan & W.E. Baggett (1987). *Ap. J.* 321, p. 459.

Reasenberg, R.D. et.al. (1988). *Astron. J.* 96, p. 1731.

Rosenblatt, F. (1971). *Icarus* 14, p. 71.

Smith, B.A. and R.J. Terrile (1984). *B.A.A.S.* 16, p. 702.

Schneider, J., and M. Chevreton (1990). *Astron. Astrophys.* 232, p. 251.

Soderblom, D.R. (1985). *Pub. Astro. Soc. Pac.* 97, p. 57.

Tarter, J.C., D.C. Black, and J. Billingham(1985).Paper IAA-85-493: Proceedings of the 36th International Astronautical Congress, held in Stockholm, October 1985

Terrile, R.J. (1988). The Formation and Evolution of Planetary Systems. Edit. by H.A. Weaver, F. Paresce, and L. Danly. Space Telescope Science Institute, Baltimore MD.

Waelkens, C., and F. Rufener (1985). *Astron. Astrophys.* 152, p. 6.

Willson, O.C., S. Gulkis, M. Janssen, H. S. Hudson, and G. A. Chapman (1981). *Science* 211, p. 700.

Young, A.T.(1974). In Methods of Experimental Physics, Vol. 12A. Edited by N. Carleton., p. 123.

Young, A.T. (1988). Second Workshop on Improvements to Photometry. NASA CP 10015.

Young, A.T., R.M. Genet, L.J. Boyd, W.J. Borucki, G.W. Lockwood, G.W. Henry, D.S. Hall, D.P. Smith, S.L. Baliunas, R.Donahue, and D.E.Epand (1991). P.A.S.P. 103, p. 221.

A DIRECT CENSUS OF THE OORT CLOUD WITH A ROBOTIC TELESCOPE

T.S. Axelrod, C. Alcock, K.H. Cook, and H.-S. Park
Lawrence Livermore National Laboratory, Livermore, CA 94550.

ABSTRACT A robotic system has been designed to search for the occultations of stars by comets in the Oort Cloud. The system employs a series of small cameras to follow the shadow of a comet as it moves along the surface of the earth. Since the expected event rate is extremely low, particular care is needed to ensure a very low rate of false alarms.

INTRODUCTION

"What is the total number of and mass of comets in our solar system?" This question is one of the most interesting unresolved questions in solar system studies. The hypotheses for the total number range from ~10^{12} through ~10^{14}; the range of estimates for the total mass extends from ~7 M_E to ~1000 M_E, where $M_E \equiv 1$ earth mass. This question is fascinating in its own right; in addition it is directly relevant to theories of solar system formation (as a diagnostic of conditions in the outer reaches of the solar system); it is directly relevant to models of biological extinction events; it is also relevant to studies of possible comet clouds around other stars.

Most of the uncertainty in the estimate of the total number of comets in the solar system results from our inability to determine the number of comets in orbits with semi-major axes between 100 A.U. and 1000 A.U. These orbits are dynamically more robust than orbits of semi-major axis > 1000 A.U., and are much less susceptible to perturbations from passing stars. In spite of this there is a growing consensus that this region of phase space is the source of new, short period comets, and further that the number of comets in this portion of phase space greatly exceeds the number in the outer region. Total numbers of comets which range up to ~ 2×10^{14} have been discussed.

The scientific importance of comets is widely acknowledged. We draw attention to a few selected issues here.

The Origin of Short Period Comets

There is growing interest in the inner portion of the Oort Cloud. This interest is motivated in large part by the realization, which has developed within the last ten years or so, that most of the comets in the solar system may have orbits with semi-major axes in the range 100 AU to 1000 AU. Some of these arguments are summarized here; a comprehensive review is given by Weissman (1986).

Fernandez (1980) suggested that a comet belt beyond Neptune might be the source of short period comets. In particular, he used Monte-Carlo simulations to demonstrate that a reservoir containing $\sim 1\ M_E$ of comets could supply all the new, short period comets, provided that some perturbers of mass $\sim 10^{24}$ g existed in the cloud to provide for dynamical evolution. Similar conclusions were reached by Duncan, Quinn and Tremaine (1988), but with a lower inferred mass for the comet belt.

Perhaps more interesting in regard to this proposal is the discussion of the whole structure of the Oort Cloud by Hills (1981). Hills argued that the so-called "inner edge" of the Oort Cloud was an artifact resulting from the dynamical robustness of comet orbits with semi-major axes $\leq 10^3$ AU. He noted that the Oort Cloud might extend inward from this boundary, and further that this hypothetical inner portion of the Oort Cloud might contain as many as 2×10^{14} comets, $\sim 100\times$ more than the better documented, outer Oort Cloud. Hills then went on to introduce the idea of comet showers, which would occur during unusually close passages by passing stars. In the event that Hills' ideas regarding this inner cloud proved correct, then during the history of the solar system more comets have entered the planetary region in these showers than from the outer Oort Cloud.

If this inner Oort Cloud extended in ≤ 50 AU it might exert a measurable influence on the orbits of Neptune and Pluto, as first noted by Whipple (1964), and further discussed by Bailey (1983). Bailey further suggested that the inner Oort Cloud might be detectable in the IRAS 100μ band.

The Great Biological Extinction Events

The most startling new developments in the science of comets originated in the claim by Alvarez *et al.* (1980) that the Cretaceous mass extinctions resulted from the impact of a large object with the earth. There is little doubt that such an impact occurred, and the association with the extinction event is persuasive to us (though not to some paleontologists). Davis, Hut and Mueller (1984) argued that the impacting object was probably a *comet,* and further suggested that impacts of this kind occurred in periodic showers induced by a companion star - this is the "Nemesis Hypothesis." We do not wish to discuss the evidence for or against periodicity in the crater record and the biological record. The suggestion that the impacting body was a comet is important for our discussion.

The primary source of comets for extinction producing showers is the inner Oort Cloud postulated by Hills. For this reason studies of the evolution of the surface of the earth, and studies of the evolution of life on earth, depend in significant ways on our knowledge of the number of comets in the inner Oort Cloud.

Comets Around Other Stars

Little is known about comets around other stars. It is reasonable to suppose that our solar system is not particularly unusual, and hence that comets are commonly present around other lower main sequence, single stars. In this respect we are studying comets around other stars when we study solar system comets.

There have been some extremely suggestive IRAS satellite observations of nearby stars reported by Aumann *et al.* (1984) and by Aumann (1984). These observations have been interpreted as evidence for comets in the inner Oort Clouds of the stars (Weissman 1984 ; Harper *et al.* 1985).

It is also known that much more massive comets clouds (i.e. more massive than the Oort Cloud) are not commonly present around other stars. Alcock *et al.* (1986) showed that the accretion of comets onto white dwarf stars would produce detectable calcium absorption lines in the spectra of these stars. The absence of the calcium absorption lines allowed Alcock *et al.* to conclude that comet clouds more massive than 3x the Oort Cloud are not common around white dwarf stars, and by extension their main sequence star progenitors.

OCCULTATION OF STARS BY COMETS (AND ASTEROIDS)

We plan to construct and use a special purpose system of monitoring cameras which will measure the rate of occultations of stars by comets in the Oort Clouds. The rate of occultations is directly proportional to the total number of comets; thus, we will be performing a direct census of the Oort Cloud. In this section we discuss the general features of this idea; the specific implementation is discussed below. This idea appears to have been around for some time. Alcock heard it first from G. McClintock (~ 1980, private communication), but it was discussed earlier by Bailey (1976). However, it appears likely that the idea has occurred to many people at different times, and abandoned as technically too difficult. We have been pursuing a practical implementation (Axelrod, Park, and Alcock, 1990).

The idea is simple to describe. One monitors the light from a sample of stars which have angular sizes smaller than the expected angular sizes of comets. An occultation is manifested by detecting the partial or total reduction in the flux from one of the stars for a brief interval. The rate of occultations is recorded over the time span of the observations. The implementation of this idea is complicated by the short expected duration of an occultation event, the large average interval between events, and by Fresnel diffraction diluting depth of the occultation. The selection of stars with small angular sizes means that relatively faint stars are used; this increases the probability of false events due to shot noise in the photons.

Consider a comet at distance r from the earth, which is occulting a distant star (of very small angular size). The photons have wavelength λ, the comet has radius a. A deep occultation (i.e. a geometrical optics occultation) can occur only if:

$$a \geq (r\lambda)^{1/2}$$

i.e. $a \geq 2.7 \text{km} \left(\dfrac{r}{100\text{AU}}\right)^{1/2} \left(\dfrac{\lambda}{5000\text{A}}\right)^{1/2}$ (1)

Equation (1) establishes the most fundamental limitation on this technique. (Smaller comets do modulate the light from a background star, but by an amount α a^{-4} for comets smaller than this limit).

The duration of the event is determined by the time the camera moves through the shadow of the comet. The earth moves at 30 km s^{-1}; with projection effects the typical relative velocity will be $\sim$ 20 km s^{-1}. For a 3 km comet, a strong occultation will have duration $\sim$ 0.2 seconds. This is the shortest duration, strong event that we anticipate due to comet occultations.

Every six months each comet "covers" a strip of sky of width (2 a/r) and length (2 AU/r). Hence in one full year the number of occultations N_O detected will be the fraction of the celestial sphere covered by this strip, multiplied by the number of detectable comets N_C, times the monitored stars N_*;

$$N_O = \dfrac{1}{4\pi} \left(\dfrac{4a.\text{AU.}}{r^2}\right) N_c N_*$$

$$N_O \approx 640 \left(\dfrac{a}{3\text{km}}\right) \left(\dfrac{100\text{AU}}{r}\right)^2 \left(\dfrac{N_c}{10^{12}}\right) \left(\dfrac{N_*}{10^3}\right)$$

(2)

It is important to stress that N_C is the number of *detectable* comets (i.e. comets that satisfy Eq.[1]). We have evaluated the equation for 10^{12} comets; there may be as many as 2 x 10^{14}.

The angular size of a 3 km object at 100 AU is 4 x 10^{-10} radians (diameter). For objects at r > 100 AU the angular size of detectable objects declines as $r^{-1/2}$ (at the diffraction limit). A typical field star has spectral type AO, radius 2 x 10^{11} cm, and absolute visual magnitude +0.7. At 330 pc this star would have angular size 4 x 10^{-10} and apparent magnitude m_v = 8.3. Hence our system should function with stars of this typical brightness.

The system we have designed will also detect occultations by asteroids. It is possible to separate ecliptic plane comets from asteroids as follows. The occultation rate and occultation duration depends on the component of relative velocity of the occulter and the camera perpendicular to the line joining them. For occulters in orbits of radius r, in au, this perpendicular velocity goes to zero at angle Θ, ahead of the meridian at midnight, where,

$$\Theta = \arccos\left(\frac{1}{\sqrt{r}}\right) \tag{3}$$

For asteroids at 2.9 au, $\Theta = 54°$; for comets at r >100au, $\Theta > 84°$. At Θ, the occultation rate goes to zero, and the duration of individual events goes to infinity. (This analysis ignores small corrections due to rotation of the earth and motion of the background star.) A feature of this nature will be prominent in the data and will serve to inform us whether we are detecting comets or asteroids.

THE ROBOTIC CAMERA SYSTEM

System Design Criteria

Our design goal is to be able to measure a rate of stellar occultations by comets of 1 per 1000 stars per year of observing time. This is comfortably lower than the rates given by Eq. (2). The duration of the occultations is widely variable, with a lower limit of about 200 msec. The proposed experiment will sample the intensities of roughly 1000 stars every 200 msec, looking for a drop in intensity caused by an occultation. Approximately $1000*3*10^7/0.2 = 1.5*10^{11}$ such observations are therefore required to detect a single event. The "false alarm" probability per observation must therefore be exceedingly small if the results of the experiment are to be interpreted with confidence. We require from our system a false alarm probability per observation of $p_{fa} \leq 10^{-13}$, resulting in a false alarm probability per year of approximately .01 for a population of 1000 stars.

False alarms can arise from a variety of causes. A major cause is the fluctuation in the number of photons detected from a light source of constant intensity due to Poisson statistics. False alarms due to this can be controlled only by detecting a sufficiently great average number of photons from the star. This clearly has implications for the telescope aperture required and the faintness of the stars that can be utilized. Extreme atmospheric fluctuations may also pose a threat of false alarms.

Additionally, false alarms can occur when a star is occulted by some nearby object. Insects, birds, aircraft, and even orbital debris have been considered as possible sources of such false alarms. In practice insects appear to be the most damaging source due to their potentially great numbers. (Orbital debris can be neglected due to the short time scale (< 1 msec) of the occultation events.) We propose to control these sources of false alarms by requiring coincident detections by at least three separated telescopes. These telescopes will be installed along an east-west line at intervals of ~5km. The distinctive signature of a comet event is that the shadow moves from east to west at a well determined rate. There is sufficient redundancy in the system that false alarms due to near field obscuration can be controlled.

Design Studies

Design studies were performed to identify the interrelated effects of optics parameters and detector characteristics on system performance. The system configuration is assumed to consist of separated telescopes for false alarm reduction, and the total system false alarm rate is kept constant at 10^{-13} per observation. Since false alarm events are assumed to be uncorrelated between any two telescopes, the system false alarm rate requirement is comfortably met when the false alarm probability $p_{fa} = 10^{-13/n}$, where n is the number of telescopes, for each telescope. The performance of the system is quantified by the number of stars that can be monitored, N_*, and by the conditional probability of detecting an occultation event, p_h.

The parameters included in the performance model are: m, the apparent magnitude of the star under observation; m_b, the diffuse background intensity (*mag arc-sec^{-2}*); d, the telescope aperture (*cm*), d_p, the pixel size (μ); F, the telescope F number; η the total system quantum efficiency; dt, the observation time (*sec*); r, the depth of the occultation modulation of the star intensity; and n_0, the sensor readout noise.

It became apparent early in the study that small aperture wide field-of-view systems are more effective in this application than more conventional large aperture systems with narrow fields. If N_* is kept constant as the field of view, Ω, is reduced, the apparent magnitude of the stars which must be monitored increases rapidly. A system with $\Omega = 1$ *sr*, achievable with focal lengths of a few centimeters, need only observe $m = 7$ stars to have 1000 available. On the other hand, a system with $\Omega = 10^{-4}$ *sr*, typical for telescopes with focal lengths of a few meters, must observe to about $m = 16$. Since the required observation time is fixed at 200 msec, the telescope aperture must also increase rapidly, and soon becomes impractically large.

Although this argument is based on the use of average sky star number statistics, the conclusion is not substantially altered even if the narrow field-of-view telescope is pointed at a local region with anomalously high stellar density such as a globular cluster. The available star numbers are now limited by field crowding, and as a result required magnitudes are still unrealistically faint. Additionally, the narrow field approach gives up the ability to measure the spatial distribution of occultations events, and therefore is unable to distinguish between Oort Cloud comets and asteroids.

Proposed System

The proposed system is a result of the design studies and an assessment of the components available at reasonable cost. The system consists of separated telescope clusters, each of which has 3 independent telescopes pointed at different areas of the sky. The clusters are aligned so that corresponding telescopes point at the same region of the sky, although high alignment accuracy is not required. All telescopes are fixed in orientation - no telescope drives are used. This system has much in common with the Explosive Transient Camera (Vanderspec, elsewhere in this volume).

Each telescope contains a Nikon 35 mm F/1.4 camera lens, a Loral Aerospace 2048 x 2048 CCD with 15 μ pixels and a frame transfer architecture, and a thermoelectric cooler to reduce dark current. The detection performance of this system is shown in Figures 1 and 2. The design parameters are as follows: $d = 2.5\ cm, d_p = 15\ \mu, F = 1.4, \eta = 0.3, dt = 0.200$ sec, and $n_0 = 50$ *electrons*. The curves in Figure 1 show the effect of photon shot noise on the conditional detection probability, p_h, versus the depth of the occultation modulation, r, for various star intensities. Each of the three telescopes in the each cluster can observe approximately 100 stars to $m = 6$, 300 to $m = 7$, 1000 to $m = 8$, and 2700 to $m = 9$, assuming sky averaged star number statistics.

We modeled atmospheric scintillation, following Dainty et al (1982), using a log-normal distribution. Figure 2 shows the probability that our system successfully detects an $r \geq 2.0$ event plotted against the probability of a false alarm, for three values of the scintillation variance. It is desirable to operate in the top left of this figure. Preliminary measurements indicate that a variance of 0.2 will be typical for our location. Clearly, a three or four camera system is viable in all except unusually poor conditions.

The predicted performance of the system is based on reasonably conservative estimates of the performance achievable from the various components. The predicted values of N_{star} and p_h are high enough to achieve the science goals of the proposal with considerable margin to spare. This gives substantial confidence that the experiment will be successful even if performance estimates for some system components turn out to be optimistic.

The data processing approach is based on technology developed by the LLNL Wide Field-of-View (WFOV) Camera Project (Axelrod et al. 1988; Park et al. 1989 ; Axelrod et al. 1989). A block diagram is shown in Fig. 3. Each incoming image is processed by an ACE (Automatic Centroid Extractor) chip, which finds stars in the CCD frames. The position centroids and intensities of these objects are calculated and passed over a serial link to an Inmos T800 Transputer, a 32 bit microprocessor. This processor monitors the intensity of each object, comparing it to the threshold stored for that object. If the intensity drops below threshold, a possible occultation event has occurred.

When a possible occultation event occurs, two actions are taken. First, the operation mode of the digital image recorder is changed from a continuous

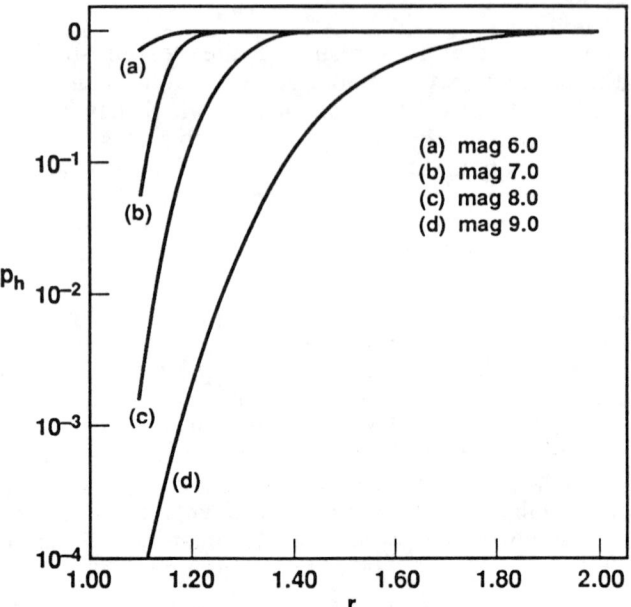

Figure 1. Probability of detection (p_h) versus modulation ratio (r).

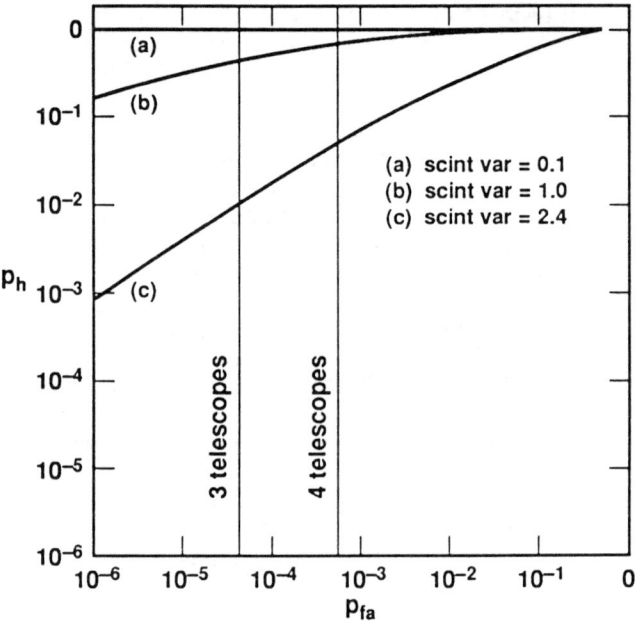

Figure 2. Effect of atmospheric scintillation when $r \geq 2.0$, p_h versus p_{fa}.

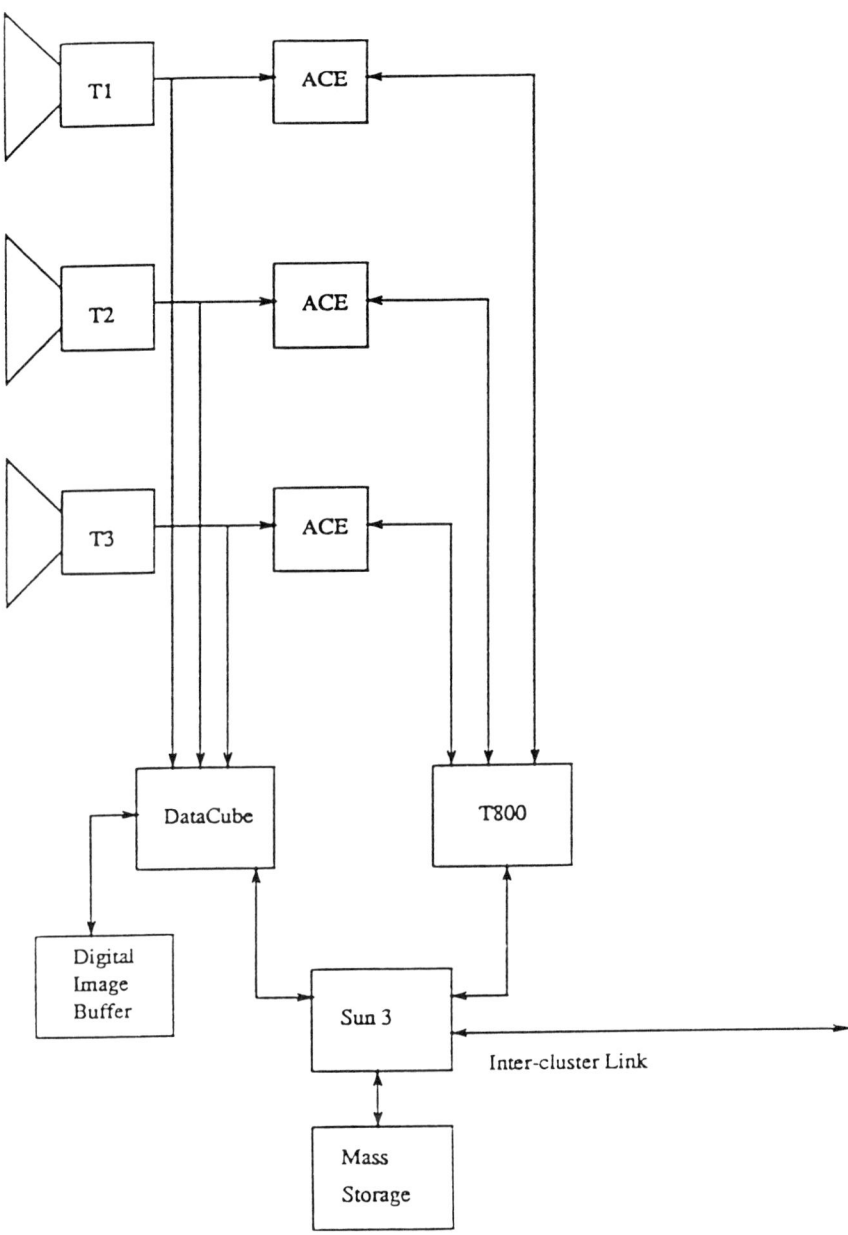

Figure 3: Block Diagram of Electronics

circular loop to a linear buffer that will be used to save approximately 100 frames centered in time around the possible event. Second, a message is sent to the other telescope clusters seeking a confirming occultation event in the same star. If confirmation is received, all saved images from the detecting telescope in each cluster are saved for offline analysis. This involves a substantial dead time while the imagery is transferred from the digital image recorder to conventional disk storage. If confirmation is not received, the event is logged as a false alarm, and operation continues without dead time.

Each telescope cluster has an attached Datacube image processor. The image processor periodically analyzes imagery from each telescope in the cluster to determine the thresholds to be used for each object, and to watch for cloud cover that requires an interruption of data collection. The Datacube, digital image buffer, Transputer, and communication link are attached to the VME bus of a Sun-3 computer. The telescope clusters take images synchronously to allow false alarm detection. Synchronization is maintained by time code receivers and the intercluster communication link.

All data processing components, with the exception of the digital image buffer, can be used unmodified in the form they were developed for the LLNL WFOV project. The digital image buffer is based on a real time video disk currently in use. The aggregate data rate from a three telescope cluster is well within the disk's capabilities, as is the image storage requirements. Some modifications must be performed, however, to accommodate the data format associated with the 2048 x 2048 CCD detectors.

CONCLUSION

We have designed a robotic system which is capable of detecting the occultations of stars by solar system comets. A system of the kind described here can determine directly the number of comets in the inner portion of the Oort Cloud, which appears to be inaccessible to all other techniques. This robotic telescope is also a sensitive detector of asteroids.

Research at Lawrence Livermore National Lab. is supported by the Department of Energy under Contract W7405-ENG-48.

REFERENCES

Alcock, C., Fristrom, C. C. and Siegelman, R., (1986) *Ap. J.* **302**, 462

Alvarez, L. W., Alvarez, W., Asara, F. and Michel H. V., (1980) *Science* **208**, 1095

Aumann, H. H., (1984) *B.A.A.S.* **16**, 483

Aumann, H. H., *et al.,* (1984) *Ap. J.* **278**, L23

Axelrod, T. S., Colella, N. J. and Ledebuhr, A. G., (1988) *Energy and Technology Review, Lawrence Livermore National Laboratory* (December 1988 issue)

Axelrod, T. S., et al., (1989) in *Real-Time Signal Processing XII*, J. P. Letellier, editor, SPIE, Bellingham, Washington.

Axelrod, T. S., Park, H. -S. and Alcock, C., (1990) *B.A.A,S.* **21**, 1154

Bailey, M. E., (1983) *Nature* **302**, 399.

Bailey, M. E., (1976) *Nature* **259**, 290.

Dainty, J. et al, (1982) appl. Optics **21**, 1196.

Davis, M., Hut, P. and Muller, R. A., (1984) *Nature* **308**, 715.

Duncan, M., Quinn, T. and Tremaine, S. D., (1988) *Ap. J. Lett.***328**, L69.

Fernandez, J. A., (1980) *M.N.R.A.S.* **192**, 481.

Harper, D. A., Lowenstein, R. F. and Davidson, J. A.,. (1985) *Ap. J.* **285**, 808.

Hills, J. G., (1981) A. J. **86**, 1730.

Park, H. -S., et al., (1989), in *Acquisition, Tracking, and Pointing III*, S. Gowrinathan, editor, SPIE Bellingham, Washington.

Weissman, P. R., (1984) *Science* **224**, 987.

Weissman, P. R., (1986), in the *The Galaxy and the Solar System*, ed. Smoluchowski, R., Bahcall, J. N. and Mathews, M. S.; Tucson: U. of Arizona Press p. 204.

Whipple, F. L., (1964) *Proc. Nat'l. Acad. Sci. USA* **51**, 711.

APPLICATION OF ROBOTIC TELESCOPES TO THE PHYSICAL INVESTIGATION OF COMETS

DAVID JEWITT
Institute for Astronomy, 2680 Woodlawn Drive, Honolulu, HI. 96822.

ABSTRACT   Robotic telescopes promise a dramatic improvement in our understanding of the physics of comets by providing a stream of linear, temporally-resolved and spatially-resolved digital photometry. We discuss the types of data expected to be of greatest value, and the physical constraints to be imposed on comets by these data.

INTRODUCTION

The purpose of this review is to provide a brief overview of those aspects of cometary science which could benefit from systematic observations by an imaging robotic telescope in the 1-m diameter class.

A broad understanding of the nature of comets is of scientific importance to both planetary and stellar astronomy. Comets are thought to be some of the first substantial bodies (the "first generation planetesimals") to form by coagulation in the pre-solar disk. The high volatile content of the comets (probably about 50% H_2O by mass) strongly suggests formation and storage in the outer solar system. Some of the original comets then accreted gravitationally to form the major planets, while still others were ejected by planets into the 50,000 AU radius Oort Cloud. The total number of comets presently bound to the sun is of order 10^{12}, and their total mass may exceed that of the planets (Weissman 1991). Since the injection of comets into the Oort Cloud is only 1% - 10% efficient, it is clear that an even larger mass of comets must have been present in the early solar system.

From a stellar perspective, the existence of the comets around the Sun implies similar populations of icy bodies around other stars. Transient red-shifted metal lines in β Pic (and other stars) have been interpreted as absorptions due to comets plunging into the stellar photospheres (Lagrange-Henri et al. 1988), although this interpretation has been questioned (Bruhweiler et al. 1991). Of equal or greater interest are submillimeter detections of large masses of dust in the vicinities of nearby pre-main-sequence stars (Weintraub et al. 1989; Beckwith et al. 1990). It is highly likely that this dust exists in circumstellar disks in the process of forming planetary systems. The lifetimes of these disks are observed to be in the 10^6 to 10^7 yr range, implying that agglomerative growth of dust particles into comets and planets is rapid.

From this brief background, it is easy to see that comets are of interest because
• They are physically and chemical primitive compared to all other solar system bodies. As such, they may preserve a record of the conditions of agglomeration in the pre-planetary disk.
• Much or most of the mass of the solar system, outside the sun, may reside in comets.
• Comets may be forming in disks around solar-mass pre-main-sequence stars. They are thus important products of circumstellar disks.

Very little is quantitatively known about the temporal variability of active comets. Most of the available time-resolved photometry has been made using purely visual estimates by observers using small telescopes. Other work has been attempted using photographic plates. The central importance of time-series imaging (as opposed to time-series one-dimensional photometry) has become apparent only in the last decade. The principal limitations in existing time-series studies are set by under-sampling and aliassing - it is difficult to obtain imaging observations over a sufficiently large period of time. For example, images of jet structures in the coma of comet P/Halley are well known, but under-sampling of the temporal variations in the jets has precluded their unambiguous interpretation. Only in recent years have linear photometers (mainly photomultipliers and charge-coupled devices) been applied to the study of cometary temporal variation. The time is ripe for a systematic study of variability in comets. This study could be efficiently undertaken using robotic, imaging telescopes.

Timescales of Variability

Temporal variations in comets are expected on a number of timescales.

- **Nuclear rotation timescale** ($\tau_n \sim 1$ day)

Rotation of the nucleus produces photometric variations in two ways. First, the known nuclei are prolate spheroids, which vary in projected cross section as they rotate about the minor axis. In comets with minimal coma, the rotation of the nucleus leads directly to a cyclic variation in reflected light. In fact, most of the available nucleus information has been obtained from measurements of this kind (Jewitt 1991).

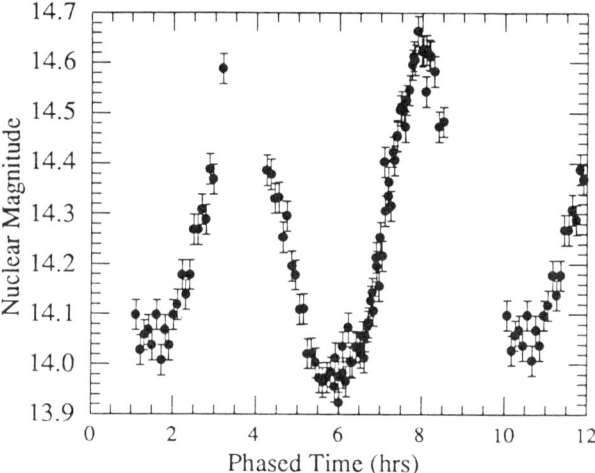

Figure 1. Rotational lightcurve of the nucleus of P/Tempel 2, from CCD observations on 4 nights in Feb. 1988. The lightcurve is due entirely to the prolate shape of the rotating nucleus (period 8.95 ± 0.01 hours). From Jewitt and Luu 1989.

Second, nucleus rotation carries localized areas of activity ("vents" in the popular parlance) into and out of sunlight. This periodic modulation of the insolation produces corresponding variations in the rate of sublimation of near-surface volatiles. The brightness of the inner coma then undergoes a forced variation driven by nucleus rotation (c.f. Figure 2).

- **Coma crossing timescale** ($\tau_c \sim$ hours)

Aperture photometry of a comet imposes a timescale on any variations which may be observed. This is because a particle released from the nucleus will contribute to the photometry for a time equal to the time needed to reach the edge of the photometry aperture.

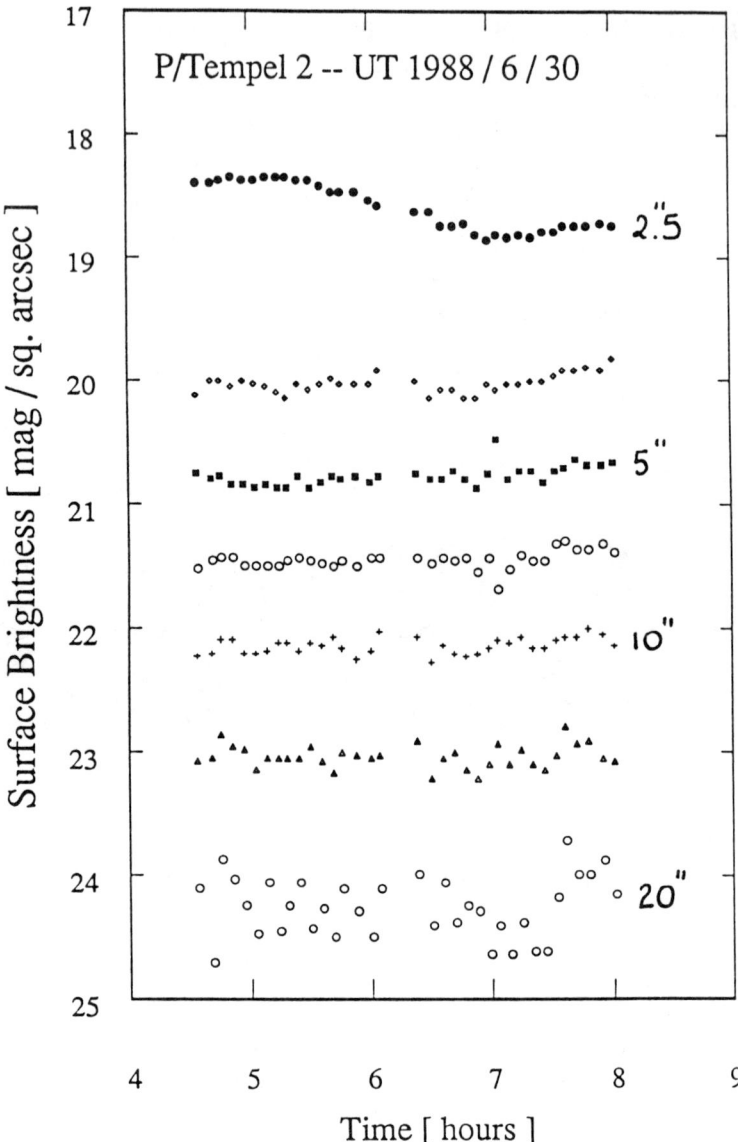

Figure 2. Example of spatially and temporally resolved photometry of P/Tempel 2. The Figure shows that temporal variations in this object are confined to the central region, and are presumably associated with the rotation of the nucleus. Note the low surface brightness of the outer coma. Aperture radii are indicated in the Figure. From Jewitt and Luu 1989.

- **Orbital/Seasonal timescale** ($\tau_{orb} \sim$ years)

Comets orbit the sun in typically eccentric orbits. The mean solar insolation varies with heliocentric distance according to the inverse square law. Near aphelion, the nucleus is cold and sublimation-driven activity is expected to be small. Near perihelion, the nucleus is hot, and sublimation will lead to the production of a strong coma. The lightcurve of the comet then reflects the long term variations in the mass loss rate from the nucleus due to the changing comet-sun distance.

The true picture is more complicated, due to the non-zero obliquity expected in cometary nuclei. This produces seasonal effects akin to those on earth. The cyclic modulation of sublimation resulting from the rotation of the nucleus will then be further modulated by the seasonal variation in the latitude of the sub-solar point on the nucleus, leading to complicated temporal variations in the total mass loss.

- **Stochastic events timescale** ($\tau_{stoch} \geq$ seconds?)

Empirically, it is known that comets exhibit short term brightness variations of a non-periodic nature, collectively (but rather misleadingly) known as "outbursts". These stochastic events might be due to exothermic heating by the phase transition of water ice from the amorphous state to the crystalline state (Klinger 1980; Smoluchowski 1981), by the rupture of embedded gas pockets in the (presumed impermeable) nucleus, and even by failure of the refractory mantle in response to gravitational and gas-drag induced stresses (Figure 3).

Questions to be Addressed Using Time-Series CCD Photometry

Numerous basic questions may be addressed using CCD photometry from a 1-m class robotic telescope.

- *What are the statistical rotational properties of cometary nuclei?*

To date, the gross rotational properties of a handful of nuclei have been determined using ground-based CCD photometry. The nuclei all seem to be prolate spheroids in singly-periodic rotation about a minor axis. There is no photometric evidence for the much-discussed precession of the nucleus, although this may reflect the short observing windows of previous studies more than the true absence of precession in all nuclei.

It would be of value to know the distributions of the rotation periods, nucleus shapes and nucleus sizes. These distributions would tell us more about the physical histories of the nuclei than a few measurements of individual nuclei.

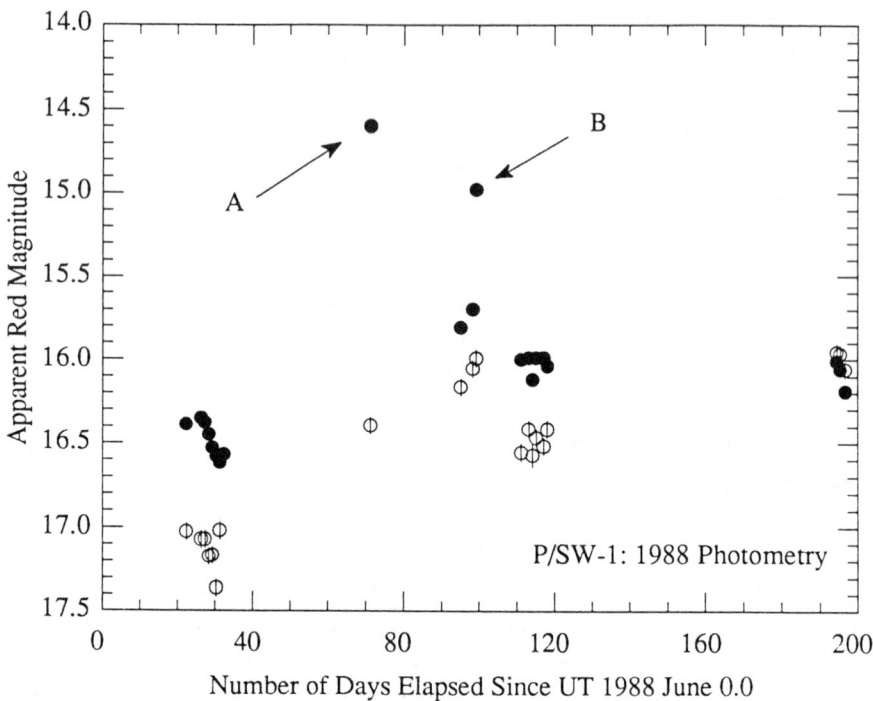

Figure 3. Outbursts (marked 'A' and 'B') in P/Schwassmann-Wachmann-1. Filled circles denote photometry from a 5" radius aperture containing both nucleus and coma; hollow circles denote a measurement of pure coma in a 5"-10" annulus. From Jewitt 1990.

• *Which of these properties are primordial?*

The rotational properties of nuclei must reflect some combination of primordial and evolutionary influences. For instance, all bodies acquire spin at the time of their formation. Comets in the hypothesized Oort cloud and Kuiper belt are likely to experience few collisions in the age of the solar system, so that their primordial angular momenta may be preserved. On the other hand, non-central mass loss may impart a torque to the nuclei of active comets, and this torque can in principle lead to precession, as well as a change in the total angular momentum of the nucleus. Unfortunately, it is impossible to predict the magnitude of the sublimation torque with confidence, since it depends on the detailed geometry of the nucleus and its outgassing areas. Thus, we are uncertain about the relative importance of primordial vs. evolutionary effects on the nucleus spin.

The same uncertainty extends to the gross physical properties of the nucleus. It has been found that the cometary nuclei, as a group, are aspherical compared to main-belt asteroids of comparable size (Jewitt and Meech 1988).

Asphericity may be a natural product of agglomerative growth of the nucleus in the pre-solar disk (Jewitt and Meech 1988; Donn 1991), but may also be produced by asymmetric mass loss in active comets.

An observational test of the origin of the nucleus physical properties could be made by comparing observations of dynamically new and old nuclei. To date, all nucleus observations have referred to dynamically old (short-period) comets, since these comets have well-known ephemerides and can sometimes be found with minimal or absent coma. Observations of dynamically new comets, probably in the presence of bright comae, might be possible.

- *What physics controls mantle growth?*

Evidence for surface mantles on comets is now overwhelming. For example, only ~10% of the surface area of the nucleus of P/Halley displayed prominent outgassing (Keller *et al.* 1987; see also Table 1 of Jewitt 1991). But the physical processes that control mantle growth are poorly understood. Simple questions that might be addressed by time-series photometry include
 • What is the timescale for mantle formation? Numerical experiments (Rickman *et al.* 1990) suggest that mantles may grow and decay on orbital timescales. If so, long term monitoring of comets should reveal secular changes in the pattern of jets in the coma, caused by changes in the positions and areas of mantle on the nucleus (c.f. Sekanina 1990).
 • Do all nuclei possess mantles? Dynamically fresh comets (i.e. comets making a close approach to the sun for the first time) might not have formed mantles. The observational signature might be the absence of collimated jets in the coma. Images of a large sample of comets of differing dynamical ages could be used to test this idea by correlating the number and strength of coma jets with cometary dynamical age.

- *What causes activity in comets at large heliocentric distances?*

Water ice sublimates negligibly at heliocentric distances $R > 6$ AU, and thus cannot be responsible for comae observed in comets at larger distances (e.g. 2060 Chiron at $R \sim 12$ AU; Luu and Jewitt 1990). It has long been speculated that distant, active comets are driven by more volatile ices such as CO or CO_2. Both of these molecules were detected in abundance in the coma of comet Halley, so it would not be surprising to find them in other comets. Protracted photometry of a set of comets as a function of R could be used to constrain the nature of the "super-volatile" which drives distant activity. For instance, pure water comets should precipitously decline in brightness near $R \sim 6$ AU, whereas the corresponding decline for CO_2 occurs near $R \sim 12$ AU, and CO is so volatile that it would be active at all heliocentric distances within the planetary region.

As a caveat to the above discussion, we note that some authors have sought to explain distant cometary activity in terms of exothermic phase transitions in near-surface water ice (c.f Klinger 1980; Smoluchowski 1981; Jewitt 1990). For instance, a recent outburst in P/Halley at $R \sim 12$ AU (Hainaut et al. 1991) has been attributed to such a transition. Systematic imaging photometry with a robotic telescope could be used to determine the rate of occurrence of such outbursts in comets as a function of heliocentric distance. This, in turn, would be a useful input to detailed thermal models of the cometary nucleus. For instance, if pre-perihelion outbursts are as frequent as post-perihelion outbursts, the phase-change explanation would be difficult to sustain.

- *How do the physical properties of the nucleus influence the structure of the coma?*

The best example of the nucleus-coma interaction is provided by the jets, which originate in active areas on the surfaces of mantled nuclei. Comets also display "sunward fans", which are cone-shaped coma extensions in the projected solar direction. These fans have been interpreted as circumpolar jets in comets with high obliquity (Sekanina 1987). The very specific predictions of this interpretation include the projected direction of the fan axis as a function of time, as well as the times of disappearance of the fan due to local sunset as seen from the active area on the nucleus. These predictions can be readily tested using time-series imaging photometry, and a robotic telescope would be ideal for this purpose.

REFERENCES

Beckwith, S. V. W., Sargent, A. J., Chini, R. S., and Güsten, R. (1990). A. J., **99**, 924.
Bruhweiler, F. C., Kondo, Y., and Grady, C. (1991). Ap. J., 371, L27.
Donn, B. (1991). "The Accumulation and Structure of Comets", <u>Comets In The Post - Halley Era</u>, eds. R. Newburn, M. Neugebauer and J. Rahe. Kluwer Academic Publishers, Netherlands. pp. 335 - 359.
Hainaut, O, Smette, A., and West, R. M. (1991). IAU Circular 5189.
Jewitt, D. C. (1990). Ap. J., 351, 277.

Jewitt, D. C. (1991). "Cometary Photometry", <u>Comets In The Post - Halley Era</u>, eds. R. Newburn, M. Neugebauer and J. Rahe. Kluwer Academic Publishers, Netherlands. pp. 19 - 65.

Jewitt, D. C., and Luu, K. J. (1989). A. J., 97, 1766.

Jewitt, D. C., and Meech, K. J. (1988). Ap. J., 328, 974.

Keller, H. U. *et al.* (1987). Astron. Ap., 187, 807.

Klinger, J. (1980). Science, 209, 271.

Lagrange-Henri, A., Vidal-Madjar, A., and Ferlet, R. (1988). Astron. Ap., 190, 275.

Luu, J. X., and Jewitt, D. C. (1990), A. J., 100, 913.

Rickman, H., Fernandez, J. A., and Gustafson, B. Å. S. (1990). Astron. Ap., 237, 524.

Sekanina, Z. (1987). "Anisotropic Emission from Comets", <u>Diversity and Similarity of Comets</u>, eds. E. Rolfe and B. Battrick, ESA SP-278, ESTEC, Noordwijk, pp. 315-322.

Sekanina, Z. (1990). A. J., 100, 1293.

Smoluchowski, R. (1981). Ap. J., 244, L31.

Weintraub, D. A., Sandell, G., and Duncan, W. D. (1989). Ap. J., 340, L69.

Weissman, P. (1991). "Dynamical History of the Oort Cloud", <u>Comets In The Post - Halley Era</u>, eds. R. Newburn, M. Neugebauer and J. Rahe. Kluwer Academic Publishers, Netherlands. pp. 463 - 486.

THE SEARCH FOR MASSIVE COMPACT HALO OBJECTS WITH A (SEMI) ROBOTIC TELESCOPE

C. Alcock, T.S. Axelrod, D. P. Bennett, K.H. Cook, and H.-S. Park
Lawrence Livermore National Laboratory, Livermore, CA 94550.

K. Griest, S.Perlmutter, and C. W. Stubbs
Center for Particle Astrophysics, University of California,
Berkeley, CA 94720

K. C. Freeman, B. A. Peterson, P. J. Quinn, and A. W. Rodgers
Mt. Stromlo and Siding Spring Observatories,
Australian National University, Weston, ACT 2611, Australia

ABSTRACT We are developing a dedicated, semi-automatic system with which we will conduct a definitive search for massive objects (such as brown dwarfs and Jupiters) which plausibly might comprise the dark matter in the halo of the Milky Way galaxy. This search will be conducted by making photometric measurements on three to ten million Magellanic Cloud (or Galactic Bulge) stars, each night for four years, in order to detect the occasional amplification of these stars by the gravitational microlens effect. Simulations have shown that this effect can be used to survey the mass range 10^{-7} - 100 $M_\odot$, with sufficient sensitivity to detect ten or more convincing amplification events if objects in this mass range make up the halo dark matter. This means that (i) if these objects make up the halo dark matter, this experiment will detect microlens effects; (ii) if these objects do not comprise the dark matter, this experiment will establish that fact.

INTRODUCTION

The evidence for dark halos surrounding the disk of the Milky Way and other spiral galaxies is now overwhelming. The most compelling case comes from rotation curves, which do not show the Keplerian fall off that would occur if the mass distribution in the outer parts of these galaxies resembled the distribution of the visible stars. In particular, the rotation curves are generally flat out to the limiting radius where measurements can be made (i.e. where the HI column density becomes too low to be detected). This limiting radius is commonly more than twice the radius of the visible disk of stars (see review by Casertano and van Albada 1990).

The rotation curve for the outer portion of our galaxy is more difficult to determine because of our location in the disk. Nevertheless, there is now compelling evidence that the rotation curve of our galaxy is approximately flat out to twice the solar radius (Fich et al, 1989). There have been many models made of the mass distribution in the halo of our galaxy; what is striking is how similar these models are, differing principally in the outer extent of the halo. The data and theory are summarized by Fich and Tremaine (1991), who also propose two extreme models: the "minimal halo" model (of mass $4 \times 10^{11} M_\odot$) and the "maximal halo" model (of mass up to $1.4 \times 10^{12} M_\odot$).

The nature of this dark matter is extremely unclear. It cannot be made of normal stars, dust, or gas, which would readily be detected. It might consist of elementary particles such as axions, massive neutrinos, or WIMPS (for discussion of these possibilities see Kolb and Turner 1990, or Primack et al, 1988). Alternatively, this dark matter could equally well be made of brown dwarfs, "Jupiters", neutron stars, or black hole remnants of primordial stars; these objects have come to be known as MACHOs (Massive Compact Halo Objects), after a suggestion by Griest (1991).

It is worth noting at this point that standard Big Bang Nucleosynthesis requires the presence in the universe of some baryonic dark matter, and that this matter could be contained in MACHOs (Silk 1991).

The project described here will use the gravitational microlens signature, as proposed by Paczynski (1986), to search for evidence of MACHOs (Alcock et al, 1990). Two other groups are attempting a similar search (Milsztajn, 1990; Paczynski, private communication). It is our intention to conduct an experimental search that will yield tens of convincing events, if MACHOs provide the dark mass in our galaxy's halo. With this expected event rate, if we see no events, we will be able to eliminate MACHOs as candidates for the dark matter. This would be a truly profound conclusion, since it would mean that some exotic elementary particle must make up the dark matter.

MICROLENSING BASICS

Gravitational microlensing refers to lensing in the case where there is significant amplification of the source, but the lensing angle is too small to be observed. Paczynski showed that the "optical depth" for microlensing by the dark halo of our galaxy is about 5×10^{-7}, so that at any given time about 1 star in 2×10^6 will be microlensed with amplification by a factor of 1.34 or larger. The fundamental unit of length for microlensing is the radius of the Einstein Ring, which is given by

$$R = 2 \left(\frac{GM x(L-x)}{c^2 L} \right)^{1/2} \tag{1}$$

where L is the distance to the source star and x is the distance to the lensing object in the halo. If R is not too small, the source star can be assumed to be a point light source, and the microlensing amplification (A) depends only on the dimensionless impact parameter ($u \equiv b/R$, where b is the physical impact parameter):

$$A(u) = \frac{u^2 + 2}{u\sqrt{u^2 + 4}} \tag{2}$$

where u = 1 corresponds to an amplification of 1.34.

Paczynski also identified the Magellanic Clouds as the optimum sources of background stars. The Clouds are far enough away to be outside most if not all of the halo, and located favorably in that the line of sight traverses much of the halo. They are close enough that individual stars may be resolved using ground based telescopes, and they provide a sufficient number of stars (>3 x 10^6 brighter than m = 19.5) to carry out a reasonable search. The most complete study of the microlensing of Magellanic Cloud stars is Griest (1991).

The Magellanic Clouds are not readily observed year round. As pointed out by Griest et al (1991), and Paczynski (1991), when the Magellanic Clouds are unobservable, one can use carefully selected Galactic Bulge fields to search for microlens events. The expected rates are lower (by about a factor of 2), but the telescope time is well used.

The time scale of a microlensing event is given by the timescale for the lensing object to traverse a distance of order R which yields

$$\Delta t \approx 2\, days \sqrt{M/M_J} \tag{3}$$

(where M_J is the mass of Jupiter) for a typical lensing object located at a distance of 10 kpc moving at a velocity of 200 km/sec. A set of characteristic amplification curves is given by Paczynski (1986).

The microlensing light curves will be distinguishable from the background of variable stars in several important ways: they are achromatic, time-symmetric, and non-repeating. These signatures should allow us to distinguish most (and hopefully all) of the variable stars from the microlensing events. In addition, the microlensing light curve is described by just 3 parameters: the maximum amplification, the time of the maximum amplification and the duration of the event. Clear evidence of this characteristic light curve will be required in order to distinguish microlensing by MACHOs from other phenomena such as variable stars.

There are two further tests which will be employed to eliminate false positives. First, follow up spectroscopy will be performed on all stars which have undergone a suspected microlens event. If the spectrum is that of a normal star, and in particular shows none of the peculiarities seen in cataclysmic variables between outbursts, our confidence will be increased. Finally, if there are many (≥10) suspected amplification events, the distribution of the maxima is known theoretically. Tests such as the one sided

Kolmogoroff-Smirnoff test will be used to compare the recorded distribution of maxima to the theoretical prediction.

There are possible complications which might tend to interfere with the characteristic microlensing light curve. These complications might result in false negatives. First, the finite size of the source star adds another parameter to the microlensing light curve, but for halo objects with $M > 0.1 M_J$, most of the red giants in the Magellanic Clouds will appear as point-like sources. Another potential complication comes from double stars. These would add two more parameters to the lightcurve fit and might make microlensing events appear chromatic, but we have found that most double star light curves can be fit very well by a single star light curve. The only real problem would be the different light curves observed in the two color bands when a member of a red giant-main sequence pair is microlensed. Fortunately, red giants are bright only for a very short portion of their lifetime, so the probability that a red giant will be paired with a main sequence companion of comparable brightness is small. One might also worry that the microlensing of double stars might be more difficult to detect because the apparent amplitude of the lightcurve will be reduced when only one member of the pair is microlensed. This is a very small effect, however, because the second member of the pair can also be lensed, and this possibility mostly compensates for the number of events lost due to the presence of double stars.

THE TELESCOPE SYSTEM

The search for gravitational microlensing by MACHOs in the halo of our galaxy requires that very large numbers of photometric measurements of Magellanic Clouds stars be made on a nightly basis, over several observing seasons. It has been decided to run the survey continuously for four years, with observations of Galactic Bulge fields being made when the Magellanic Clouds are inaccessible. The system described here is not fully robotic, though it possesses a high degree of automation. The data taking and reduction are fully automated. Telescope scheduling and operations are automated, but we will need an operator to monitor the system, to respond to inclement weather, and to initiate or terminate a night of data taking.

The MACHO survey will be performed using the 1.27m telescope at Mt. Stromlo (the "Great Melbourne Telescope").

The Magellanic Clouds cover tens of square degrees. There are more than $\sim 3 \times 10^6$ stars brighter than $m = 19.5$. With a 1.27m telescope and a CCD detector, a signal to noise of better than 2% at $m = 19.5$ during full moon requires an exposure time of 300 seconds. With time to read out the CCD (expected to be ~50 seconds) and some overhead for moving the telescope from field to field, we expect to be able to take ≥ 8 exposures per hour. It is clear that we need an uncommonly wide field of view in order to expose on sufficient stars in one night to perform a convincing survey for MACHOs.

The first step toward a wide field of view is obtaining a fast beam. This is achieved by working at prime focus; the 1.27m primary mirror is f/4.47. The telescope is undergoing considerable modification for this purpose (previously the prime focus lay outside the dome!). In addition, entirely new drives are being installed, which we expect will provide smooth, accurate tracking during the 300 second exposures. These modifications are underway at present, with completion expected by November 1, 1991.

A parabolic mirror does not have good off axis performance. A system of corrector lenses has been designed by E. H. Richardson and I. Lewis which reduces the off axis coma and astigmatism to acceptable levels. Specifically, the image of a point source always falls within a 15 micron diameter circle throughout the field of view, which has a diameter of 1°. The corrector cell reduces the f ratio to f/3.88 (giving an image scale of 41 arcsec/mm). In addition, a dichroic filter is used to split the beam into two channels (which we have chosen to be 4000A -6400A, and 6400 - 8100A) to give simultaneous imaging in two filter bands. The corrector cell is being fabricated by OCA Applied Optics of Garden Grove, California. Delivery is expected by August 30, 1991.

In order to image these large fields of view, two exceptionally large CCD cameras are being fabricated (for the "red" and the "blue" channels). Each will contain 2 x 2 mosaics of 2048 x 2048 CCD imagers, fabricated by Loral Aerospace (formerly Ford Aerospace) at their Newport Beach, California plant. These CCDs were designed by John Geary; each has 15 micron pixels and two readout amplifiers. The pixel size corresponds to 0.6" on the sky. Working at 50 khz, the CCDs may be read out in 42 seconds. Under no circumstances do we expect readout noise to impact significantly our measurements; sky background and crowding effects dominate our errors. Most of the electronics for the cameras has been acquired from Photometrics, Inc. of Tucson, Arizona.

These cameras will produce image data at a prodigious rate. Approximately every six minutes, 32 million pixels are read out, and digitized into two byte words: 64 MBytes total! It is necessary to process each night's data within 24 hours in order to keep up; our ultimate goal is real time processing, with each image reduced to photometry during the exposure of the next image.

The primary data processing will be performed by a four processor Solbourne computer. The architecture is shown in Figure 1. Raw image data from the cameras passes down a fiber optic line to a "descrambler" board, which separates the sixteen channels, and is written onto 64 MBytes of buffer memory, which sits on the VME Bus. The images are then written onto the magnetic disks. The photometry is performed by the CPUs independent of the state of the cameras. The photometric results will be stored on magnetic disks (9 GBytes capacity). This computer also controls the telescope (issuing commands to the control VAX over the ethernet), and the cameras (via an RS422 link to the camera controller).

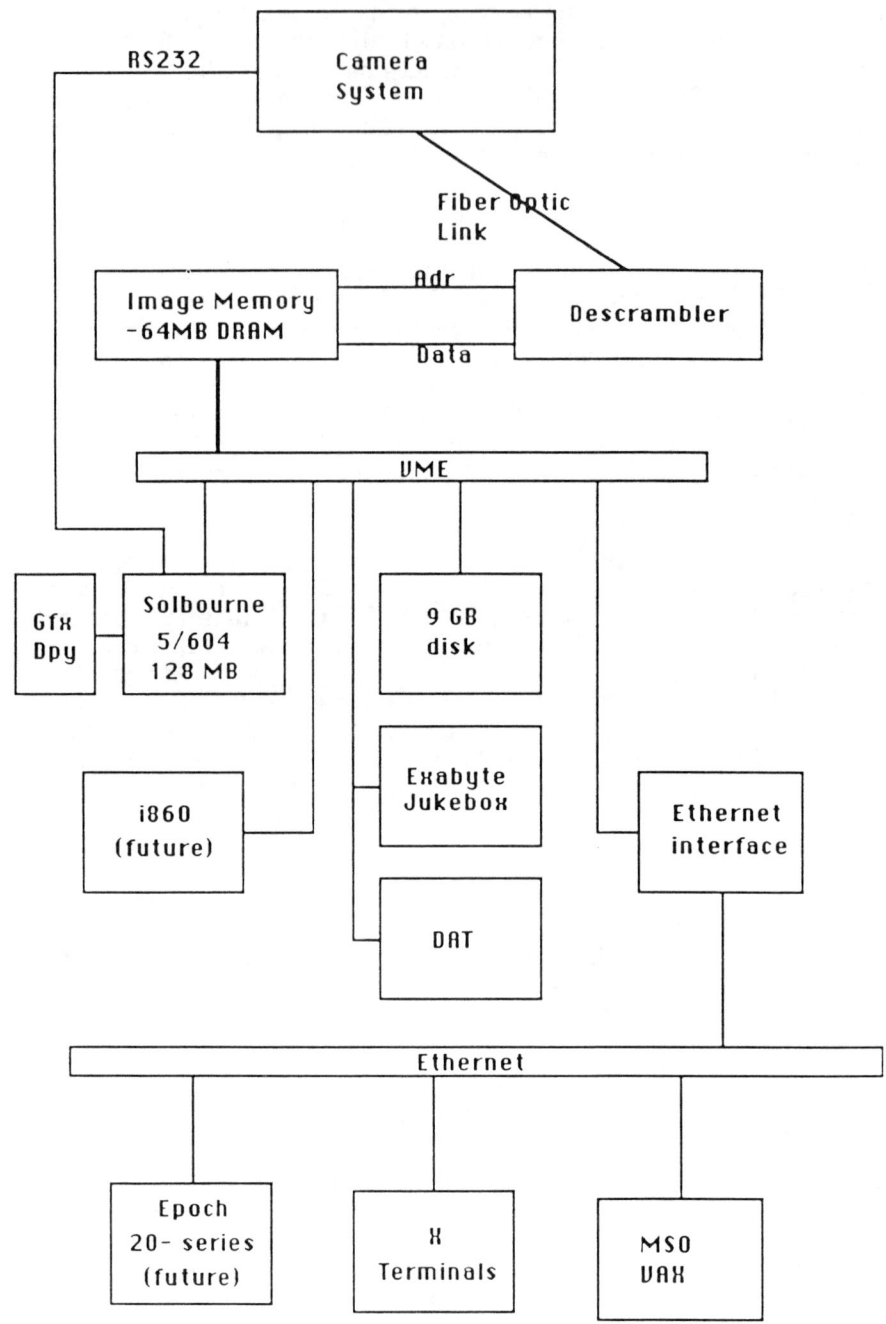

FIGURE 1: Computer Architecture

The photometry is performed by a code "Sodophot" that is derived from Dophot. Sodophot determines an analytic fit to the point spread function for the star images using a preselected set of bright, relatively isolated stars, and uses this fit plus a template of the field to fit the star brightness and the sky for each object in the template. Preliminary tests performed on data obtained for selected Magellanic Cloud fields indicates that Sodophot can process a full night's data in less than 24 hours, on the four processor Solbourne.

A database containing the entire photometric histories of all the stars in the templates will be compiled. Much of this database will reside on the disks and on a jukebox of Exabyte tapes. These histories will be analyzed for variations characteristic of a MACHO event.

In addition to the photometric results, we will record all the bias subtracted, flat fielded images on Exabyte Tapes. Each night will generate up to 4 GByte of image data. This will require 1 high density Exabyte tape per night. We believe that this step is important for three reasons (i) it enables us to co-add frames in order to work with fainter stars, (ii) it enables us (or anyone else) to check or reanalyze a given photometric reduction, (iii) these data will almost certainly have uses we have yet to think up!

Also shown in Figure 1 are two desirable upgrades to this system. Coprocessor boards based on the Intel i860 chip might increase the performance of the system to the point where real time processing of the images becomes possible. Real time analysis would allow us to respond to interesting photometric variations by altering the observing schedule, perhaps to obtain extra images of that field. The other upgrade outlined is a mass storage device, which would provide an online database containing all the images. This would greatly facilitate detailed analysis of the candidate MACHO events.

EXPECTED SYSTEM PERFORMANCE

The expected performance of the system depends on details of the observing strategy, and on the nature and frequency of interruptions due to bad weather, etc. We have made a conservative estimate of the system performance by evaluating a naive observing strategy, and simulating the performance of this system using 8 years of continuous weather records taken at Mt. Stromlo. (Our precise observational strategy has not been decided yet.) In the naive strategy, we will survey stars in the LMC, the SMC and the galactic bulge, to a limiting magnitude of about $m < 19.5$ with 5 minute exposures (at full moon). The estimated star densities ($m < 19.5$) in these fields range from about 380,000 stars per square degree in the center of the LMC to about 85,000 stars per square degree 3 degrees away from the center of the LMC, to 30,000 stars 2 degrees from the center of the SMC. Our analysis of CCD images of several regions in the Magellanic Clouds taken in different seeing conditions using Sodophot shows that we can obtain 6% (or better) photometry on 350,000 stars per square degree with 2.1" seeing, but only 96,000 stars per square degree with 3.4" seeing. We can obtain 10% (or better) photometry on 450,000 stars per square degree in 2.1" seeing, and

210,000 stars per square degree in 3.4" seeing. Note that we are picking up stars with m > 19.5 to achieve these numbers. With this performance of the data analysis, we should be able to obtain measurements on up to 10 million stars in a good night!

The naive strategy is most sensitive to the mass range $M_J < M < 100 M_J$. which spans the range for Jupiters and brown dwarfs. The time scale for these events ranges from a couple days to a few weeks, so a reasonable strategy would be to observe as many different stars as possible each night. As many of the same stars as possible will be observed on successive nights so as to give daily sampling of the light curves for as many stars as possible. In order to search for halo objects of larger mass (black holes?), it will be preferable to sample more stars less frequently. One relatively easy way to do this will be to co-add data from successive nights in order to get accurate photometry on fainter stars.

Smaller mass dark matter candidates ($M < 0.1$ M_J) pose several different problems. First, the lensing light curves for red giants (which make up two thirds of the stars to M < 19.5) in the clouds become more complicated because the point-like source approximation begins to fail. For impact parameters comparable to the stellar radius, this results in a large increase in the amplification, but when the impact parameter is smaller than a stellar radius, the amplification is reduced. The other difficulty with these small mass objects is that they tend to have lightcurves which last about 24 hours, which makes it difficult to obtain complete light curves. The detection of microlensing in this mass range will probably require the use of at least two different observing sites at different longitude.

The table shows the average number of events "detected" using 8 years of weather data from Mt. Stromlo for different masses of Halo objects. The two detection criteria used are (1) to require 2 points with amplification A>1.2, on each of the rising and falling parts of the light curve and (2) at least 5 points on the light curve with at least 2 on each of the rising and falling portions of the curve with A>1.2, and 1 point with A>2. Note that each "point" on the light curve refers to 2 simultaneous images in different color bands.

TABLE I. Microlens Event Rates

Mass	4 pts. with A > 1.2	5 pts. with A > 1.2 & 1 with A > 2
$0.1 M_J$	2.3 events/yr.	0.8 events/yr.
$0.3 M_J$	7.2 events/yr.	2.8 events/yr.
M_J	15.9 events/yr.	6.6 events/yr.
$3 M_J$	23.7 events/yr.	10.8 events/yr.
$10 M_J$	24.2 events/yr.	11.4 events/yr.
$30 M_J$	19.3 events/yr.	8.9 events/yr.
$100 M_J$	12.8 events/yr.	5.7 events/yr.

"NON MACHO" ASTRONOMY

During the course of this experiment, we will make $>10^{10}$ photometric measurements. These data are being gathered for the primary purpose of searching for microlensing events, but the database will be a treasure trove of information regarding temporal variability in stars. We draw attention to some of the possibilities here.

The experiment, and the first season observing strategy, is especially well suited to the study of Cepheid variables. We will reliably detect all the Cepheids within our fields, but in addition we will have well sampled light curves for each of them! This will greatly increase the store of information about one of the most important distance indicators in astronomy.

We will also gather similarly useful data on the RR Lyrae stars. However, the naive strategy described above, with one measurement per night on stars down to m ~19.5, is not well suited to the study of RR Lyraes: their periods are of order hours, and at minimum light they are fainter than m = 19.5. In subsequent observing seasons we will change our strategy to perform more frequent sampling of smaller numbers of stars, and hence will be able to perform a survey of RR Lyrae stars that will produce more light curves than all previous surveys combined.

The experiment will also be particularly well suited to the study of transient phenomena, such as cataclysmic variable stars and foreground flare stars. We will pay particular attention to these objects because they potentially contribute a significant astronomical background to the microlens search. It is worth stressing that all known transient variables can be eliminated as potential microlens candidates because of the asymmetry of their light curves, and because of significant color variation during the out burst.

CONCLUSION

We have designed and are constructing a semi-robotic telescope that can perform millions of photometric measurements per night. We expect to bring this system into operation by the end of 1991. This system will perform a sensitive search for gravitational microlensing by MACHOs, thus determining whether or not they comprise the dark matter in the halo of our galaxy.

Work performed at LLNL is supported by the DOE under contract W7405-ENG-48. Work performed at CfPA is supported in part by the Office of Science and Technology Centers of NSF under cooperative agreement AST-8809616.

REFERENCES

Alcock, C., Axelrod, T. S., and Park, H. -S., 1990, B.A.A.S., **21**, 1149.

Casertano, S., and van Albada, T. S., 1990, in "Baryonic Dark Matter",eds. D. Lynden-Bell and G. Gilmore (Dordrecht: Kluwer), 159.

Fich, M., Blitz, L., and Stark, A. A., 1989, Ap.J., **342**, 272.

Fich, M., and Tremaine, S., 1991, Ann. Rev. Astron. Astrophys., **29**, in press.

Griest, K., 1991, Ap.J., **366**, 412.

Griest, K., et al., 1991, Ap.J. (Letters), **372**, L79.

Kolb, E. W., and Turner, M. S., 1990, "The Early Universe", (Addison Wesley: New York).

Milsztajn, A., 1990 in Proceedings of the XXVth Rencontres de Moriond, ed. O. Fackler and J. Tran Thanh Van (Editions Frontieres: Gif-sur-Yvette).

Paczynski, B., 1986, Ap.J., **304**, 1.

Paczynski, B., 1991, Ap.J. (Letters), **371**, L63.

Primack, J. R., Sadoulet, B., and Seckel, D., 1988, Ann. Rev. Nucl.PartSci., **B38**, 751.

Silk, J., 1991, Science, **251**, 537.

2MASS: THE 2 μm ALL SKY SURVEY

S. G. KLEINMANN
Department of Physics and Astronomy, University of Massachusetts
Amherst, MA

ABSTRACT   This paper describes a new sky survey to be carried out in three wavebands, J(1.25 μm), H(1.65 μm), and K(2.2 μm). The limiting sensitivity of the survey, 10 σ detection of point sources with K $\leq$ 14 mag, coupled with its all-sky coverage, were selected primarily to support studies of the large-scale structure of the Milky Way and the Local Universe. The survey requires construction of a pair of observing facilities, one each for the northern and southern hemispheres. Operations are scheduled to begin in 1995. The data will begin becoming publicly available soon thereafter.

INTRODUCTION

This paper describes the 2 Micron All-Sky Survey project (hereafter, 2MASS), which is designed to survey the entire sky to a sensitivity that will allow the detection of point sources brighter than K = 14 mag at the 10σ level.

More than 2 decades ago, Neugebauer and Leighton (1969) carried out the *Two Micron Sky Survey* (hereafter, *TMSS*), which covered 77% of the sky and resulted in the detection of 5612 stars brighter than K = 3 mag. In terms of average surface density of objects (but not the type of objects that were detected) the *TMSS* is similar to the *Yale Bright Star Catalog* (Hoffliet and Jaschek 1982): there are on average 0.25 *Bright Stars* per square degree, and 0.20 *TMSS* sources per sq. deg. The 2MASS project aims to cover $\geq$95% of the sky with 25,000 times greater sensitivity than the *TMSS*. One can with confidence predict that 2MASS will provide a powerful means of addressing problems ranging from the structure of the Local Universe to the detection of brown dwarfs. Yet with such a leap in sensitivity, its serendipitous potential can hardly be overestimated.

This new survey is made possible by advances in detector technology and in data processing. The rapid developments in infrared array detector technology have resulted in chips with suitably large formats, high read-out rates, and low noise. The improvements are so dramatic that it can fairly be said that the infrared window on the Universe has effectively been re-opened. At the same time, data processing technology has produced nearly 100-fold increases in the speeds of microprocessors in the past decade, along with corresponding drops in the cost per MByte of memory. An additional and important consideration is that infrared astronomy has a mature data processing center—the Infrared Processing and Analysis Center, managed by JPL—with a demonstrated ability to produce quality catalogs and atlases on time and within projected budgets.

This project is not a demonstration of robotic telescopes, *per se*. That is, automation that enables general purpose use of the telescope is not an objective of the project. Rather, the 2MASS telescopes will operate in a simple, restricted pattern, optimized for the execution of the survey. Automation of the steps involved in the survey operation will be implemented wherever possible.

SCIENTIFIC GOALS

Large Scale Structures

The major objective of the survey is to enable the study of large-scale structures, exploiting a wavelength range that is relatively insensitive to interstellar dust, yet sensitive to a wide range of stellar masses.

To quantify these advantages, note that a cloud producing 1 mag of extinction in the blue—the effective wavelength of the galaxy catalogs used for modern, large-scale redshift studies—causes only 0.07 mag of extinction at 2.2 μm. This has a dramatic effect on the ability to make a census of galaxies in the Local Universe: whereas the "Zone of Avoidance", the belt of galactic latitudes within which any survey for galaxies becomes incomplete due to extinction by foreground dust—covers 40% of the sky at blue wavelengths, it covers only 2.8% at 2.2 μm. By minimizing the mask imposed by interstellar dust extinction, it is possible to seek much larger structures in the Local Universe than the "Great Wall" already discovered from redshift studies of blue-selected galaxies.

Likewise, the insensitivity of the near-infrared to dust obscuration reopens the exploration of our own galaxy: At optical wavelengths, the gradient of dust extinction in the galactic plane is near 2 mag/kpc. Thus, the diffuse clouds in the Milky Way become optically thick in only 0.5 kpc. One can see 10 times as far at 2.2 μm before the optical depth of diffuse clouds exceeds unity. Observation of the Milky Way stars at a wavelength where the transparency of the interstellar medium is high should not only reduce the biases in our current determinations of fundamental parameters such as the scale length of the disk, it will also revitalize studies of the stellar content of the bulge, and enable studies of other large structures that are important for understanding the dynamical evolution of the Galaxy, such as a stellar bar and/or warp.

The second advantage of the near-infrared for large-scale structure studies is its sensitivity to low-mass stars. When these stars evolve to the giant branch, they become excellent tracers of the gravitational potential, because they are luminous, common, and old. Such stars emit most of their energy in the near-infrared, and dominate the total visual and near-infrared luminosity of our own and other galaxies. The photographic surveys of the past were most sensitive to hot, massive stars, and even the *IRAS* survey of galaxies was most sensitive to hot, massive stars, in that *IRAS* detected dust heated by such stars.

Searches for Rare Objects

The second objective of the survey is to seek certain elusive classes of objects that emit strongly in the near-infrared. An obvious class of objects expected to have this characteristic are the so-called "brown dwarfs," stars with such low masses that they are unable to sustain hydrogen-burning. Most brown dwarfs in the solar neighborhood should have near-infrared color temperatures ≤ 1000K, and should not have significant mid-infrared excesses (such as the colors expected of an evolved star). The I-band detection limit of the new POSS plates is 18.3 mag (10σ) making 1000K $0.1R_\odot$ blackbodies elusive unless they lay within 0.6 pc of the Sun. The 2MASS survey would reveal such objects out to 7 pc, sampling a volume of space 1000 times larger. Thus, the sensitivity and all-sky coverage of the 2MASS project should provide a hard test of the current expectation that there may be as many as 1 brown dwarf per pc^3 in the solar neighborhood.

Regardless of the outcome of the search for brown dwarfs, 2MASS should detect thousands of the coolest dwarf stars. To distinguish these stars from cool giants, follow up observations will be needed. But the search list can be made tractable: faint, red stars at high latitudes will be the best candidates.

Another, perhaps less obvious class of objects which should be strong infrared emitters are quasars. It is estimated that 2MASS may detect more than 10,000 quasars and active galaxies. The real benefit, however, will not be in the number of quasars found, but in obtaining a census of quasars that is (at least) differently biased than current, optical surveys. Studies of the quasars found in the *IRAS* survey provided strong evidence of the difference between the properties of far-infrared selected quasars differ and those found optically.

Setting the Context for Studies of Selected Regions

In order to take maximum advantage of the new infrared array detector technology, it is essential to develop a framework within which one can interpret new classes of objects that are likely to be discovered in studies of selected areas and medium- or small-scale targets. A uniform sky survey does just that.

Studies of star forming regions illustrate the point. These studies often focus on deeply embedded regions of active star formation. The extraction of luminosity functions for these forming clusters requires a means of distinguishing cluster members from background stars, which in most cases will be highly reddened by foreground cloud material and not readily distinguished from stars still enshrouded in their circumstellar envelopes. Without sensitive, high resolution spectroscopy one must use statistical techniques to determine the fraction of sources which belong to the background. Reliable determination of the background population requires the evaluation of star counts over an extended area surrounding the cluster of interest, a task requiring an order of magnitude greater observational effort than the directed observations of the cluster itself. 2MASS will provide a characterization of the galactic stellar population along any particular line of sight, ultimately providing a general model of infrared galactic star counts.

Perhaps the most important aspect of the survey is to provide the context for the interpretation of the serendipitous discoveries that will result from the survey itself. Analogies can be made to the *IRAS* discoveries of infrared cirrus, the "Vega phenomenon" and the ubiquity of galaxies with luminosities rivalling quasars, or to the *TMSS* discovery of stars surrounded by multi-solar-mass dust envelopes. The number and type of serendipitous discoveries that might emerge from the 2MASS project can not be foretold. However, the sheer number of volume elements being sampled—some appropriate product of the sensitivity and spatial resolution—suggests rich rewards.

SCIENTIFIC REQUIREMENTS

The baseline parameters for the survey are given in the Table below:

Table 1. 2MASS Science Requirements

Sensitivity (10 σ detection)	$K \leq 14$ mag (point sources)
Wavebands	J, H, K
Sky Coverage	$\geq 95\%$ (goal of 100%)
Pixel Size	2.5" x 2.5"
Photometric Accuracy	5% for point sources w/ $K \leq 12.5$ mag
Positional Accuracy	1"
Reliability	≥ 0.9999
Bright Source Limit	$K = +2$ mag
Completeness	$\geq 98\%$
Overall Time Limit	Complete by 2000
Redundancy	Seconds, Minutes, Years

The sensitivity limit was chosen to enable the detection of a number of galaxies that significantly exceeds the number listed in the Zwicky catalog. Based on an uncertain extrapolation from existing optical surveys, we estimate that as many as $\geq 500{,}000$ galaxies smaller than 2' will be detected with 2MASS. This is an order of magnitude more galaxies than are listed in the Zwicky catalog. Such sensitivity will permit the detection of supergiants in galaxies in the Local Group, as well as giants and supergiants in the Magellanic Clouds. This latter step will provide measurements on the near-infrared luminosity functions for these stars, which is crucial for interpreting 2MASS results in terms of the structure of the Milky Way.

The sky coverage goal of 100%, except for confused areas, is essential to the studies of large scale structures. Any significant deviation from that number introduces aliases into such studies, whose removal requires *a priori* knowledge of the structures we are trying to examine! Nearly complete sky coverage also demands uniformity—substantial non-uniformity would also introduce aliases.

The choice of wavebands was also driven by a number of scientific and practical concerns. Our main considerations were to (1) minimize sensitivity to interstellar dust obscuration, (2) maximize the sensitivity to the photospheres of cool stars and to normal galaxies even in unreddened fields, and (3) minimize effects of moonlight. Also, three wavebands are required to help distinguish reddening from temperature effects. At wavelengths much longer than 2 μm, the thermal emission from the telescope severely limits one's sensitivity to normal stars and galaxies, while at wavelengths shorter than ~ 1 μm, interstellar reddening and moonlight present severe limitations. The three ground-based "windows" accessed through traditional J, H, and K filters satisfy our constraints.

Choosing the pixel size is a complicated issue. It is important to note, first of all, that the speed with which one completes a background-limited survey is independent of pixel size, since with smaller pixels, the sensitivity goal can be reached in shorter integration times, so that one can compensate for the reduced sky coverage of a single exposure with more exposures per unit time. This does not mean the pixel size is unconstrained. The arguments favoring larger pixels are: Larger pixels favor the detection of more extended objects (e.g., galaxies), and are less sensitive to the vagaries of seeing. On the other hand, smaller pixels are can be used to distinguish faint galaxies from stars, reduce susceptibility to confusion, and can, in principle, provide a means of obtaining more accurate photometry and positions. Without *a priori* knowledge of the expected surface brightness distribution of a near-infrared selected sample of galaxies, it is impossible to make a demonstrably optimum choice. The upper limit of pixel size is dictated by optical constraints of the telescope/camera system.

We have adopted a pixel size (2.5″) near the minimum necessary to remain insensitive to seeing, but also adopted a survey strategy that overcomes limitations in positional and photometric accuracy that would otherwise result from using large pixels (see below). One of the most important objectives of our observational studies during the next year or two will be to determine whether this choice of pixel size and survey strategy will satisfy the survey's major science goals.

One more issue regarding pixel size deserves mention. Current models of the expected surface density of near-infrared sources indicate that the survey will certainly be limited by confusion to the detection of stars brighter than K = 11 mag within an area $\leq$ 300 sq. deg. near the Galactic Center. However, this has little impact on the science goals of the survey, because (1) heavy interstellar dust obscuration would prevent the detection of galaxies in this direction independently of the pixel size, (2) estimates of the surface densities of rare objects like brown dwarfs and quasars will be best obtained from high latitude studies, where the surface density of common objects is minimized, and (3) studies of the structure of the Milky Way will employ giants, whose luminosities are so great that they will outshine the confusion limit even near the Galactic Center. The giants near the Galactic Center will readily distinguish themselves (on a statistical basis) from other, foreground stars, by their redder colors.

The remaining scientific requirements are included in Table 1 for completeness, but space limitations prevent more detailed explanation here.

THE BASELINE SURVEY DESIGN

The survey mapping strategy has been driven by the following considerations: With a pixel size of 2.5″, the field of view of a single 256 x 256 array is 10.6′. There are 1.3×10^6 such fields in the sky. The integration time per position can be calculated with the following assumptions:
- there will be two survey telescopes, one each in the northern and southern hemispheres,
- 30% of the nights at each site are useable,
- each telescope covers 55% of the sky,
- a typical night has 10 hours, and
- the sky is to be surveyed in 2 years. Then the maximum integration time per field is just under 11 s.

If the survey were to operate in a simple, "step-stare" mode, then the telescope would have to move a large fraction ($\geq 50\%$, say) of the field of view of the camera, and stop in a time much less than 11 s. This puts a tight constraint on telescope dynamic performace. The problem is actually much more severe than that: to minimize vagaries in the sensitivity of the survey due to variations in sensitivity across a detector array, and mitigate effects of dead pixels, it is necessary to obtain several much shorter exposures, say, 1 s, rather than a single long exposure. This means that the telescope must move several arc minutes (i.e., to a different region of the array, since "bad" pixels tend to be clustered) and stop *dead* on a time scale much less than 1 s!

An alternative scanning mode places far fewer constraints on the telescope drive system. It uses a "ratchetting secondary" mirror to freeze the image in the focal plane for a time equal to the short integration time (about 1 s) while the telescope scans at a constant rate in a direction opposite to the motion of the secondary. This method has a large dynamic advantage over the step-stare method, because the secondary has much less mass than the telescope.

The survey scanning mode that has been adopted is a slight variation on the "ratchtetting secondary" mode described above: we will obtain 2 sets of 5 1 s integrations on each field. The second set of integrations will be obtained about 15 minutes after the first set. Comparison of the two sets will provide a further check on the photometric accuracy of the observations, and allow the identification of moving such as asteroids.

To carry out this scanning strategy efficiently, the survey camera will incorporate 2 arrays for each waveband. The arrays will be offset in right ascension. (They will also be offset in declination, but this has little effect on the outcome.) At the end of each declination scan, the telescope will be offset in right ascension by an amount almost equal to one array width (about 10 '). Thus, the second set of 5 1 s integrations will be obtained when the second array scans the same declination path as had been previously scanned by the first array.

One last complication constrains the survey scanning strategy, and camera design. The large size of the pixels, compared to the average expected point spread function, could reduce photometric accuracy due to intra-pixel variations in sensitivity. Our strategy is to reduce this problem by sampling the set of images for each star at different relative pixel locations. To do this, we will tilt the array slightly to the scan path, and sub-sample in the scan direction.

The preliminary design of the survey camera thus calls for 6 arrays, 2 each for J, H, and K bands. Fig. 1 shows the drawing of the focal plane developed by Infrared Laboratories, Inc.

In order to achieve the required sensitivity goal (10 σ detection of 2.2 μm point sources) with the selected pixel size and integration time, it is necessary to use a moderate size low-background telescope. This telescope must be carefully designed for the survey in order to support the survey strategy described above. The requirements are listed in Table 2.

Table 2. 2MASS Telescope Requirements

Aperture:	~1 meter diameter
Field of View:	40'
Image Size:	≤ 1".5
Secondary Mirror:	Articulated
Pointing Jitter:	≤ 0".5 in 2 minutes

A fully dedicated facility is required to insure the speedy completion of the survey and to minimize instrument failures. To meet these requirements and those in Table 2, it will be necessary to construct a pair of telescopes, one each for the Northern and Southern hemispheres.

The survey facility must be complemented by a sophisticated on-site data system capable of (1) controlling the telescope and camera, (2) handling the enormous data volume (131 KBytes per 1 s integration per array times 36000 s per night, or 28 GBytes per night per site), and (3) providing sufficient quality checking to support efficient survey planning. Our studies show that present technology is able to support the development of a system matching these requirements at reasonable cost. The final design will implement expected advances in high volume data storage in the next couple of years.

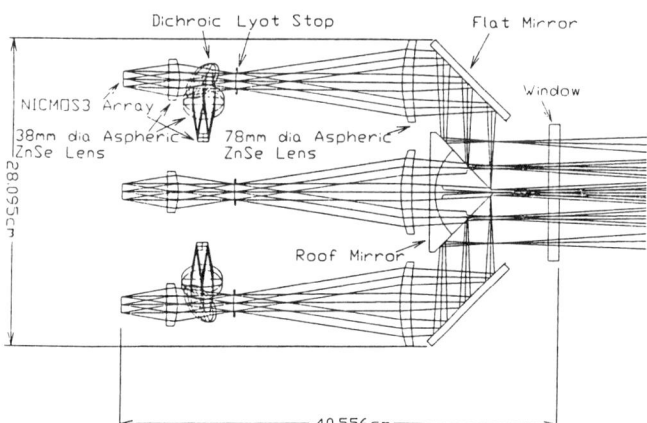

Figure 1. Side view of the camera and the beam path. The two channels bent by the dichroic mirrors are tilted from the plane of the figure by 45°. The four flat mirrors adjust each channel axis by a small angle to point the line-of-sight toward the center of the secondary.

OUTPUT FROM THE SURVEY

The survey data will be processed at IPAC. The primary goal of the processing will be to produce a catalog of discrete sources, i.e., point sources and extended sources smaller up to a maximum size of $\leq 2'$. This catalog will contain discrete sources, including galaxies with near-infrared sizes below the upper limit, brighter than 14 mag at 2.2 μm.

In support of archival research on extended objects, there will also be a facility for producing survey "co-adds", similar to the *IRAS* Super Sky-Flux data product, which attempts to normalize the sky background between frames to a common level. It is not yet clear whether it will be feasible to do this for the whole sky. In any case, the software will be developed to enable coadding of selected regions.

Access to the entire database of raw images will also be provided to archival researchers. This will allow users to apply alternative processing algorithms in search of phenomena not optimally detected by the discrete source processor used to generate the catalog described above.

The project will, of course, provide an explanatory supplement to describe the data taking process, the main properties of the discrete source catalog, and known anomalies.

A preliminary hardware/software system design shows that the processing of the survey data can be done with a small "farm" of *present day* workstations, keeping up with the rate at which survey data is obtained. Again, expected advances in computing hardware and software are expected to ease this task.

It is a major objective of the survey to provide broad access to the survey data as soon as possible. That is, once the integrity of the data and the data processing have been validated, which we aim to accomplish within a few months of the start of formal survey operations, the data will become available to interested users. No proprietary access period for the science team has been reserved.

PROJECT MANAGEMENT

The design of the survey has been being studied by a team of 20 astronomers, listed in Table 3. Various members of the core team are responsible for developing specific aspects of the survey. Both the core team and the science advisory group will take part in the various stages of formal reviews. The first such review is the Conceptual Design Review, scheduled for this coming Fall.

Table 3. 2MASS Team

Principal Investigator
S. G. Kleinmann (UMASS)

Core Team

C. A. Beichman (IPAC)	G. L. Kopan (IPAC)
T. J. Chester (IPAC)	F. J. Low (UofA)
J. H. Elias (NOAO)	P. A. Seitzer (UMich)
F. C. Gillett (NOAO)	M. F. Skrutskie (UMASS)

Science Advisory Group

G. G. Fazio (CfA)	S. D. Price (AF/PL)
J. P. Huchra (CfA)	M. M. Shara (STScI)
C. J. Lonsdale (IPAC)	B. T. Soifer (CIT)
K. Y. Matthews (CIT)	C. G. Wynn-Williams (IfA)
G. Neugebauer (CIT)	

In order to maintain our goal of beginning survey operations early in calendar year 1996, construction of the survey telescopes must begin early in 1993.

European astronomers are also interested in surveying the southern sky in the near-infrared, but have selected a very different approach than the 2MASS project as described here. Discussions are ongoing to explore a basis for collaboration between these projects.

The survey team is currently focussed on developing a prototype camera, whose use will provide far more realistic estimates of the achievable photometric and positional accuracy of the survey, the variations in the OH background, and the cost/time required for processing of the data. Data obtained with this camera will also be used to test current models of the infrared sky, and to obtain an initial census of the infrared colors and shapes of various objectively selected samples of galaxies.

We plan to begin development of the survey facilities next year. This will involve construction of two telescopes, one each in the northern and southern hemispheres. This period will also see continued use of the prototype camera, improvements to the data system associated with the prototype camera, and, eventually, the construction of the two 6-array survey cameras. The data obtained with the prototype camera will be used as the basis for developing an optimized pipeline processing method at IPAC.

ACKNOWLEDGEMENTS

Support for this project has been obtained from the National Aeronautics and Space Administration, the National Science Foundation, the Air Force, and the University of Massachusetts.

REFERENCES

Hoffleit, D., and Jaschek, C. 1982, *The Bright Star Catalog*, 4th Rev. ed. (New Haven: Yale Univ. Observatory).
Neugebauer, G., and Leighton, R. B. 1969, *Two Micron Sky Survey* (NASA SP-3047).

AUTOMATION OF INTERFEROMETRIC OBSERVATIONS

M. BESTER, C. G. DEGIACOMI, W. C. DANCHI, L. J. GREENHILL, AND
C. H. TOWNES

Space Sciences Laboratory and Physics Department, University of California, Berkeley, CA 94720

ABSTRACT The Infrared Spatial Interferometer (ISI) is a heterodyne interferometer that operates in the 9 – 12 μm atmospheric window. It is located at Mount Wilson and consists of two 1.65 m Pfund-type telescopes. Presently baselines range up to 35 m. Lately the performance of the ISI was improved significantly, providing higher quality interferometric data. The improvements include all-reflective front-end optics, larger bandwidth and higher quantum efficiency heterodyne detectors, a fringe calibration system, a CCD autoguiding system and a more advanced computer control system. The newly developed control software allows the observations to be largely automated.

INTRODUCTION

In the mid-infrared region, at a wavelength of 11 μm, the diffraction limited beamwidth of a single filled-aperture telescope with a typical diameter of 3 m is approximately 0.9 arc sec. For substantially higher angular resolution, telescopes constructed from discrete multiple apertures are required. The ISI has been designed for high-resolution aperture synthesis imaging, using the variation of the baseline provided by the rotation of the Earth, a technique that is well known from radio interferometry (Thompson et al. 1986).

Each of the two novel-design Pfund-type telescopes (Fig. 1) consists of a 2 m diameter flat mirror siderostat in an alt-az fork mount and a fixed f/3.14 parabolic mirror with 1.65 m diameter (Townes 1985). The flat mirror reflects the signal from the sky onto the parabola which in turn focuses it onto an optics table behind the flat mirror. An advantage of this particular geometry is that it has no support struts for a secondary mirror which usually give rise to diffraction effects. Another advantage of the Pfund geometry is the absence of the singularity in the zenith that is common for conventional alt-az mounts. A disadvantage is limited sky coverage but this can be eliminated by rotating the trailers 180°. This compact design allows the mirror mounts plus all additional optics and electronics to be contained in one standard-size semi-trailer per telescope. Each system weighs less than 20 metric tons, which is relatively light weight for its size.

The ISI array is laid out in seven stations with baselines between 4 and 35 m in various directions, yielding lobe spacings of 0.5 and 0.06 arc seconds, respectively (Bester et al. 1989, 1990). In the observing mode all mirror mounts are resting on three kinematic supports, and are mechanically decoupled from the trailers. This allows for stress-free expansion or contraction of the mounts with temperature changes and provides excellent

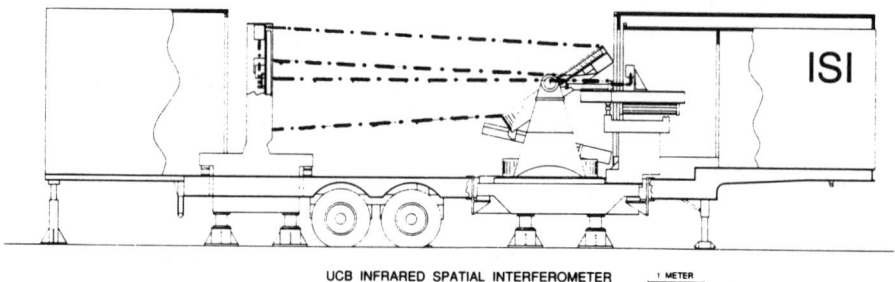

Fig. 1. ISI semi-trailer. The dash-dotted lines are beams of HeNe laser distance interferometers used for measuring length changes and atmospheric fluctuations within the telescope optics.

mechanical stability. The mirror mounts are lifted by the trailers, when moving from one station to another, in order to change the baseline.

ISI: RADIO INTERFEROMETRY AT 11 MICRONS

The Infrared Spatial Interferometer is similar to a radio interferometer, but it operates at a wavelength of 11 µm, much shorter than generally thought of in terms of radio techniques. Like other interferometers, the ISI consists of two apertures separated by the length of the baseline (Fig. 2). The infrared signal received by each telescope is down-converted into the microwave intermediate frequency (IF) range using a cooled photo-diode mixer and a local oscillator, and then processed using standard radio techniques. Correlation between the two signal paths is maintained with a delay line in one of the signal paths. Two CO_2 lasers are used as local oscillators, one at each telescope, and they are phase-locked to each other. The phase and frequency offsets are adjustable so that the fringes appearing at the output of the correlator are kept at a fixed frequency of 10 Hz. Without such a measure the fringes would change in frequency due to the Earth's rotation (the "natural" fringe rate) and the fringe signal would have to be analyzed in a more complex way. The mixers have a quantum efficiency of about 40% and the IF bandwidth is 2 GHz. A more detailed description of the detection system is given in Danchi et al. (1988). A fringe spectrum obtained under average seeing conditions is shown in Fig. 3. The sidebands around the central spike are due to phase modulation in the atmosphere and are part of the stellar signal rather than noise. The visibility of an astronomical object is obtained by comparing the integrated power in the fringe spectrum with that obtained from a point source. Visibilities are calibrated using the total infrared power measured on either telescope simultaneously with the fringe observations (Danchi et al. 1990a).

ARCHITECTURE OF THE COMPUTER CONTROL SYSTEM

An overview of the architecture of the computer control system is shown in Fig. 4. The host development system and user interface is a Sun (*UNIX 4.2 BSD*) workstation. This workstation is linked via Ethernet with three dedicated VMEbus computers (*MVME 147S*, Motorola) running *VxWorks* (Wind River Systems) as their real-time operating system. VME0 is the data acquisition system, and VME1 and VME2 are the system controllers

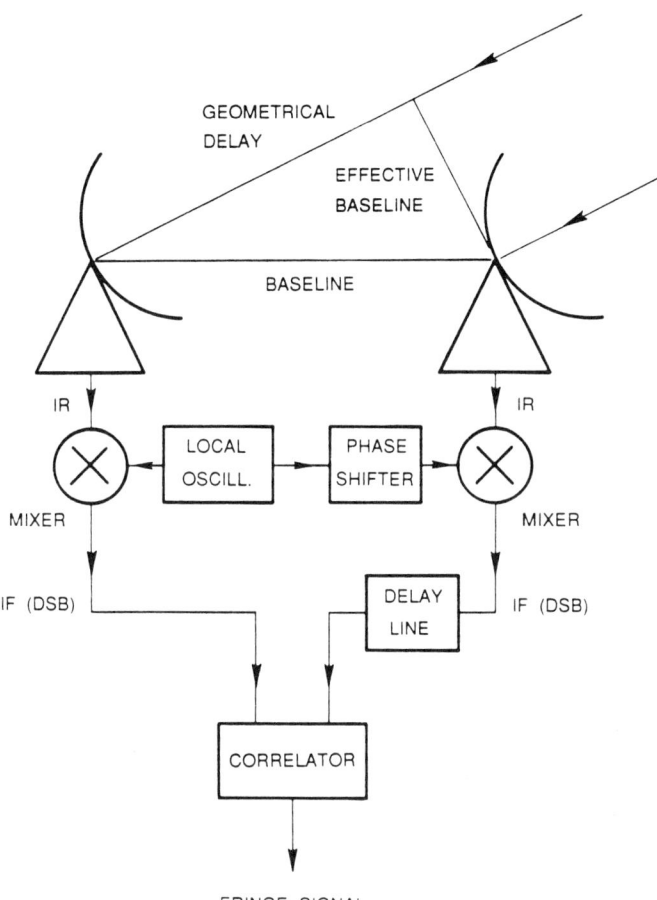

Fig. 2. Schematic of the infrared heterodyne interferometer.

for telescope #1 and #2, respectively. Time information is received from WWVB and is distributed to all real-time systems in form of the IRIG-B time code. Custom electronics developed in-house are interfaced through parallel I/O's from the VMEbus crates. The Sun workstation is also the disk and tape server for the real-time computers. Each of the VMEbus systems runs its own set of tracking calculations once per second and monitors system parameters at a rate of 1 kHz. The communication between the *UNIX* workstation and the *VxWorks* real-time computers is accomplished via remote procedure calls. All ISI software has been written in 'C' code.

A key feature of this modular system architecture is that it allows for future expansion of the interferometer with respect to more telescopes, as well as more workstations for on-line data analysis.

A block diagram of the servo controllers is shown in Fig. 5. The trackball is also linked to the autoguider on either telescope, which allows guiding on the visible part of

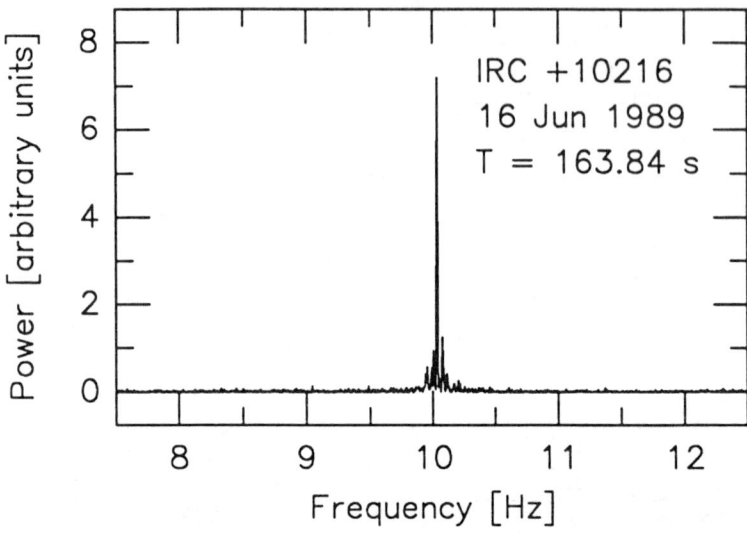

Fig. 3. Fringe spectrum of IRC +10216 taken on June 16, 1989, in a 164 s integration time. The small sidebands around the central spike are due to atmospheric fluctuations and are part of the signal rather than noise.

the stellar spectrum down to $m_V = 13$ in a 1 second integration time. The autoguiders consist of a PC controlled CCD camera (*Lynxx PC Plus System*, SpectraSource Instruments) with customized software for centroid calculations and star tracking (Degiacomi and Bester, 1990). The feedback loop is closed by feeding artificial trackball pulses into the servo microprocessors. The telescopes are equipped with conventional optical encoders having a resolution of 0.06 arc sec in azimuth and 0.07 arc sec in altitude on the sky.

The interconnection scheme between the two telescopes is shown in Fig. 6. All components shown to the left of the dashed line are installed in telescope #1, while the others are in telescope #2. Ultimately, fiber optic links will be used to transmit all necessary analog and digital signals between the two telescopes. These signals comprise RS 232 data lines and the Ethernet, the video signal and some analog signals, which are monitored in the control room in trailer #1, as well as the line that carries the microwave intermediate frequency band from the detection system on telescope #2. Fiber optic links are advantageous as they provide electrical decoupling and are safer in the event of a lightning strike. There is also no signal loss or cross-talk between signals for the longer baselines. However, at this time the fiber optic links are not yet fully implemented.

THE USER INTERFACE

The user interface has been written in the *X Window System* environment with the *OSF/Motif* widget set and is completely menu/mouse driven. A hardcopy of the *xobserve* window in Fig. 7 shows a typical set-up of an observing session. One would start with a 5-point beam map, for example on a strong source like α Ori, in order to align the boresight of the infrared detectors with the optical boresight as seen by the autoguiding

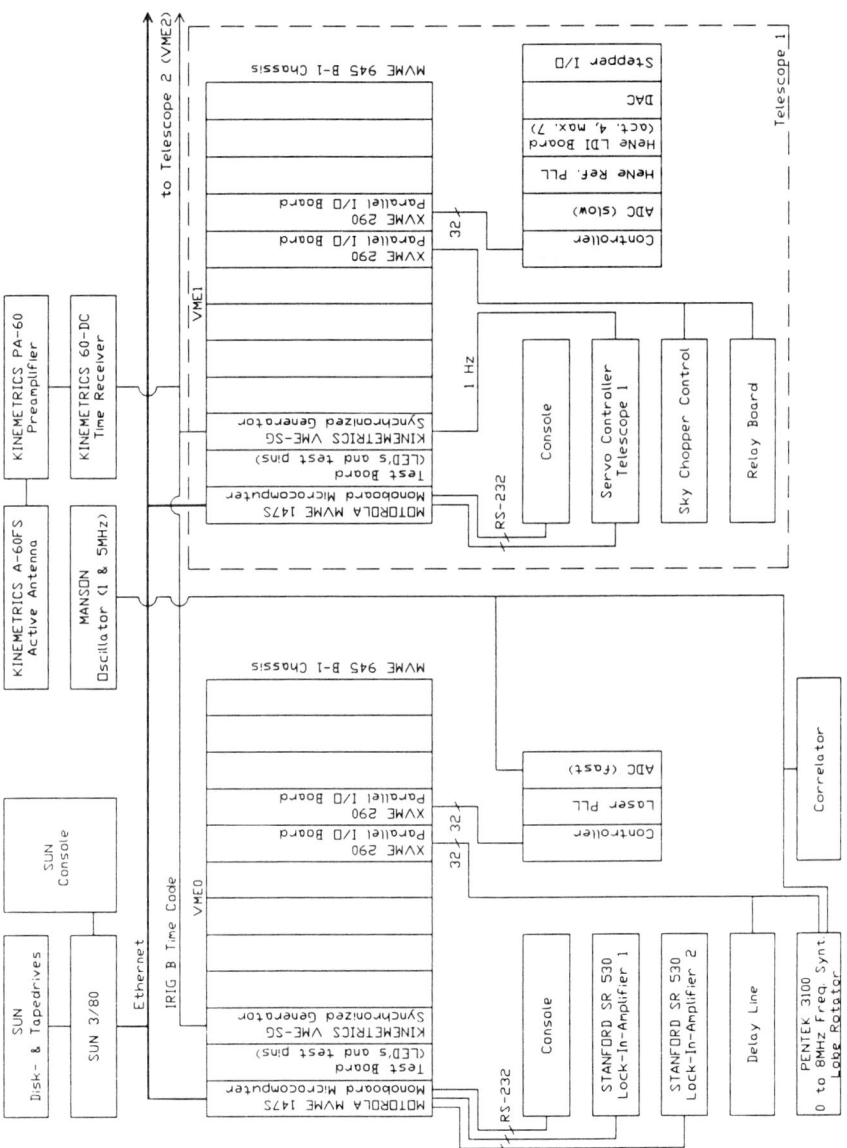

Fig. 4. ISI computer control system. The dashed box includes all control electronics for telescope #1. An identical unit, not shown in the figure, is installed on telescope #2.

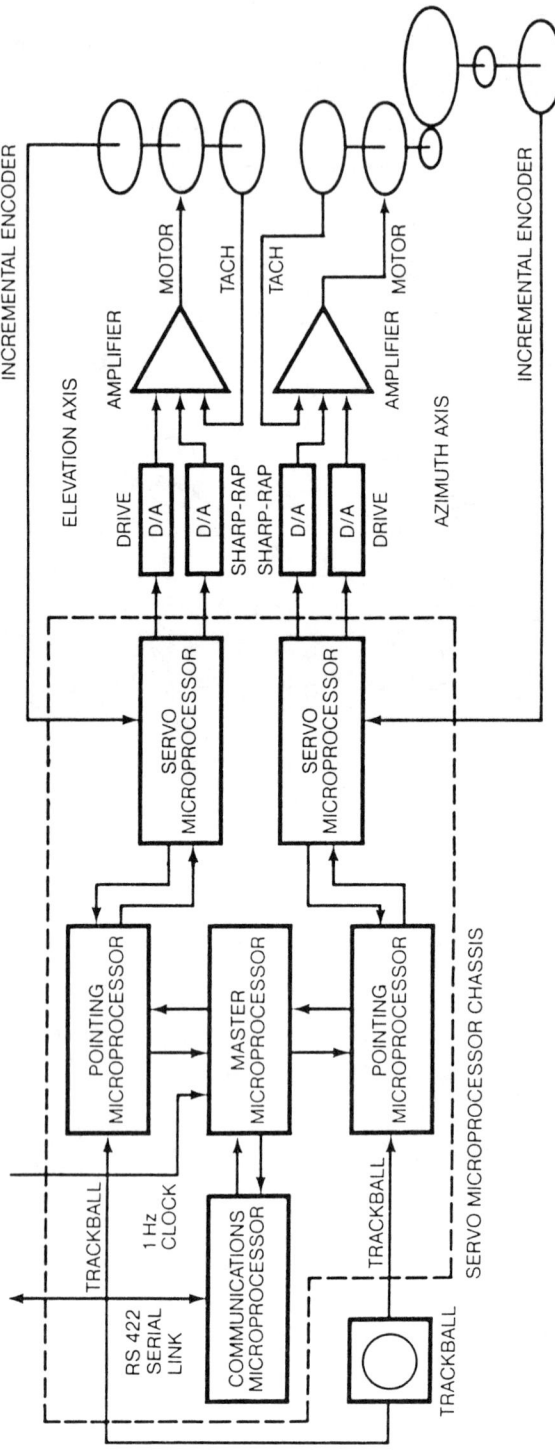

Fig. 5. ISI servo control system. The autoguider, not shown in this figure, is connected to the trackball.

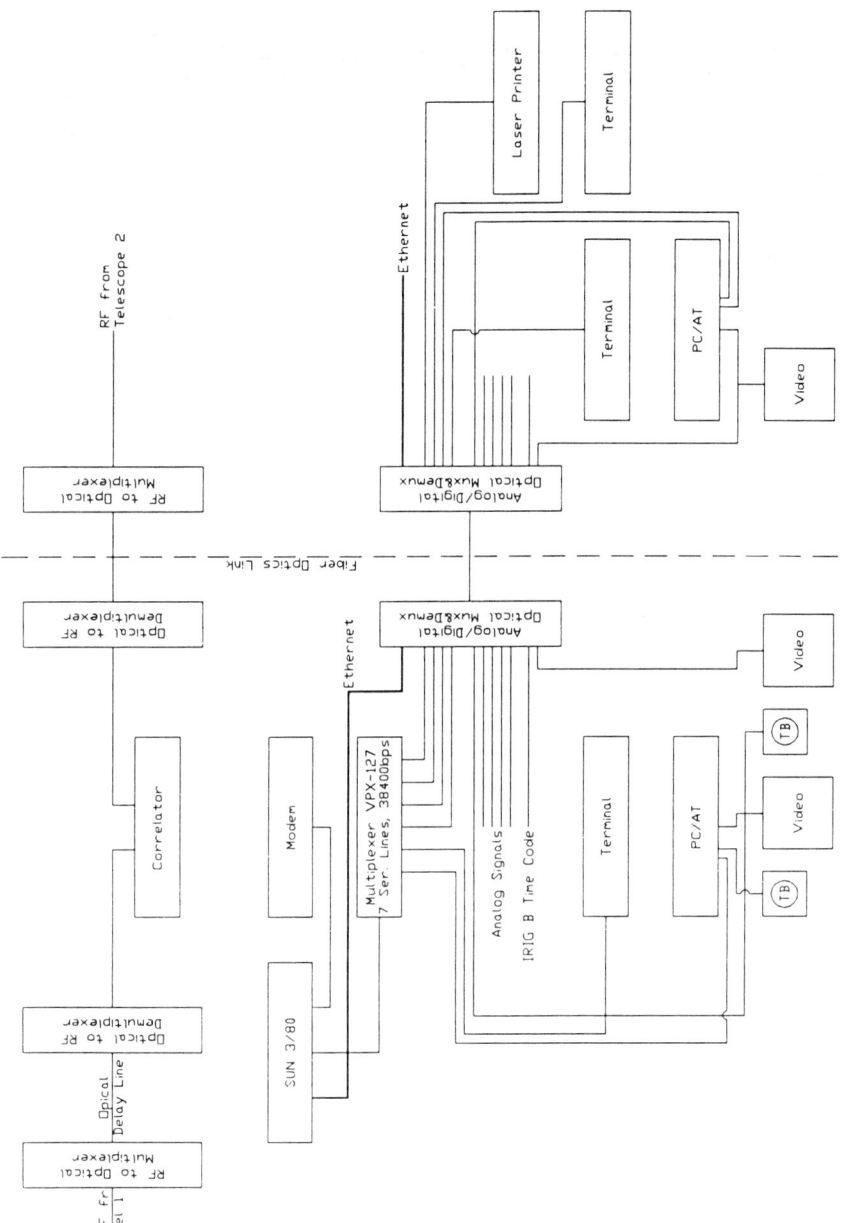

Fig. 6. Interconnection diagram between the ISI telescopes. All components to the left of the dashed line are installed in trailer #1, while those to the right are in trailer #2.

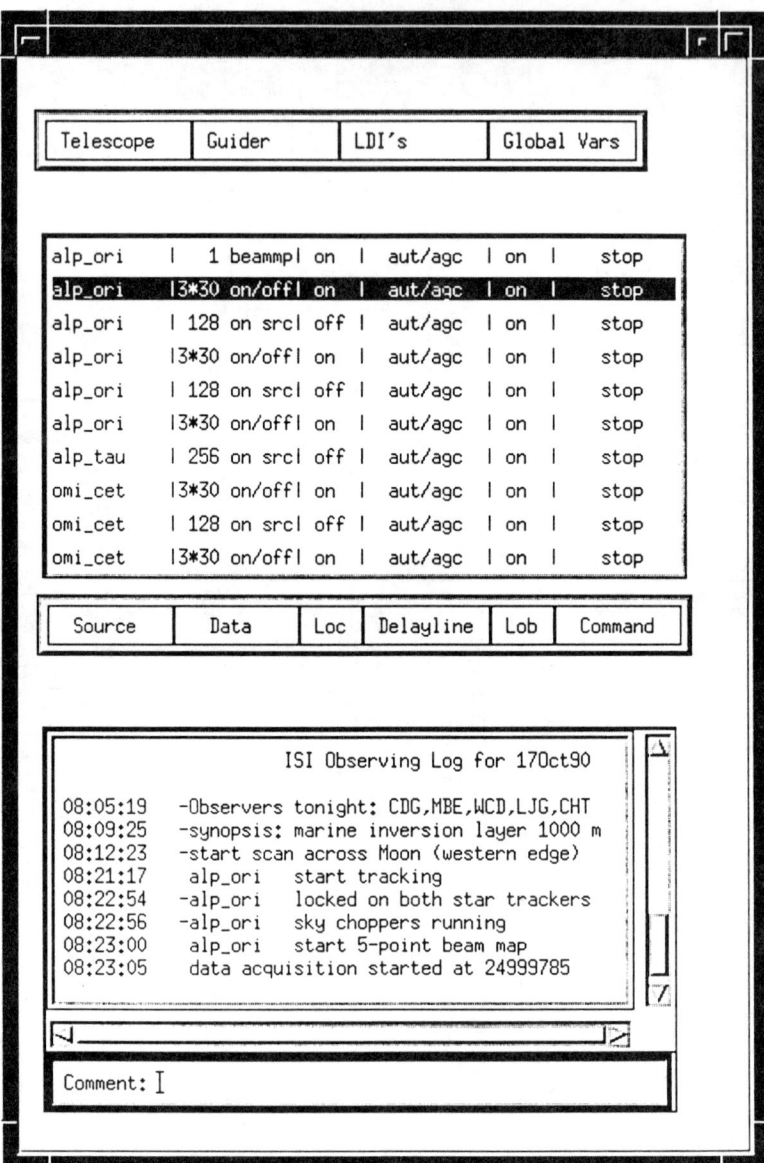

Fig. 7. Control window of program *xobserve*. The buttons above and below the panel that shows a pre-programmed observing sequence activate several submenus. The most recent entries in the log book are shown in the lower panel. Entries beginning with a hyphen are comments entered by the observer.

cameras. Then a sequence of measurements on and off-position is performed with the sky choppers running to measure the total infrared flux simultaneously with the coherent fringe signal, and to obtain the cold sky offset. Interleaved with these measurements are measurements with the sky choppers turned off, in order to get a stronger fringe signal. An observation of α Tau would follow to perform a point source calibration of the visibility. Then a similar sequence would follow for another source, e.g. o Cet. The four buttons above and the six below the window that shows the observing sequence activate several layers of sub-menus. These allow the observer to select a new source from the on-line J2000.0 star catalog and to set all other parameters required for an observing session, and finally to start the pre-programmed observing sequence. This user interface allows each night's observations to be automated and observing time to be used optimally. All actions are recorded in a log book displayed in the bottom panel. A comment can be entered manually. Error messages due to system error conditions or inappropriate user commands are shown in a foreground window. The observer has to acknowledge each message and take corrective action, before being allowed to proceed. Other windows on the Sun console are used to monitor the tracking and data acquisition parameters on the real-time systems, and to perform on-line Fourier transformations to check the quality of the recorded fringe data.

ILLUSTRATIVE SCIENTIFIC FINDINGS AND FUTURE PLANS

Between 1988 and 1991 we obtained a large number of interferometric data on a variety of sources, mainly well known late-type stars. For IRC +10216 and o Cet the dust was found closer to the star (3-5 stellar radii) and at a higher temperature (~ 1200 K) than indicated by previous measurements (Danchi et al. 1990b, Bester et al. 1991). Visibilities of IRC +10216 and o Cet change with luminosity phase and new dust appears to form at still smaller radii during minimum luminosity. Near maximum luminosity, the central star of IRC +10216 produces a visibility of 0.02 at 11 μm, from which a stellar diameter of 0.061 arc sec and a temperature of 1900 K are deduced. In contrast α Ori has a very different dust distribution, with a dust shell of substantially larger inner radius and very little dust close to the star (Bester et al. 1991). This indicates a sporadic production of dust and little or no recent dust formation. There appears to be a discrete thin shell at a radius of 0.95 arc sec and velocity of 11 km s^{-1} (Bernat et al. 1979), probably emitted during the unusual activity of the late 1920's. The distance to α Ori would hence be 125 pc. Data on other late-type stars like VY CMa, α Leo, α Her, α Sco and χ Cyg are presently analyzed.

The 11 micron interferometer described above is sensitive enough for investigation of several thousand IRAS sources with high angular resolution. In particular, the use of a heterodyne detection system is well suited for studying the distribution of certain molecules like NH_3, SiH_4 and C_2H_2 around evolved stars like IRC +10216. Other work will include astrometry and the investigation of atmospheric turbulence.

ACKNOWLEDGMENTS

This work was supported by the United States Office of the Chief of Naval Research Grant N00014-89-J-1583 and grants from the California Space Institute and the Perkin-Elmer Corporation. W. C. D. has been supported in part by the L. W. Frohlich Research Fellowship of the New York Academy of Sciences, C. G. D. by the Swiss National Science Foundation, and L. J. G. by a Miller Research Fellowship of the University of California at Berkeley. We thank Walt Fitelson and Carl Lionberger for their continuous technical support.

REFERENCES

Bernat, A. P., Hall, D. N. B., Hinkle, K. H., and Ridgway, S. T. 1979, *Ap. J. (Letters)*, **233**, L135.

Bester, M., Danchi, W. C., McCullough, P. R., and Townes, C. H. 1989, in *The Physics and Chemistry of Interstellar Molecular Clouds*, eds. G. Winnewisser and J. T. Armstrong, *Lecture Notes in Physics*, Vol. **331**, p. 396.

Bester, M., Danchi, W. C., and Townes, C. H. 1990, in *Amplitude and Intensity Spatial Interferometry*, ed. J. B. Breckinridge, *Proc. SPIE*, Vol. **1237**, p. 40.

Bester, M., Danchi, W. C., Degiacomi, C. G., Townes, C. H., and Geballe, T. R. 1991, *Ap. J. (Letters)*, **367**, L27.

Danchi, W. C., Bester M., and Townes, C. H. 1988, in *High-Resolution Imaging by Interferometry*, ed. F. Merkle, *ESO Conference and Workshop Proceedings* **29**, p. 867.

Danchi, W. C., Bester, M., Degiacomi, C. G., and Townes, C. H. 1990a, in *Amplitude and Intensity Spatial Interferometry*, ed. J. B. Breckinridge, *Proc. SPIE*, Vol. **1237**, p. 327.

Danchi, W. C., Bester, M., Degiacomi, C. G., McCullough, P. R., and Townes, C. H. 1990b, *Ap. J. (Letters)*, **359**, L59.

Degiacomi, C. G., and Bester, M. 1990, *Test of a SpectraSource Instruments Lynxx CCD Imaging System*, Technical Report, Space Sciences Laboratory, University of California at Berkeley.

Thompson, A. R., Moran, J. M., and Swenson, G. W., Jr. 1986, *Interferometry and Synthesis in Radio Astronomy*, John Wiley & Sons, New York.

Townes, C. H. 1984, *J. Astrophys. Astron.*, **5**, 111.

PROSPECTS FOR AUTOMATED RADIAL-VELOCITY OBSERVATIONS

R. F. GRIFFIN
The Observatories, Madingley Road, Cambridge CB3 0HA, England

ABSTRACT Measurements of stellar radial velocities, especially those of late-type stars, have become substantially automated since the development of direct cross-correlation methods some 25 years ago. What is needed now, to automate the complete process, is to combine a modern radial-velocity spectrometer with an automatic telescope. So far the author has been unsuccessful in attracting support for such a system; he would be interested in collaborating with anyone who is in a position to contribute the telescope side of the equipment.

Andy Fraknoi was at pains to stress yesterday that this Society is a national society. He might almost have claimed international status for it — after all, there are at this meeting three speakers who even come from the Eastern Hemisphere (counting myself, since my institution is 22 seconds of time to the East of the Greenwich meridian!). I have been a member of the ASP for more than 25 years, and I regret to say that this is the first time I have managed to get to a meeting. It is certainly a pleasure to be here, especially now that one's operating system is coming to terms with living in a partial vacuum!

Since I perceive that many of you are photometrists, I hope that you will not *all* be insulted if I begin by outlining why radial-velocity measurements are significant. They bear on the large subject of dynamical astronomy — the study of the movements of celestial objects. All assemblies of objects must be moving under the influence of their own, and/or external, gravitational fields, and the movements of the objects are of interest both individually and for the purposes of charting the gravitational potentials of the assemblages to which they belong. Examples of such assemblages are the solar system, binary stars, star clusters, the Galaxy as a whole, external galaxies, and clusters of galaxies.

There are two basic techniques of dynamical astronomy, and they are complementary. One refers to the crosswise or tangential movement of objects across the sky, and involves angular measurements either of absolute or of relative positions. Basically you measure the positions of the objects of interest now, and then again some time in the future. Before you can determine the rate of movement you have to wait for the movement to become perceptible. Some stars show perceptible movement in years, some in decades, but most have not shown any so far; galaxies and quasars are altogether beyond the reach of the method. Radial velocities, on the other hand — the motions of objects along the line of sight — are determined directly by the Doppler shift in the spectrum; no waiting is involved, and there is no dependence of the ease or accuracy of the measurement upon the distance of the object observed.

In the last quarter of a century considerable steps have been taken to automate radial-velocity measurements. So far those steps have related only to the measuring instruments and not to the telescopes, but that is all the more reason why advantage should now be taken of the independent development of automated telescopes to combine the advances in both instrumental fields to produce equipment of unprecedented potency.

The classical way to measure radial velocities was to photograph the spectrum of the object of interest between spectra of a laboratory reference source (often an iron arc) whose purpose was to provide a velocity zero-point. After the photograph was processed, the astronomical and laboratory spectra on it had to be measured in a measuring machine, normally line by line, and a more or less labor-intensive reduction procedure had to be followed in order to obtain the final result. The method had some fundamental inefficiencies, not the least of which was the need to spread out the light from the object into a spectrum and in effect to measure the intensity of that spectrum at each one of hundreds (if not thousands) of independent wavelengths, each of which had to have sufficient light to register on the plate, when all that was wanted was one single quantity — the Doppler shift. In the early 1960s, therefore, I developed a method — I did not invent it, I only made it work — whereby the Doppler shift in the spectra of cool stars could be measured directly without any need to record the spectrum. The method works by cross-correlation, a technique that was then novel but is now in almost universal vogue for the measurement of radial velocities. The original instrument, which I regret to say is still in use at Cambridge, performs the cross-correlation without the user having to know that this is what it is doing. It is basically a high-resolution spectrograph in which the spectrum is received not on a detector but on a mask which is mostly opaque but has transparent slots in it — hundreds of them — at places corresponding to absorption lines in the star spectrum. The principle of the measurement is to scan the mask along the spectrum, monitoring all the while the amount of light it is transmitting, to locate the position at which it is in register with the spectrum. That position is revealed by the sudden diminution of the light transmission when every aperture in the mask is occupied by the corresponding dark line in the star spectrum.

The method that I have just described, and analogous procedures in which the cross-correlation is performed digitally upon spectra which need only be observed with extremely low signal-to-noise ratio, is in use around the world and has enabled radial velocities of high accuracy to be measured quickly and routinely at magnitudes a thousand times fainter than were accessible before. But ambition grows at least as fast as ability, and now there comes a cogent reason for wanting the radial velocities of far more stars than have been measured in the past, to complement the crosswise motions that are being measured by the satellite Hipparcos. Hipparcos, which cost, I believe, about 400 million dollars, is measuring the positions and — in the cases where they are perceptible — the proper motions and parallaxes of some 100,000 stars. The scientific case for measuring the radial velocities of the same stars is overwhelming. To understand the Galactic orbits of the stars concerned, one needs to know the radial as well as the transverse motions. The radial velocities of about two-thirds of the stars — those that are of spectral types later than about early F — could be measured by the spectrometer method I have described; not only can they be measured by ground-based equipment costing a mere one-half percent of the cost of Hipparcos while practically doubling the value of the results, but they will be more accurate than the transverse motions for all but a handful of the stars. The accuracy of the transverse motions falls off rapidly with increasing distance, both because of the diminishing proper motions and also because of the diminishing accuracy of parallaxes as the distance increases.

The hardware proposed for the project consists of twin systems, one in the northern hemisphere and one in the southern, consisting of compact radial-velocity spectrometers mounted on 40-inch automated Cassegrain telescopes. There are compelling technical reasons, notably the need to keep atmospheric dispersion parallel to the length of the spectrometer slit, for wanting the telescopes to be on altazimuth mountings. The spectrometers could follow the general pattern of the very successful "Coravel" instruments designed by a consortium led by Dr. Mayor of the Geneva Observatory. On the very conservative assumption that each telescope would observe 200 stars a night (a number that can often be achieved in "manual" observations with the existing Coravel at ESO) on 200 nights a year, the total number of observations would be 80,000 a year — more than enough to include all the Hipparcos stars that are of late enough types for measurement. All those stars, therefore — more by far than the total number of stars whose radial velocities have ever been measured — could be measured *every year* until the constancy or otherwise of their velocities was sufficiently established. Even at the start of the project there would appear to be power to spare for other programs, and as time goes on the extraordinary power of the system could obviously be progressively released for all sorts of individual research programs and for further ambitious survey work.

Unfortunately, despite substantial efforts over the past four years, I have not succeeded in obtaining funding for this project. I am now reduced to trying to fund a northern-hemisphere system only, since I understand that a consortium similar to that which built the original Coravels is now building an automated system to be placed at ESO. However, I shall very shortly be scrapping the old Cambridge spectrometer in favor of an updated Coravel-type instrument. Although I am intending at present to use the new instrument for hands-on observing at Cambridge, it is to all intents and purposes an automated instrument, and in the last resort I would be prepared to subscribe it to a fully automated system at a good site. If there is anyone here, or anyone who reads the written version of this harangue, who would be ready to enter into a collaboration on this project on the basis of subscribing the necessary telescope, I would be only too pleased to discuss the matter.

ROBOTIC TELESCOPES WITH FIBER-COUPLED SPECTROGRAPHS

LAWRENCE W. RAMSEY
Department of Astronomy and Astrophysics, The Pennsylvania State University, University Park, PA 16802

ABSTRACT We address the scientific need for a Robotic Spectroscopic Telescope (RST) and discuss some of the issues that must be considered in the design of the telescope, the fiber couplings, and the spectroscopic instrumentation. Performance modeling based on experience with fiber coupled spectrographs indicates that a one-meter class telescope can provide the capability to carry out a wide variety of scientific programs, as well as most economically demonstrate the feasibility of the RST concept.

INTRODUCTION

As other papers in this volume have shown, robotic telescopes have been quite successful in carrying out photoelectric photometry. There is also tremendous utility in attempting to do robotic spectroscopy. Spectroscopic instruments are somewhat more complicated, however, and the data more cumbersome to handle. The development of fiber-coupled spectrographs during the last decade can help here, as a fiber cable allows the instrument to be placed in a suitable environment off the back of the telescope. This simplifies both the telescope and spectrograph design problem. We will discuss below some of the requirements for such a system as well as the anticipated performance.

SPECTROSCOPIC SCIENCE WITH ROBOTIC TELESCOPES

The most important question to ask is why one should build a robotic telescope dedicated, at least in some large fraction, to spectroscopy. Robotics is best utilized for repetitive, relatively simple tasks. This leads to two prerequisites for robotic spectroscopy. First, the scientific program must be able to be carried out with no real-time participation by the observer. Second, the instrumental configuration must be fixed or tightly constrained. Three types of observational programs are most suitable to robotic observations: surveys, time-domain astrophysics, and ground-based/space coordination.

Surveys are a fundamental resource in astrophysics. Several recent (IRAS, GRO, and ROSAT) and upcoming (EUVE) space missions have a major survey component. Ground-based surveys utilizing the dramatic developments in detector technology have been rare. Notable exceptions that come to mind are John McGraw's transit telescope at Arizona and the highly productive CfA redshift survey. The 2 micron survey discussed at this conference will certainly prove even more productive. In principle, surveys are ideal programs for robotic telescopes as they involve repeated measurements of a large number of objects with a set instrumental configuration. Surveys fall into two broad categories: (i) all sky, and (ii) concentrated area (including pencil beam). All-sky surveys tend to be object class specific. Examples may be galaxies detected by IRAS, dwarf novae, or RS CVn binary systems. Robotics are well suited for this type of survey as the telescope can efficiently carry out observations all night with a set configuration. Limited-area surveys, such as obtaining spectra of all ROSAT sources around the ecliptic pole, are not as well matched to a robotic telescope, unless one is willing to let part of the night be unused or blend with another survey, perhaps requiring an instrument change.

The time domain is relatively unexplored in astrophysics. This is particularly true for phenomena with time scales between a few days to a few months. These time scales are likely to be an area in which robotic observatories can make the biggest impact. Phenomena with time scales of minutes to hours are well served, at least potentially, by extant facilities. It is going to be hard to address a wide range of problems in extragalactic astronomy with moderate size telescopes. Here, 2-m class instruments are needed. One area that stands out is monitoring variability in AGNs. Shields (1990) discussed this problem recently. There is the possibility of doing some of the brightest objects with a 1-m class instrument, as we will see below, but at least a 2-m class telescope will be needed for major advances in this area and other areas of extragalactic astronomy. On the other hand, there is a wide range of problems in stellar astrophysics that will benefit from time series spectroscopy with small (1 m) aperture telescopes. Wade (1990) summarizes many of these in a recent workshop article. Another relatively obscure publication that describes such science is the SHIRSOG workshop. The *Proceedings* of this workshop are published by NOAO and edited by Mark Giampapa at NOAO.

Ground–Space coordinated observing is another area in which robotic telescopes with versatile scheduling can make a large impact. Such opportunities often arise on short notice and cannot be accommodated by the national observatory. Even university observatories cannot always respond due to over-subscription of telescope time as well as scientist time.

TELESCOPE REQUIREMENTS

Spectroscopy imposes a different set of requirements on a robotic telescope than PMT photometry. The very nature of spectroscopy (i.e., dividing the light into narrow bandpasses) requires a larger aperture. One can make reasonable arguments that a 2–3 meter class telescope would serve the broadest range of science. However, nobody to my knowledge has a full-time RST (Robotic Spectroscopic Telescope). Thus, it seems prudent to demonstrate the instrumental concept as well as scientific utility by starting out with a 1-m class telescope. Such an aperture can, as I will demonstrate below, carry out a large number of interesting science programs. Two-meter class telescopes present truly exciting possibilities, but are realistically farther away.

One-meter telescopes are now reasonably economical to build. A particular advantage for a fiber-fed RST is that we can locate the acquisition camera and fiber at either prime or Cassegrain focus. In either case the on-telescope instrument package consists only of an acquisition and guiding camera and the fiber feed itself. A simple set of calibration lamps might also be resident in this assembly. Using the prime focus has some throughput advantage but, as we will see below, would require a longer focal length primary — an advantage in making a mirror, with attendant longer telescope structure, but a disadvantage in making a stiff structure with good pointing characteristics.

The most critical parameter to consider for feeding fiber is Focal Ratio Degradation (FRD). Ramsey (1988) discusses this and concludes that the transmission is optimized for f-ratios around $f/5$ to $f/7$. Barden (1989) reaches the same conclusion, in addition to showing that the slower f-numbers yield better scrambling. Adopting $f/6$ for feeding the fiber, the scale for a 1-m telescope is $34.4''\,\text{mm}^{-1}$. This yields, for a 200 μm fiber, a subtended angle of $6.87''$. As Ramsey (1988) discusses, there is good reason not to use smaller fibers. The implication for the telescope image quality is obvious; 1–$2''$ images are fine.

The slew rate of the telescope need not be very rapid. The goal is always to minimize dead time. But the slew time can be used to advantage by taking flat fields or wavelength calibrations. Time to reach a new object should be less than 1 minute. Currently, commercial hardware can easily handle this. Accurate pointing and tracking are essential. For a variety of reasons, mostly economical, the field of view (FoV) will be reasonably small (about $10'$).

Given the scale of about $0.5'\,\text{mm}^{-1}$, we can gauge the required pointing accuracy by how close we need to put the object within the field of the acquisition CCD. Very economical CCD systems that are quite adequate for acquisition can be used if the pointing is within 10–$15''$. Clearly something like a half-inch CCD will give a very respectable $6'$ square FoV, but smaller CCDs are acceptable. Richmond and Filippenko (1991) have discussed the guiding problem for the case of an automatic imaging telescope. The problem for spectroscopy is almost identical with one additional step. Once a guide star has been acquired, one wants to peak the program star on the fiber. This is done by rastering or spiraling the program star around the fiber aperture while monitoring the output of the fiber with a PMT. We routinely perform this operation by hand at our local observatory to peak both position and focus, and automation is straightforward. As Richmond and Filippenko (1991) point out, the tracking will need to be accurate to $0.3''$ for times of seconds, as some of the guide stars might require integration of the guide CCD for that period. It is actually the guiding camera that imposes the telescope image quality requirements.

SPECTROGRAPH/FIBER SYSTEM

I will limit my discussion here to the type of instrument one might consider for the 1-m class telescope I have been discussing. I will also assume only single-object spectroscopy. Efficient multi-object spectroscopy requires objects with a high surface density on the sky. This implies faint objects which are, for the most part, out of reach of the 1-m class telescope under discussion.

Given the type of programs discussed above, there appear to be two classes of spectrographs one might like to see on a RST. The first would have low resolution with spectrophotometric capability. Such an instrument might prove valuable in ground-based/space coordinated programs or time variability studies at the limit of the reach of our 1-m class telescope. The second instrument would be employed in radial velocity and line profile/intensity variability studies. This instrument should have a resolution of at least 5000 and perhaps up to 10,000. In reality both instruments could be constructed on one optical bench and both be "on" the telescope simultaneously, with separate fiber cables going to the focal plane terminating in separate holes in the decker. Thus, switching instruments becomes the relatively simple matter of exactly where in the focal plane you place your objects. This allows more versatility for carrying out limited-area surveys without an extra mechanical (and failure prone) switching system. Two detectors systems would be required, however.

Designing Spectrographs with Fiber Feeds

There are several factors to consider when designing a spectrograph for use with a fiber optic feed. The most important is to consider FRD. This effect, due to the imperfect nature of fibers (Nelson 1988; Ramsey 1988), changes how we match a spectrograph to a telescope. Normally the collimator and telescope have the same f-ratio, and the collimator (grating height) is sized according to the resolution desired. Because of FRD, the collimator should always be a bit faster than the telescope. This represents a loss in the throughput-resolution product (Schroeder 1970).

This brings up what I call *Myth I* concerning fibers. Quite a few astronomers avoid fibers because of Myth I, which says that fibers are lossy. Through most of the visible and near-UV range (360 to 900 nm), fibers of moderate length (less than 30 m) have throughput as high as, or better than, a single aluminum reflection. Fibers *do* degrade the throughput-resolution product in a spectrographic application. Just how much they degrade that product depends on how well the telescope f-ratio is matched to the properties of a given fiber. The plots in Ramsey (1988) provide some guidance here. Two general rules of thumb can be applied. First, bigger fibers are better. This is because FRD is due, in large measure, to imperfections in the core cladding interface that tend to be independent of the size of the core. This represents a large perturbation for smaller fibers. The second rule of thumb is that fibers work best at about $f/6$–$f/8$. Three factors must be considered here. First, the transmission is maximized for f-ratios greater than 3. Second, FRD is minimized for f-ratios less than 8. Last, scrambling is maximized for larger f-ratios.

Low Resolution Spectrograph

We will consider low resolution to be in the range $\lambda/\Delta\lambda = 100$–$500$. The first conflict one has is in specifying the aperture or slit (fiber) size. Spectrophotometric capability and minimizing the effect of atmospheric dispersion argue for a large aperture, while resolution and minimizing sky noise push in the opposite direction. Another factor is the need to provide additional fibers to allow sky subtraction. The question is how many. The best sky subtractions are done with long-slit spectrographs. Ideally one would like to have a large number of fibers to maximize the signal-to-noise ratio in the sky spectrum before subtracting it from the object spectrum. Clearly one fiber is a minimum. For the type of science to be done with a 1-m class telescope this is not a critical question, but three would be a nice compromise. What is important is to address another myth concerning fibers. This myth, which I call *Myth II*, comes from experience with multi-object spectrographs. Myth II says that fibers have poor sky subtraction properties (e.g., Ellis and Parry 1988).

It is true that sky subtraction is a problem, but it is more a problem with the object spectra than the sky spectra. The myth and the problem arise because fibers are not perfect scramblers. In particular, a fiber does transmit some memory of the radial distribution of the illuminating light, whereas the azimuthal information is well scrambled. The sky uniformly illuminates the fiber, but an object may underfill it, or if it overfills, the centroid of the illumination will vary with seeing and guiding errors. Heacox (1987) discusses the scrambling properties theoretically, and Hunter and Ramsey (1991) explore the effects experimentally. Hunter and Ramsey show that the most important effect here is a change in distribution of the light over the collimator, and hence over the grating and camera. This is a small effect at the few percent level, but it can lead to subtle mismatches of object and sky spectra. One benefit of bad seeing is to overfill, or more uniformly illuminate, the fiber — there is always a silver lining.

Moderate Resolution Spectroscopy
Moderate resolution spectroscopy ($\lambda/\Delta\lambda$ = 3000–10,000) will have a different set of problems. Particularly important here is optimizing the throughput-resolution product with a good choice of fiber size and telescope f-ratio. The moderate resolution instrument should have as its goal the production of highly repeatable, high-quality spectra for radial velocity or line variation studies. A fiber-coupled spectrograph constructed on an optical bench is well suited for this. Ramsey and Huenemoerder (1986) constructed such an instrument in 1985; it has been in regular use at Kitt Peak since 1987.

Again the issue of fiber scrambling arises in what I call *Myth III*. This myth says that fibers are perfect scramblers and thus provide excellent radial velocity stability. Myths II and III are obviously related. The fact is that azimuthal scrambling in fibers is nearly perfect but, as mentioned above, the radial scrambling is not complete. However, the radial scrambling is very good and much better than in a conventional slit spectrograph. Heacox (1988) gives an excellent review of the problems of wavelength precise spectroscopy. Without resorting to extraordinary means, one can expect a fiber to be an adequate scrambler to a few hundredths of a pixel, and with care to a hundredth of a pixel. Indeed, our experience with our Fiber Optic Echelle (FOE) at Kitt Peak has shown this to be the case. In some applications, such as stellar oscillations and planetary detection, this is not enough as Brown *et al.* (1991) discuss. The ultimate in scrambling requires some additional effort. Brown (1990) describes a device that we constructed at Penn State. Hunter and Ramsey (1991) show that a single fiber with this scrambler inserted has the nearly perfect radial and azimuthal scrambling required for the most exacting applications.

A critically important aspect of any modern spectrograph system, as well as fiber-coupled systems, is the "flat field" calibration. Suffice it to say that the best results will be obtained when the beam from the calibration source matches that from the telescope. This has been the practice at most observatories since the advent of CCDs. In a fiber spectrograph what this yields is a lamp spectrum that illuminates the CCD in the same manner as the star or galaxy spectrum.

If one is going to achieve resolution on the order of 5000 and maintain good spectral coverage, a cross-dispersed echelle like the FOE will be required. This format with a fiber feed yields 15–40 orders, each only a few pixels wide, on the CCD. This makes the lamp spectra critically important. With the FOE at Kitt Peak, we have found that the best results are obtained when we obtain lamp spectra every two hours. This is because liquid nitrogen dewars drift with time. With narrow orders, a drift component perpendicular to the spectral dispersion of more than 0.1 pixel between an object spectrum and a lamp calibration spectrum reduces the quality of the observation. Some liquid nitrogen dewar designs are better than others, but we have found that thermoelectric device (TEC) cooling provides a much more stable dewar than liquid nitrogen. One would like to minimize the number of lamp spectra during the night, and TEC dewars offer this potential.

CCD Detector System Requirements

Varying sky conditions can yield vastly different signal for constant exposure on one object. A wide dynamic range may be the most important property for the detector CCD in a robotic spectroscopy application. Another highly desirable property is low dark current. The low dark current allows a TEC to be used instead of expendables like liquid nitrogen. At Penn State we have developed TEC-cooled dewars that can operate at -70 C. The current generation of Tektronix CCDs seem to have the requisite property of large dynamic range, low readout noise, low dark rate, and high quantum efficiency. Their biggest drawback is the self-induced "cosmic ray" rate, which amounts to a hundred or so per hour.

EXPECTED PERFORMANCE

In the spirit of asking what kind of science can be done with an RST, it is useful to estimate the kind of performance one might expect from the 1-m baseline telescope with fiber-fed spectrographs. We have constructed and used several instruments at Penn State ranging in resolution from 200 to 12,000 including the FOE which is now at KPNO. The performance of the FOE has been extensively evaluated at Kitt Peak (Barden *et al.* 1987). I have used our experience with the FOE and other instruments at Penn State to construct a performance model for a generic fiber-couple spectrograph on a 1-m telescope. The results are in Figure 1, which shows the signal-to-noise ratio per pixel for an exposure time of 10 minutes. They assume a CCD with the properties of a Tektronics 512×512 pixel, anti-reflection (AR) coated device, and a spectral resolution element of 3×3 pixels. A dark sky is assumed in all cases. It is clear that such a system would allow a wide variety of interesting science to be done with a robotic telescope of quite modest aperture.

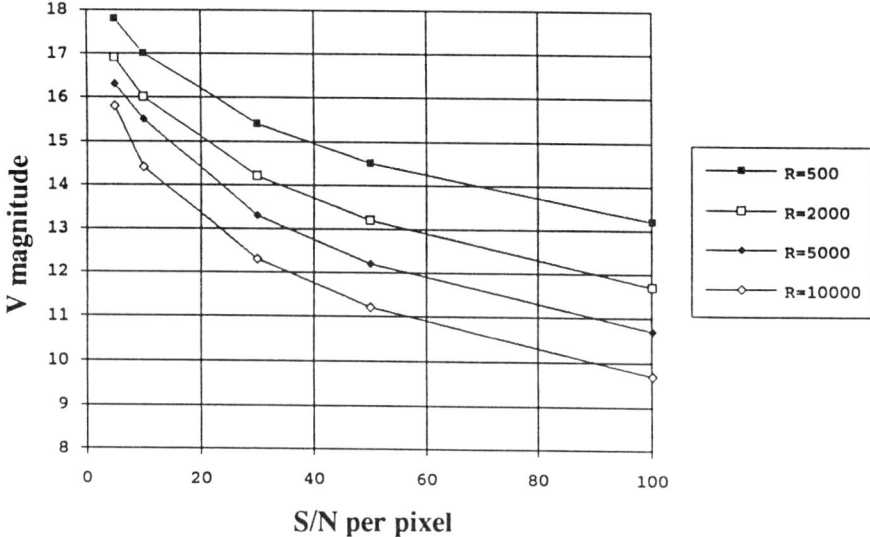

Fig. 1. Limiting V magnitude at a given signal-to-noise (S/N) ratio per pixel around 5500 Å in a 10 minute integration time is given for spectral resolutions of 500, 2000, 5000, and 10,000. All calculations assume a 1-m telescope feeding a 200 μm fiber at a Cassegrain focus. CCD properties are typical for current Tektronix AR coated chips.

CONCLUSIONS

To date, ground-based robotic astronomy has proven to be an effective means of obtaining scientifically exciting photoelectric photometry. Telescopes for robotic CCD imaging are under development (Richmond and Filippenko 1991). There do not appear to be any obstacles to the development of a robotic spectroscopic telescope. The scientific need is there, as is a well defined technological path that benefits greatly from coupling the instrumentation to the telescope with optical fibers. However, since this path has not been tread before, it seems prudent to start out with a modest size telescope in the one meter class. Such a telescope can address a wide range of scientific problems. I would like to end on one cautionary note. An RST will produce a large amount of spectroscopic data particularly from the echelle formats. One should not underestimate the effort required to turn such a data set into scientific results.

ACKNOWLEDGEMENTS

The author would like to acknowledge the support of the National Science Foundation, particularly the REU program, which has facilitated our continuing studies of fiber optics in astronomy. He also thanks Alan Welty for assistance in preparing this document for publication.

REFERENCES

Barden, S. C. 1989, private communication.
Barden, S. C., Ramsey, L. W., Huenemoerder, D. L., and Buzasi, D. 1987, *Bull. A.A.S.*, **19**, 1099.
Brown, T. M. 1989, in *CCDs in Astronomy*, ed. G. H. Jacoby (San Francisco: Astron. Soc. Pacific), p. 335.
Brown, T. M., Gilliland, R. L., Noyes, R. W., and Ramsey, L. W. 1991, *Ap. J.*, **368**, 599.
Ellis, R. S., and Parry, I. R. 1988, in *Instrumentation for Ground-Based Optical Astronomy*, ed. L. B. Robinson (New York: Springer-Verlag), p. 192.
Heacox, W. D. 1987, *Journal of Opt. Soc. of America A*, **4**, 488.
Heacox, W. D. 1988, in *Fiber Optics in Astronomy*, ed. S. C. Barden (San Francisco: Astron. Soc. Pacific), p. 204.
Hunter, T. R., and Ramsey, L. W. 1991, *Pub. A.S.P.*, submitted.
Nelson, G. W. 1988, in *Fiber Optics in Astronomy*, ed. S. C. Barden (San Francisco: Astron. Soc. Pacific), p. 2.
Ramsey, L. W. 1988, in *Fiber Optics in Astronomy*, ed. S. C. Barden (San Francisco: Astron. Soc. Pacific), p. 26.
Ramsey, L. W., and Huenemoerder, D. L. 1986, *Proc. S.P.I.E.*, **627**, 282.
Richmond, M. W., and Filippenko, A. V. 1991, In *Advances in Robotic Telescopes*, ed. M. A. Seeds and J. L. Richard (Mesa: Fairborn Press), p. 81.
Schroeder, D. J. 1970, *Pub. A.S.P.*, **82**, 1253.
Shields, G. A. 1990, in *The Spectroscopic Survey Telescope*, proceedings of a Workshop held at Penn State in October 1990, ed. F. A. Córdova, p. 49.
Wade, R. A. 1990, in *The Spectroscopic Survey Telescope*, proceedings of a Workshop held at Penn State in October 1990, ed. F. A. Córdova, p. 53.

GLOBAL NETWORKING: THE AUSTRALIAN CONNECTION

JOHN J. VAN-VEGCHEL
The Southern Automatic Patrol Telescope Service, Nanango Astronomical Observatory, P.O. Box 66, Nanango 4615, Queensland, Australia

ABSTRACT   An Automatic Photoelectric Telescope Service will be operating from the Nanango Astronomical Observatory in Australia, providing much needed southern sky exposure to northern hemisphere astronomers.

INTRODUCTION

The Southern Automatic Patrol Telescope Service located at the Nanango Astronomical Observatory (NAO) will provide the first Automatic Photometric Telescope Service outside of the United States. The placement of this facility in the southern hemisphere will provide the obvious advantage of exposing southern skies to northern astronomers, graduate/undergraduate students, and educators, while allowing sufficient overlap of the northern skies to permit real-time sharing and collaboration between other observatories (robotic and manned) in suitable global longitudinal distribution.

The NAO project is the only observatory in Australia dedicated to robotic telescopes with an unlimited capability of expansion. Auxiliary observatories will provide astrometric mapping and solar Hα photography. Other professional observatories will be moving onto location to take advantage of the sub-arcsecond resolution and lack of light pollution, not to mention security.

APT SERVICE

The NAO will provide the only Automatic Photoelectric Telescope Service (Van-Vegchel 1990) in the southern hemisphere. This service is based on the existing Automatic Photoelectric Telescope Service (APTS) operating at the Fairborn Observatory on Mt. Hopkins, Arizona, USA (Genet et al. 1987). Fees for the placement of telescopes and charges for observational requests will be identical to those of the APT Service at Fairborn and will not be described here. The important aspect is that compatible instrumentation and operating systems will allow a standardized service to operate. Real science for less dollars is the aim of both northern and southern hemisphere services.

OBSTACLES ENCOUNTERED IN ORDER TO ACHIEVE REALIZATION

In the words of Albert Einstein, "Great spirits have always encountered violent opposition from mediocre minds."

The NAO project was originally time-tabled for completion in early 1990. This estimate was based on the approvals given by Australian financial institutions in mid 1989. Unfortunately, the approvals were conditional on the banks entering into a joint partnership arrangement which they later dishonored (for no better reason than the current financial climate). The search was then commenced for an underwriter to secure collateral so that the 3 million dollar project could continue.

This search was successfully achieved by approaching United States financial houses who found scientific and educational programs a necessity for the future development of science. With securities ready for lodgement with a lending institution, we discovered another obstacle, Saddam Hussein. His consequent invasion of Kuwait led to the freezing of our securities until such time as the conflict ended. I am indeed pleased to announce that the day after the liberation of Kuwait the collateral suppliers to the NAO project continued to process the loans which are now closing. Casualties of war can take many forms!

GLOBAL NETWORK

By far the most powerful aspect of the NAO will be its contribution to the Global Network of Automatic Telescopes (Baliunas et al. 1989). Collaboration and cooperation between Mt. Hopkins and Nanango will allow observational requests to be transferred via high-speed modems, thus creating an environment which will permit time sharing with other robotic telescope users for stars within the northern and southern overlap windows made possible by NAO's latitude. Real-time photoelectric and imaging response between telescopes at the NAO and Fairborn will not be available until satellite (see COMMUNICATIONS section) links are explored between Australia and the United States.

The Global Network is not exclusively the realm of robotic telescopes. The NAO will fill an important longitudinal window in the Whole Earth Telescope (Henry 1989) concept of continuous observation by providing an additional site from which new data can be collected.

QUALITY CONTROL

The NAO is dedicated to quality science. Photometric data collected will be substantiated by separate standard star observations from another robotic telescope 0.45 m in diameter. Integrity of observation will be provided by strip chart recordings. Full weather details will also be forwarded to observers along with their data. An astronomer will be on duty at the observatory at all times. All telescope functions will be overseen to make sure that sudden changes in criteria (e.g., weather) can be monitored with suitable notations logged and passed onto the user. Contingencies for the observation of sudden stellar phenomena (supernovae, etc.) will be in place so that a change in scheduled observations can take place at short notice at the request of the users.

INSTRUMENTATION

The Southern Automatic Patrol Telescope Service will initially dedicate two AutoScope Robotic Telescopes to the APT Service. In the original publications concerning the NAO, a 0.80-m robotic telescope was proposed. After reviewing the optical capabilities of the slumped meniscus mirror (80% of the light in 3.0 arcsecond diameter), it was decided that, in order to achieve greater sensitivity in CCD photometry and imaging, not to mention versatility in future instrument changes, a smaller, more exacting optical specification was necessary. Therefore, a 0.60-m Ritchey-Chrétien optical train (80% of the light within 0.50 arcsecond diameter) was agreed upon (Genet 1991).

Telescope #1 is a Model 600, 0.60-m telescope with a Model 200 slow-scan thermoelectrically cooled CCD camera and RP–1000 precision photometer. This telescope will be housed in a 6-m dome or equivalent Multiple Mirror Telescope (MMT) type enclosure.

Telescope #2 will be a Model 406, 0.45-m telescope with RP–500 PMT photometer and CCD–100 camera. Once again, this instrument will consist of Ritchey-Chrétien optics of similar precision to the 0.60-m telescope. This instrument will be housed in a roll-off roof observatory which will have the capability of accommodating 10 telescopes of 1.0-m diameter. This extra floor space is available for telescopes owned and operated by other institutions and maintained by the NAO.

Complimenting the robotic telescopes will be a solar observatory with a 7-inch refractor used for Hα photography and sunspot monitoring. Numerous field telescopes ranging from 0.5 m to 0.32 m diameter will be available for student access and are incorporated within an education complex which will cater to upper high-school and University students.

SITE

The NAO is located in Queensland, Australia at a latitude of 26°37'00" South and a longitude of 151°58'30" East. The site experiences sub-arcsecond resolution made possible by low atmospheric turbulence, low particle scattering (no dust, aerosols, or other pollutants), *no* light pollution, and a dry stable environment. The heavily forested region enables cool temperate conditions to prevail. With an altitude of 425 m, the NAO is well elevated by Australian standards.

COMMUNICATIONS

Initially, high-speed modems and optical fiber links will allow for data transfer abroad. This method is inadequate for imaging transfer, which requires at least 6 Mb per second transfer. Current existing communications networks being used by the Australia Telescope (the only high resolution, high sensitivity aperture synthesis radio instrument with frequency diversity and spectral-line capability in the southern hemisphere) are being investigated for their possible use in real-time operation between robotic telescopes in Australia and in the United States. An AUSSAT (Australia Satellite) communications dish at the NAO can be linked into the Australia Telescope which in turn has a microwave link to the NASA Deep Space Tracking Station at Tidbinbilla, Canberra. Transmission would then proceed via the mid-Pacific satellite to the VLA site in New Mexico, USA, and finally via microwave link to Mt. Hopkins. If the proposal is accepted, the expense needed to construct a communications link between Mt. Hopkins and Nanango would be reduced.

A second option being investigated will require an Earth station at the NAO and another at the Fairborn Observatory to access and communicate via the intelsat mid-Pacific satellite. Currently a proposal is being prepared in response to a NASA Research Announcement (NRA 91-OSSA-13) which will outline the advantages of direct communication between APT Services in the United States and Australia as an innovative research program.

SECURITY

The 120 acre site has only one access road and is surrounded by dense forest and bush. The observatories will be permanently supervised and patrolled. Other professional astronomers and institutions are invited to take advantage of the facilities and relocate their observatories and instruments to the NAO. Observatories can be left unattended without fear of vandalism or theft.

FUTURE

The NAO plans to install during the construction phase suitable utilities so that future expansion can proceed without delay brought on by inadequate power distribution or communications. Individually surveyed parcels of land are available for lease by local or international institutions. Other nations are invited to take advantage of southern skies at this prime location in a politically stable country.

Conference and accommodation facilities at the site provide for symposia, conventions, and general meetings. I would like to propose for consideration of this gathering the possibility of holding the 105th Annual Meeting of the ASP at the NAO. This indeed would be a first for the society, as well as an excellent opportunity to see the operation of your southern cousin.

CONCLUSION

The Nanango Astronomical Observatory is a high-technology, professional astronomical facility offering astronomers worldwide access to high-resolution southern skies via the Southern Automatic Patrol Telescope Service. International cooperation and collaboration to produce quality science and better science per dollar are our goals. The Global Network of Automatic Photoelectric Telescopes will no longer be an ambition but instead a reality.

REFERENCES

Baliunas, S. L., Cornell, J., and Genet, R. M. 1989, in *Automatic Small Telescopes*, ed. D. S. Hayes and R. M. Genet (Mesa: Fairborn Press), p. 125.
Genet, D. R. 1991, *Robotic Observatories 1991 Catalog*, (Mesa: AutoScope Corporation).
Genet, R. M., Boyd, L. J., Kissell, K. E., Crawford, D. L., Hall, D. S., Hayes, D. S., and Baliunas, S. L. 1987, *Pub. A.S.P.*, **99**, 660.
Henry, G. W. 1989, in *Remote Access Automatic Telescopes*, ed. D. S. Hayes and R. M. Genet (Mesa: Fairborn Press), p. 41.
Van-Vegchel, J. J. 1990, *IAPPP*, **42**, 63.

ROBOTIC TELESCOPE NETWORKS

RUSSELL M. GENET
AutoScope Corporation, and the Fairborn Observatory, 3435 East Edgewood Ave., Mesa, AZ 85204

ABSTRACT Various types of robotic telescope networks are described. These include local site, regional, global, and Earth–space networks. Two ways of managing networks are also discussed. Networks of automatic telescopes offer great promise for observations of astronomical objects.

INTRODUCTION

Fully automatic telescopes are naturally disposed towards networking; of necessity, their inputs, outputs, and behavior are completely and logically defined. Furthermore, many automatic telescopes are already accessible via modem and telephone or computer network.

Many types of robotic telescope networks are possible. In this paper, local site networks between two or more telescopes at the same site will be discussed. Regional networks, where several telescopes in the same region are networked together, will also be included, as will global networks. Finally, Earth–space networks, in which telescopes on Earth and in space work cooperatively together, will be considered.

Besides the extent of network coverage, two major types of networks, from the network management viewpoint, will be discussed. These are (1) dedicated networks where all the telescopes in the network "belong" to the network coordinating agency, and (2) voluntary or "democratic" networks where the participating telescopes independently belong to different institutions.

LOCAL NETWORKS

Local networks of robotic telescopes consist of interconnected telescopes at the same site. One simple reason we have run across at the APT Service for a local network is "telescope avoidance." We have four 30-inch telescopes placed in an area only 16 feet on a side. While the telescopes are very compact, they could actually run into each other if their hour angle excursions were not limited. Furthermore, even before physical contact, the telescopes start vignetting the light from each other — something very undesirable in precision photometry.

We solved this problem by limiting hour angle coverage of each of the telescopes, but we have seriously considered a much more elegant but simple solution, and that is a local network. To simplify the situation consider two telescopes (instead of the actual four telescopes). Assume that there is a "zone of conflict" where the two telescopes could either hit each other, or at least block each other's light. At the beginning of the night (or on any resumed operations), the first of the two telescopes to "claim" the zone of conflict could operate in it, locking the other telescope out of the zone. When the telescope operating in the zone moves out of the zone, it would give up its claim for the zone, and at this point the other telescope could claim the zone for operations if it so desired. As the zone of conflict is a small portion of the operable sky for the two telescopes, they should have no problem in sharing with the simple rules outlined above. If there were a problem, one could add another rule establishing time limits in the zone if another telescope wanted to use it.

Another local network of interest is one that can do simultaneous photometry of two stars. While not fully automated, J. Donald Fernie has had such a two-telescope network in operation at David Dunlap Observatory for a number of years. In Fernie's network, one telescope observes the variable star, while the other telescope observes the comparison star, which is located very close to the variable star in the sky. Thus, any thin cirrus clouds lurking about are likely to affect both stars similarly, and the differential photometry of the variable star will be largely preserved due to the simultaneous observations of two very close stars.

As large-format CCD cameras that can simultaneously observe a number of stars come into greater use, and CCD photometry is more fully automated, single telescopes will be able to do simultaneous multiple star photometry. In fact, the first such system, described at this meeting by Kent Honeycutt, is already in operation. Nevertheless, Fernie has developed and operated for a number of years a most interesting local network that overcame serious problems with thin cirrus clouds.

Another interesting possibility for a local network is the use of multiple telescopes to provide light to a single, common instrument. This has been discussed for some years in the realm of spectroscopy, where a fiber from each of several telescopes would converge to a single automated spectrograph. In this case, economics is the main driver, the thought being that, for instance, four 2-m telescopes might cost less than one 4-m telescope, although the areas (and hence equivalent aperture for spectroscopy) would be the same.

An additional possible advantage beyond economics for such a local network would be the ability, in the example of four 2-m telescopes mentioned above, for the telescopes to operate, at times, as four independent 2-m telescopes, each doing photometry, imaging, or some other task. On nights of high photometric quality, such a local network might concentrate on photometry, operating as four 2-m telescopes, while on nights with thin cirrus clouds, the system might concentrate on spectroscopy as a single 4-m telescope.

Finally, there are a number of specialized applications where a local network of robotic telescopes would be appropriate. During the meeting, for example, Roland Vanderspek described a two-telescope local network for observing transient events, where one telescope is used to confirm the results of the other telescope.

REGIONAL NETWORKS

Regional networks consist of two or more telescopes that are in the same regional area, but are not at the same site. An example of a regional network, to be implemented in the near future, is a two-telescope network planned by Tennessee State University, the Smithsonian Institution, Vanderbilt University, and the Mount Wilson Research Institute. In this regional network, an existing fully automatic telescope at the APT Service on Mt. Hopkins will be supplemented with a second, fully automatic telescope to be placed at Mt. Wilson in southern California. While these two telescopes would only be about 400 miles apart, the weather patterns on the California coast and in southern Arizona are close to complementary of each other. Arizona tends to have long periods of clear skies in the winter, but the monsoon season in the summer makes observations essentially impossible during July and August. (In fact, the APT Service simply shuts down for these two months.) Mt. Wilson, on the other hand, has generally clear summers, with many months of good weather, while the winter nights are often cloudy.

By simply placing the same stars on the programs at both locations, one can be assured of avoiding the summer gaps that are present in the current data obtained only from Mt. Hopkins. However, with even modest communications between the two sites, one can readily improve efficiency further. For instance, if it is clear at both sites, there may be no reason to observe the same star at both sites, and it may be more efficient to instead observe a larger number of different stars in a coordinated manner between the two sites. A simple way of doing this might be to have primary and secondary lists of stars to be observed. If one site is clouded out (or the equipment is down, etc.), then the other site, if not also clouded out, could observe the primary list. If both sites were clear, then one site would handle the primary list, and the other site could handle the secondary list. A single phone call between the control computers at the two sites at the beginning of each night might suffice to implement such a strategy in a reasonably efficient manner. Additional phone calls (or computer network contacts) throughout the night could further increase efficiency by taking into account those nights where the weather changed during the course of the night itself.

Another interesting example of a regional network of fully automatic photoelectric telescopes is being established by Kelly Cowles and David Erickson at the Jet Propulsion Laboratory. The goal of this local network is to demonstrate that reliable optical communications would be possible with a remote (deep space) spacecraft directly from a local area on Earth. As with the previous example, JPL has chosen the complementary weather patterns of southern Arizona and the California coast; it is placing APTs on Mt. Lemmon (Arizona) and Table Mountain (California). A third APT will be placed at an additional location (such as Texas). It is hoped that between the three systems, at least one will have sufficiently clear skies for deep space optical communications well over 90% of the time.

Yet one more example of a possible regional network was discussed during the course of this meeting by Roland Vanderspek and Louis Boyd. The thought here was to move one of the two transient optical event telescopes from Kitt Peak to Mt. Hopkins, which are about 50 miles apart. Such a separation would allow easy discrimination against Earth-orbit satellites, leaving the desired transient events located much farther away.

Freeman Dyson and others have suggested placing a line (or lines) of small automatic telescopes over a distance of $\sim$ 100 miles to detect comets in the Oort cloud. Many other specialized and dedicated networks are certainly possible. Robert Millis, and others, have used networks of non-automatic telescopes for a number of years to observe asteroids when they pass in front of a star. The multiple locations allow various chords across the asteroid to be observed, and from these observations the actual shape of the asteroid can be deduced.

GLOBAL NETWORKS

Global networks of robotic telescopes have been considered by many over the years, and at least one such network is being implemented. As with regional networks, global networks can overcome to some extent the vagaries of local weather and equipment failures. In addition, if the robotic telescopes are appropriately spaced in longitude, they can overcome the day/night cycle that interrupts observations on a daily basis.

An excellent example of what can be done has been provided by the non-robotic but global network called the "Whole Earth Telescope." Pioneered by the University of Texas in conjunction with many participating observatories around the world, the network was used to keep stars under constant observation for extended periods of time, allowing long unbroken records of their slight pulsations to be built up. In records where there are frequent large gaps, aliases caused by the gaps make interpretation of fine details impossible. However, when the record is unbroken for a long period of time (weeks or months), the fine structure of the variations becomes apparent and can be used to improve our astrophysical understanding in such areas as the interiors of stars.

Underway, but not fully implemented yet, is the GONG world-wide network of small, fully robotic solar telescopes. By providing long, unbroken records of velocity variations on the surface of the Sun, the fine structures of the Sun's "astro-seismological" behavior will be revealed, greatly increasing our understanding of the interior and heretofore "invisible" inner regions of this nearby star.

As a greater number of independent robotic telescopes come into operation at various locations on Earth, one can envision them exchanging time on a voluntary basis to obtain round-the-clock coverage of selected objects. Such a network would be a voluntary or "democratic" network, as opposed to a dedicated, centrally managed network. Such a voluntary network will be discussed in greater detail later in this paper.

It might be noted that there are at least two alternatives to global networks for obtaining round-the-clock coverage of selected objects. One is operation at, or near, one of Earth's poles (such as the South Pole). In principle, one could obtain unbroken records approaching six months in duration, although in practice this is precluded by weather interruptions and also, for some types of observations, by severe auroral disturbances. Robert Wilson and others have noted that there is a large plateau (elevation 18,000 feet) not far from the South Pole that may make a good astronomical site, not to mention a possible test site for Lunar Outpost shelters and activities.

Whether multiple Antarctic locations could provide protection against bad weather and bright auroras remains to be seen. If this turns out to be the case, then perhaps three properly chosen locations in Antarctica could provide near-continuous observations of the southern skies for almost six months of the year.

Besides a polar location on Earth, the other alterative to a global network for providing continuous observations is to place telescopes in space, such as in Earth-synchronous orbit, or in the vicinity of the poles of the Moon. Here one does not need to worry about interruptions by poor weather or auroral displays, but one does, of course, have other things to worry about, including high costs.

EARTH-SPACE NETWORKS

Ground and space telescopes have been cooperating for years. One interesting area of cooperative observations involves the International Ultraviolet Explorer (*IUE*) and the "Phoenix-10" APT on Mt. Hopkins. Much of the time on the Phoenix-10 APT has been used to make photometric observations prior to, during, and after UV spectroscopic observations of the same object by *IUE*. The more extended photometric observations allowed the spectroscopic observations to be placed in the appropriate astrophysical (variable star) context at low cost.

A number of APT users have arranged to have simultaneous observations with space telescope observations, or with observations by other Earth-based telescopes, such as much larger telescopes making spectroscopic observations.

The Space Sciences Division at the Lawrence Berkeley Laboratory is planning on developing an automated interface between a number of space telescopes and fully automatic APTs on Earth. The thought here is that the space telescopes would request Earth-based backup (simultaneous photometric observations), and Earth-based telescopes would be automatically asked to provide these observations. Of course, many details will need to be worked out before this becomes a routine capability.

Another ground/space connection being discussed is the testing here on Earth of space telescope prototypes (or actual space telescopes), and the use of such prototypes as a training and demonstration ground for subsequent operations in space. Currently, space telescopes require sometimes sizeable ground crews for operation. Consideration is being given by NASA-Ames Research Center and AutoScope Corporation to automating all, or at least major portions of, such ground operations. However, it may be prudent to first demonstrate on Earth that the elimination of these ground crews is feasible before placing such systems in space.

Also, some thought has been given to the possible advantages of operating space telescopes on Earth prior to operating them in space. This might allow some of the more obvious problems (such as the optical problems encountered with the Hubble Space Telescope) to be worked out here on Earth prior to sending the telescopes into space. While there are many obvious (and some not so obvious) differences between operations on Earth and in space, the similarities probably outweigh the differences in at least some cases, making "Earth-first" operation a reasonable possibility.

There are some "Moon like" locations on Earth. The 18,000 foot plateau near the South Pole mentioned earlier has many lunar-like environmental features. A location that really looks like the Moon has been pointed out by

Harlan Smith (who visited and took pictures of it). It is a 20,000 foot volcanic peak in the middle of the northern Chilean desert. Mountain ranges to both the east and west block moisture-laden air, with the result that there is only a slight rain every ten years or so. The terrain is totally barren, with no living things whatsoever. There is a good road to within 400 feet of the summit — the highest road in the world. The road is in constant use to transport sulfur from a pure outcropping near the summit to the world's largest copper mine, located about 100 miles away in the middle of this otherwise totally uninhabited area of the world.

This mountain top may be the best astronomical location on Earth. It is certainly one of the highest, driest, and most cloud-free locations. It may also be the most lunar-like location. If totally unmanned telescopes were operated by astronomers from many different institutions, via computer network and synchronous satellite relay, for a number of years from this location, using solar power, batteries, and low-power components, then it would not seem implausible to operate comparable systems from the lunar outpost in a similar manner.

INTEGRATED EARTH–SPACE NETWORKS

Looking ahead well into the future, we might expect there to be many telescopes in space, including a large contingent of telescopes on the Moon. Essentially all of these telescopes will be robotic. Also, we might expect that an increasing number of Earth-based telescopes will be fully automatic. It would seem likely that astronomers might tend towards similar operation of all telescopes, be they on Earth or in space. Remote is remote and, from the viewpoint of using astronomers, it may not be crucial whether the telescopes are a few hundred miles away, a continent away, or located on the Moon.

If logic prevailed, one could envision that the interface between astronomers and their remotely located telescopes might become standardized, and thus operation of remote telescopes, be they on Earth or in space, might be similar or even essentially identical. The development of truly user-friendly operation of remote telescopes and instruments is not a simple undertaking, and it might be hoped that some standardization might evolve to avoid having to reinvent the wheel too many times.

Assuming that common user interfaces between Earth-based and space-based telescopes do evolve, then forming networks, temporary or permanent, out of a number of different telescopes would probably become a fairly frequent occurrence. Space telescopes could be used to complement Earth-based telescopes in portions of the spectrum not accessible from Earth. Earth telescopes could, in many cases, provide low-cost backup and supplemental observations to space telescopes.

DEDICATED AND VOLUNTARY NETWORKS

From a management viewpoint, networks can be classified as dedicated or voluntary. Each type of network has its area of application and advantages.

Many of the networks discussed earlier are dedicated networks. Typically they are owned by one or a few cooperative agencies, and they have either a single mission or several closely related missions. One of the advantages of such dedicated networks is that standards, interfaces, instruments, etc., can all be "dictated" without the need for forming a consensus among many independent elements in the network.

Voluntary networks may, at least initially, be slower in forming, but in the long run they might come to predominate. For voluntary networks to be effective, a fair number of robotic telescopes must already be in place. While independently owned and operated, these various telescopes must be capable of "speaking" a common language, at least in terms of how requests for observations are made and how results are returned. They must also have sufficiently similar instruments to allow calibrations across various telescopes to be effectively conducted.

The basic operational mode across independently owned telescopes may be either trading time between telescopes to achieve with multiple telescopes what cannot be achieved with single telescopes, or selling time at some prearranged rate to users that need network time. There have already been discussions between early owners of fully automatic telescopes about trading time to start getting some networking advantages.

At least in the area of photometry, standards have been set for both requesting observations and receiving results. The de facto standard is the Automatic Telescope Instruction Set (ATIS). ATIS is a published, agreed-on standard that is currently being used by some three or four dozen users. Plans are being made to extend ATIS into automated imaging/CCD area photometry and to automated spectroscopy.

So far, the various users of automated telescopes, particularly in the area of photometry, have shown the ability to organize themselves, exchange information, maintain standards, share software, etc. If this is a sign of things to come, it bodes well for user self-organization, voluntary networks, and a cooperative but democratic approach to automated astronomy.

Perhaps it will not be far in the future when networks for all sorts of different projects will be organized, accomplish their missions, and be replaced with other networks, many of them very transitory. It would not be surprising to see many independently owned automatic telescopes participating, all at nearly the same time, in a number of different network projects as well as making purely local observations on projects that do not require networking. As networking grows, however, purely local projects may become fewer and smaller as users take advantage of the many benefits of networked observations. Someday the idea that observations are made locally at a single telescope may seem strange and, well, old fashioned and quaint.

ACKNOWLEDGEMENTS

As mentioned previously, many others have considered networked automatic telescopes for a number of years. I have greatly benefitted from discussions on networking with David L. Crawford, David R. Genet, Donald S. Hayes, Sallie L. Baliunas, and Louis J. Boyd.

Robotic Telescopes in the 1990s
ASP Conference Series, Vol. 34, 1992
Alex V. Filippenko (ed.)

APTs FOR INDIAN UNIVERSITIES

RANJAN GUPTA
Inter-University Centre for Astronomy and Astrophysics (IUCAA), Post Bag 4, Ganeshkhind, Pune 411 007, India

ABSTRACT We are in the process of fabricating small Automated Photoelectric Telescopes (APTs; 14-inch diameter) to be used for various astronomical observations by interested Indian University Groups. The major design features for these APTs are discussed.

AIM

We are developing small Automated Photoelectric Telescopes (APTs) for Indian Universities. The aim is to have interested University groups involved in the design and fabrication of small automated telescopes ($\sim$ 35 cm, or 14 inches, in diameter) to be used for astronomical observations from their respective locations.

The expected outcome of this effort is a group of active workers who will be exposed to the technology of observations with automatic equipment, PCs, software, digital communications, CCD cameras, image processing, etc. This approach would generate more interest in experimental astronomy within the Indian University environment.

WHY APTs?

The present state of APTs has reached a mature level where one can easily introduce such telescopes for educational purposes. While one can do a lot of useful astronomy with small manual telescopes, APTs have many more advantages over them (Genet and Hayes 1989); these need not be highlighted here. It is the sophisticated instrumentation that goes with the APTs which gave us the impetus to introduce them for the Universities.

DESIGN ASPECTS

Figure 1 shows a sketch of an APT for illustration purpose. The APTs would use commercial optical tube assembly (Celestron 14 OTA), thereby allowing us to use the readily available compact OTA. All the mechanical mounting, drive and control electronics, etc., will be developed at IUCAA.

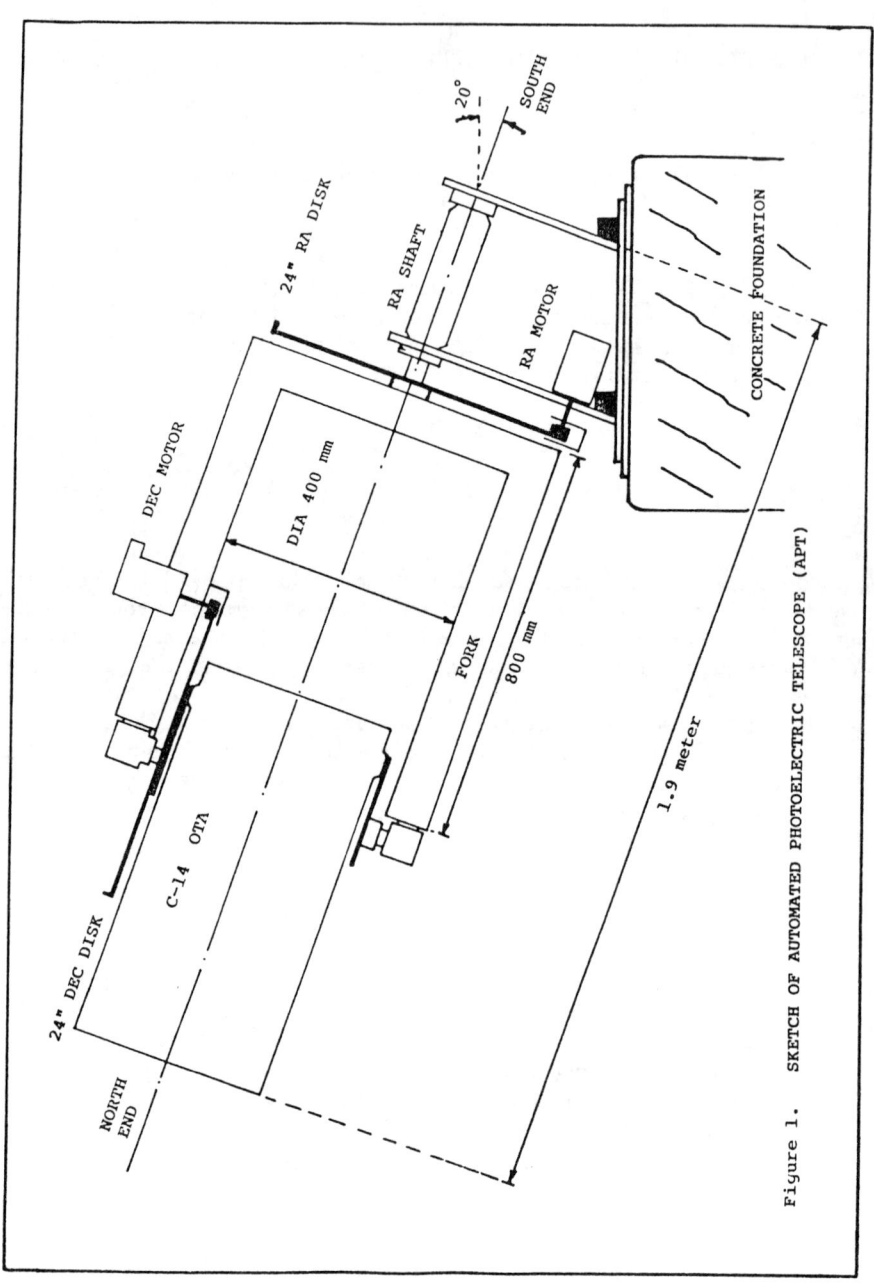

Figure 1. SKETCH OF AUTOMATED PHOTOELECTRIC TELESCOPE (APT)

The fork mount design is chosen since it is best suited for this class of telescopes; moreover, one is required to insert an eyepiece or frequently change various backend instrumentation.

Friction gears will be used for backlash-free motion using 24-inch diameter disks with 1-inch rollers for both right ascension and declination motions. The 1-inch rollers are pressed by 100 kg springs against the disks. The rollers and disks are machined from mild steel, with the contact surfaces ground and hard-chrome coated to 50 micron thickness.

Microstepping motors with 25,000 steps per revolution are used in conjunction with the 1:24 friction gear ratio to render an ultimate resolution of about 2 arc seconds per step. This value is comparable to the seeing conditions at the sites. The motors are AEROTECH Model 310 SM (holding torque 2.6 N-m) with DM 8010 stepping translator and Cybernetics controller model CY 545.

The mechanical structure design has been constrained such that the pointing accuracy will be better than 3 arc minutes without any external guidance. This value was obtained after considering the mechanical flexures and fabrication accuracies.

Real-time guidance of the APT will be done by using a CCD star tracker. The CCD star tracker of SBIG (Model ST-4) make is an attractive option since it allows one to go down to fainter stars apart from its other advantages. This tracker can track stars up to $m = 13$ mag (for a 35 cm telescope) every 3 seconds with an accuracy of one arc second. The error signals from the detector are used to generate the appropriate number of steps to drive the stepping motors under a servo loop. The main Cassegrain beam will be used for tracking as well as actual observations by using a beam splitter.

Figure 2 gives a block diagram illustrating the servo loop controlling the telescope operation.

An estimate of the natural frequency of vibration for the telescope's mechanical structure was made and found to be more than 6 Hz. This value is much larger than the required bandwidth of the servo loop (less than 1 Hz). The effects of wind-generated torques were also made for an upper limit of 40 km per hour (gusts, etc.) by considering a tube of 40 cm diameter and 80 cm length. This gives approximately 40 N force, with a leverage of 0.5 m and hence a 20 N-m torque, which is much larger than the friction and the acceleration torques needed. This defines the required driving torques of the stepping motors.

The present design is limited to photometric observations using lightweight back-end instrumentation such as a Solid State Photometer (SSP-3 of OPTEC make) or a CCD Camera (ST-4 of SBIG make).

The total weight of the telescope is expected to be about 150 kg. The nominal right ascension axis elevation will be 20°, but it can be adapted for the local latitude range from 5° N to 35° N by providing a suitably tilted concrete platform.

The mechanical drawings of the APT have been made and the mechanical fabrication should be complete in 3–4 months. Simultaneous development of the control electronics is ongoing. It is expected that the first tests with the prototype APT can be done in about a year. In the initial phases, the APT will be used in the presence of an operator who will take care of starting and stopping the operations.

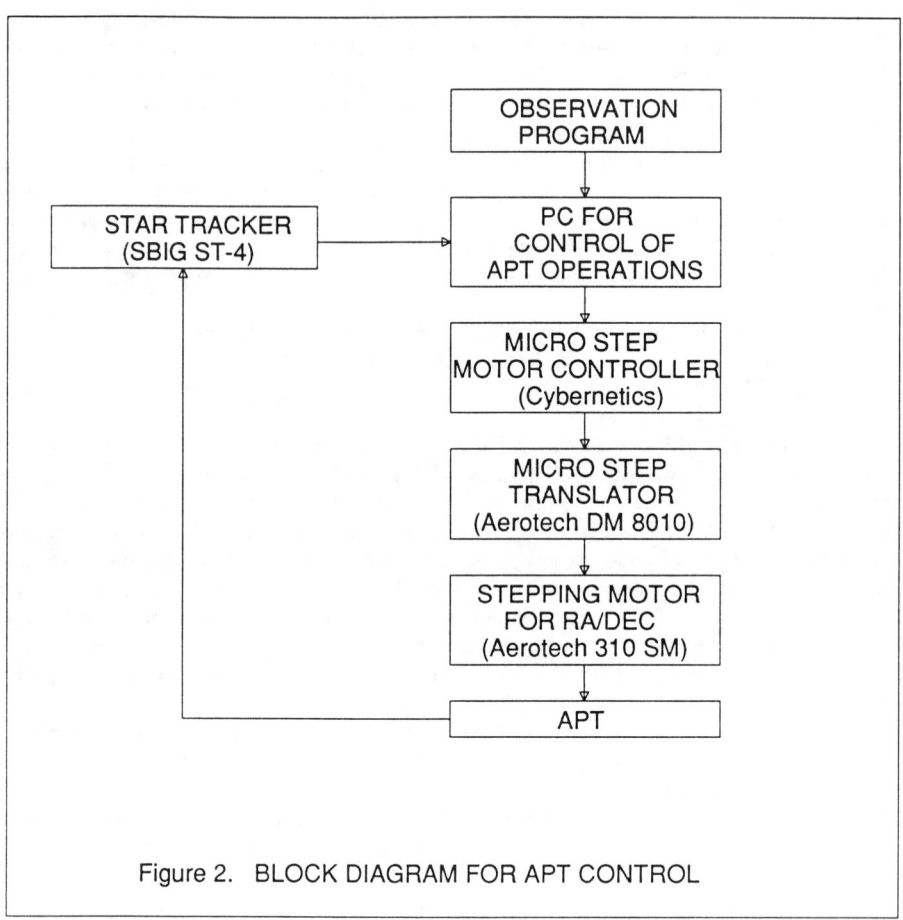

Figure 2. BLOCK DIAGRAM FOR APT CONTROL

ACKNOWLEDGEMENTS

I wish to acknowledge the suggestions and fruitful discussions I had with Professor S. N. Tandon and encouragement from Mr. Russell M. Genet.

REFERENCES

Genet, R. M., and Hayes, D. S. 1989, *Robotic Observatories* (Mesa: AutoScope Corporation).

THE UCSB REMOTE ACCESS ASTRONOMY PROJECT

PHILIP LUBIN
Physics Dept., University of California, Santa Barbara, CA 93106, USA

JANET VAN DER VEEN
Physics Dept., Adolfo Camarillo High School, Camarillo, CA 93012, and University of California, Santa Barbara

ABSTRACT The remote access astronomy project is a computerized telescope and data distribution system that has the potential to substantively change the way astronomical concepts are taught to undergraduate students, and to improve the teaching of astronomy, earth science, and physics at the high school and undergraduate levels. In addition, particularly at the secondary school level, it can serve as a forum for the distribution of curriculum materials and data among teachers, and as an educational network between high schools, universities, and professional astronomers. The system is undergoing testing at the undergraduate level, with several high schools, and a local museum.

INTRODUCTION

The UCSB Remote Access Astronomy Project (RAAP) grew out of experiences in teaching undergraduate astronomy several years ago. It quickly became clear that most students were less than impressed with the usual roof top observation sessions with small telescopes. In addition to having to come back to campus when it is cold and dark, the students could see very little more (qualitatively) in the eyepiece than with the unaided eye. This is particularly disappointing when compared to the beautiful pictures presented in the textbook. Clearly, for them, astronomy should be done from the textbook and not with a telescope! With the current need for vast improvements in science education at all levels of pre-university and university teaching, it seemed desirable to extend the use of high quality astronomical images to high school and undergraduate physics and astronomy students and teachers through the use of a dial-in data distribution system connected to the local area network. In this way, images from our telescope, as well as images from other sources, could be made available for high school students and teachers as well as to UCSB students.

From a modern perspective of photon counting devices, some comparisons of the eye, emulsions and photon integrators (CCD, etc.) are very instructive. The major difficulty with the eye as a light detector is not "read noise", "dark current" or quantum efficiency, but rather integration time. The eye has an effective integration time of only about 50 ms. Though near single photon sensitivity is possible, typical dark-adapted sensitivity is 10-100 photons on a receptor in an integration time. Emulsion grain sensitivities are typically of order 100 photons per grain.

Film has excellent spatial resolution, but suffers from low quantum efficiency and dynamic range. The UCSB Remote Access Telescope (RAT) uses a cooled CCD as the photoreceptor. It has an average quantum efficiency of 40% and read noise of 12 electrons and a dark current of about 1 electron per second. Pixel capacity (well depth) is about 150,000 electrons. For a comparison, the UCSB RAT, which has a 14-inch aperture with its CCD camera is roughly as sensitive as the Lick 120 inch was with film 40 years ago. A large amount of astronomy was done then, and can still be done now at this sensitivity, which corresponds to a detection of a 20th magnitude point source in a few minutes.

LIMITING BACKGROUNDS AND SENSITIVITY

A number of issues enter the calculation of a signal to noise ratio. These include background flux from airglow, back-scattered terrestrial light, zodiacal light and diffuse galactic and extragalactic light as well as optical throughput and detector noise (read noise and dark current). The fundamental limiting backgrounds from the sky are summarized in Table I.

TABLE I Signal to Noise Considerations

Fundamental Limiting Backgrounds

Source	S_{10} Equivalent # 10th Mag Obj per sq deg
Airglow	50
zodiacal light(away from ecliptic)	150
Starlight (mean)	100
Galaxies	1
Total	~ 300

The total is equivalent to 10^{-2} photons/$cm^2 - sec$ per sq arc second.

$$1 \, (\text{deg})^2 = 10^{7.113} \, (\text{arc sec})^2 \, (= 3600^2)$$
$$1 \, S_{10} = 6.86 \times 10^{-6} \, erg - cm^{-2} - s^{-1} - st_2^{-1}$$
$$1 \, S_{10} = 1 \, 10^{th} \, \text{mag obj}/(\text{deg})^2 = 10^{-7.113} \, 10^{th} \, \text{mag obj}/(\text{arc sec})^2$$
$$= 27.78 \, \text{mag}/(\text{arc sec})^2$$
$$300 \, S_{10} = 21.59 \, \text{mag}/(\text{arc sec})^2$$

N_R - readout noise e^-

i_{DC} - dark current e^-/s

Q_e - Quantum eff

F - point source signal flux on telescope $\gamma/s - cm^2$

F_β - background from sky $\gamma/s - cm^2 - arcsec^2$

Ω - pixel size $arcsec^2$ (assume > seeing)

ϵ - telescope optical efficiency

τ - integration time   sec

A - effective telescope area cm^2

$$\text{Signal} \quad S = F\tau A\epsilon Q_e \quad e^-$$

$$\text{Noise} \quad N = \left(N_R^2 + \tau(FA\epsilon Q_e + i_{DC} + F_\beta A\epsilon Q_e \Omega)\right)^{\frac{1}{2}}$$

let $A_\epsilon = A\epsilon Q_e \equiv$ Effective area

let $N_T = FA_\epsilon + i_{DC} + F_\beta A_\epsilon \Omega$

$$S/N = FA_\epsilon \sqrt{\tau} / \left[\frac{N_R^2}{\tau} + FA_\epsilon + i_{DC} + F_\beta A_\epsilon \Omega\right]^{\frac{1}{2}}$$

$$= FA_\epsilon \sqrt{\tau} / \left[\frac{N_R^2}{\tau} + N_T\right]^{\frac{1}{2}} = FA_\epsilon \tau / \left[N_R^2 + \tau N_T\right]^{\frac{1}{2}}$$

I. for small τ $S/N \sim \tau$ N_R Limited

II. for large τ $S/N \sim \sqrt{\tau}$ N_T Limited

Time to measure a given S/N:

$$S_N = S/N = FA_\epsilon \tau / \left[N_R^2 + \tau N_T\right]^{\frac{1}{2}}$$

$$\tau = \frac{S_N^2 N_T \pm \sqrt{S_N^4 N_T^2 + 4F^2 A_\epsilon^2 S_N^2 N_R^2}}{2F^2 A_\epsilon^2}$$

$$= \frac{S_N^2 N_T}{2F^2 A_\epsilon^2} \left[1 + \sqrt{1 + \frac{4F^2 A_\epsilon^2 N_R^2}{S_N^2 N_T^2}}\right]$$

As can be seen, this ideally is about 21.6 magnitudes per square arc second equivalent. Most easily accessible ground-based sites are worse that this. Given the background flux, we can compute the signal to noise ratio (S/N) given the object flux (magnitude), detector (CCD) characteristics and optical throughput.

Figure 1 shows the results of this calculation for objects of magnitude m=15, 17.5 and 20 given our system characteristics with $1\times$ and $10\times$ ideal dark sky conditions. A couple of caveats are in order here. First, we have assumed all the light is dumped on one pixel (2" pixels) i.e., this is for point sources. Second, we have assumed perfect guiding during integration. This is not trivial for the longer integrations. In any case, it is an instructive calculation.

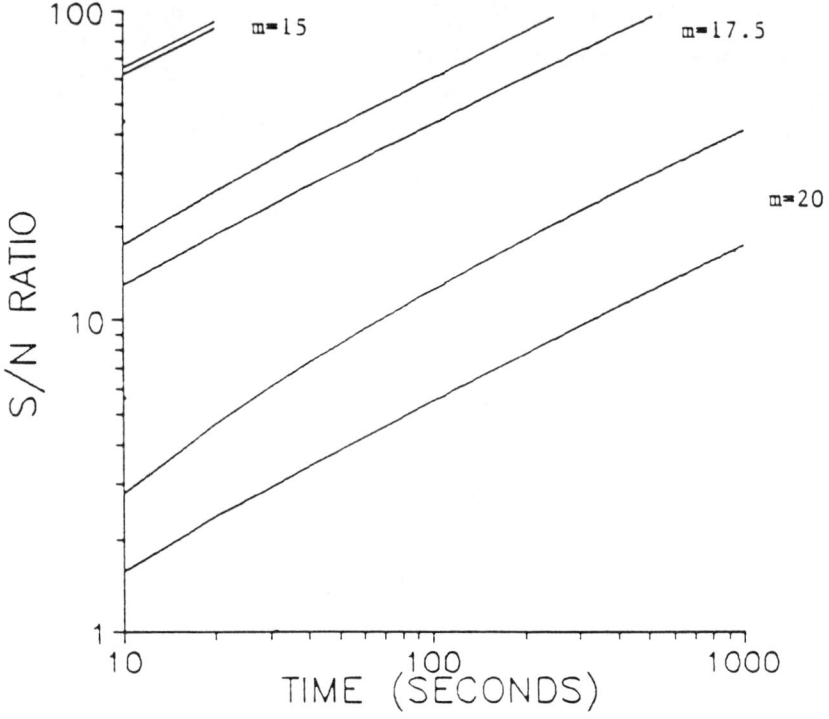

Figure 1:   Signal to Noise Ratio vs. Integration Time

CURRENT SYSTEM

The system consists of a 14-inch diameter aperture telescope, computer controlled with micro step-driven stepper motors, a thermoelectrically-cooled CCD camera, 16-position filter wheel (dual 8-position wheels), and a servoed focus. The telescope is housed in a computer-controlled, weatherized enclosure. A computerized weather station and infrared sky monitor determine meteorological conditions before opening the "dome". This part of the system communicates via a microwave link to the base of the system, which consists of an ethernet-connected network of workstations for local undergraduate use and a high speed modem interface for remote users (high schools and other colleges). The dial-in interface uses V.32/V.42 modem technology for effective throughput of 1-1.5 KBs (Kilo Bytes per second) for compressed image files, 2 KBs for binary executable files and 3-4 KBs for ASCII text files. Our current Thompson CCD has a format of 576×384 pixels, and after image compression yields image sizes of 90-300 KB, depending on the image complexity. Typical image transfer times over the dial in link are 1-4 minutes per image at 9600 baud. Figure 2 shows the current system.

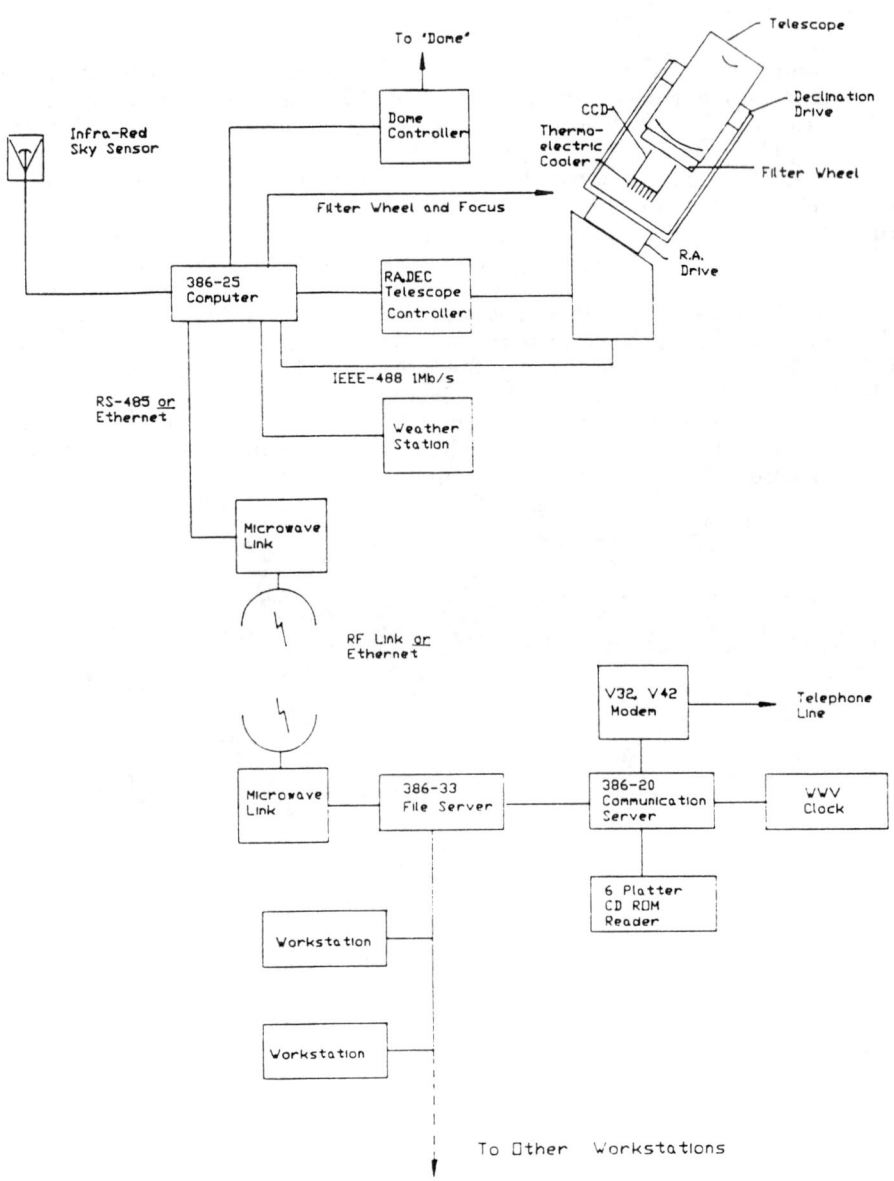

Figure 2: System Diagram

Undergraduate and high school users can submit target requests to the system during the day, or have their computer automatically call in at night when costs are lower. The next day they can download their images and analyze them during normal class hours. This eliminates the need for evening classes which, for the majority of students and teachers, is not practical on a daily basis.

Typical costs for night-time image transfer over the phone lines are 10-40 cents per image (for long distance calls). Eventually, the system may be put on one of the university networks (internet, etc.), but at present, very few high schools have access to such a network. We are currently operating on a commercial telephone number, but may go to an 800 number to centralize costs.

The workstation for image analysis can vary considerably, though at present we are targeting 32-bit IBM-PC class machines (80386 or 80486 CPU). This class of machine currently has from 1.5 MIPS (16 MHz 386SX) to 20 MIPS (33 MHz 486) and has enough power to handle the image manipulation. Cost for our current minimum-recommended workstation is about \$2000, including the Super VGA display. A V.32/V.42 modem adds about \$500. As all these costs are constantly dropping, it is anticipated that within a few years costs will drop again by a factor of 2. Other computers and image processing programs (e.g., Apple-MacIntosh) can also be used with our data, as all our images are stored in standard "FITS" (Flexible Image Transport) format.

In addition to the telescope, we have an extensive library of images on CD-ROM and magnetic disks which include NASA images of Jupiter and Saturn and their satellites, Uranus, Neptune, Mars, Venus, Earth and the Moon. We also have radio and infrared sky maps, x-ray survey maps, images from the Keck Telescope and Hubble Space Telescope, and daily solar images. We currently have about 50,000 images available comprising over 20 GB of data.

These images are an excellent aid for teaching high school astronomy, physics, and earth science courses. Teachers and students can call our bulletin board and request any of these images to be loaded into their directories for use in class assignments or individual research projects. We are also developing activities and teaching units to accompany some of the images, which will be available for distribution beginning in the fall of 1991.

Some of the units that we are developing are: demonstrating Kepler's Laws by plotting the orbits of some of the Jovian and Saturnian satellites; comparing craters on our moon, the Saturnian moons Dione, Callisto, and Mimas, and known or suspected craters on Earth to study history of the solar system, erosion processes on different planets; estimating the masses of bodies that caused observed craters; observing sun spots using daily solar images; calculating trajectories of volcanic debris from the images of volcanic eruptions on the Jovian moon Io; using the radio, infrared, and x-ray maps of the sky to learn about how the universe might "look" if we could "see" at wavelengths other than the visible.

Our bulletin board (The "Astro-RAAP") is designed to serve as a forum for the exchange of information between high schools, colleges, and universities. We have a variety of programs available for downloading, which include educational software for the IBM-compatible PC, image processing programs, general utilities, and text files. E-mail can be exchanged between schools in different parts of the country, and between high schools and the university. As already demonstrated in two California high schools, student response is extremely positive, and students are excited about coming to class and looking for their e-mail messages!

In the longer term we may be able to support one hundred or more schools on our system. It is hoped that participating schools will work towards developing common curricula, thus allowing a larger number of teachers to participate, and making it easier for new teachers to implement this technology in their classrooms. Our system provides a very natural way to distribute curricula and images together.

ACKNOWLEDGEMENTS

This project has been supported by the University of California, the National Science Foundation Center for Particle Astrophysics and the National Aeronautics and Space Administration. To a large extent, this project has been made possible by the efforts of the undergraduate Physics majors at UC Santa Barbara. We greatly appreciate their enthusiasm and dedication to this project.

PHOTOMETRY FROM SPACE

M. J. NELSON, R. C. BLESS, and J. W. PERCIVAL
Space Astronomy Lab, University of Wisconsin, 1150 University Avenue, Madison, WI 53706

R. L. WHITE
Space Telescope Science Institute, 3700 San Martin Drive, Baltimore, MD 21218

ABSTRACT A brief description of the High Speed Photometer (HSP) of the Hubble Space Telescope is given, in particular the HSP light path, detectors, entrance apertures, and filters. The status of HSP testing to date is reported, and problems encountered with the bright Earth and the telescope pointing system are described.
   The calibration effort for the HSP is well underway. Results of internal (instrument coordinate) aperture locations good to $0''\!\!.05$ and external (telescope coordinate) locations good to $0''\!\!.02$ are shown. The effects of spacecraft pointing and jitter on HSP photometry are detailed, and a preliminary measurement of spacecraft jitter with HSP is shown. The aperture calibration effort is verified by accurate pointing of a star to different HSP $1''\!\!.0$ entrance apertures, and photometric performance of the instrument is shown to be accurate to the 2% photon noise of the observations. Future science verification and guaranteed observing time programs are listed. Finally, suggestions are made for future space-based photometers.

THE HIGH SPEED PHOTOMETER (HSP)

The HSP of the Hubble Space Telescope (HST) is a photometer designed to work in the visual through ultraviolet wavelengths available from HST. The photometer consists of 5 detectors with different photocathodes, each capable of taking measurements at rates as high as about 10^5 Hz. The primary detectors are two ITT Image Dissector Tubes (IDTs), with multialkali photocathodes sensitive in the range 1600–6500 Å (the so-called VIS and POL detectors), and two solar-blind CsTe IDTs with photocathodes sensitive in the range 1200–3200 Å (the UV1 and UV2 detectors). The fifth detector is a photomultiplier tube in a 1P21 configuration with a GaAs photocathode (1600–8500 Å).

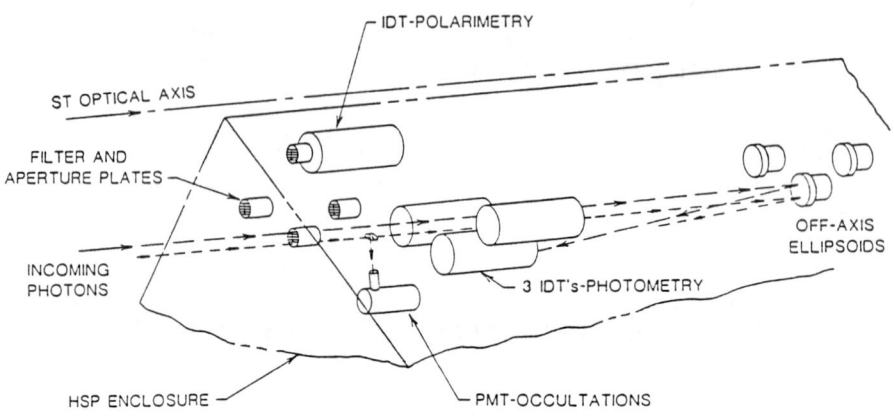

Fig. 1. The light path of the HSP (from the HSP Handbook).

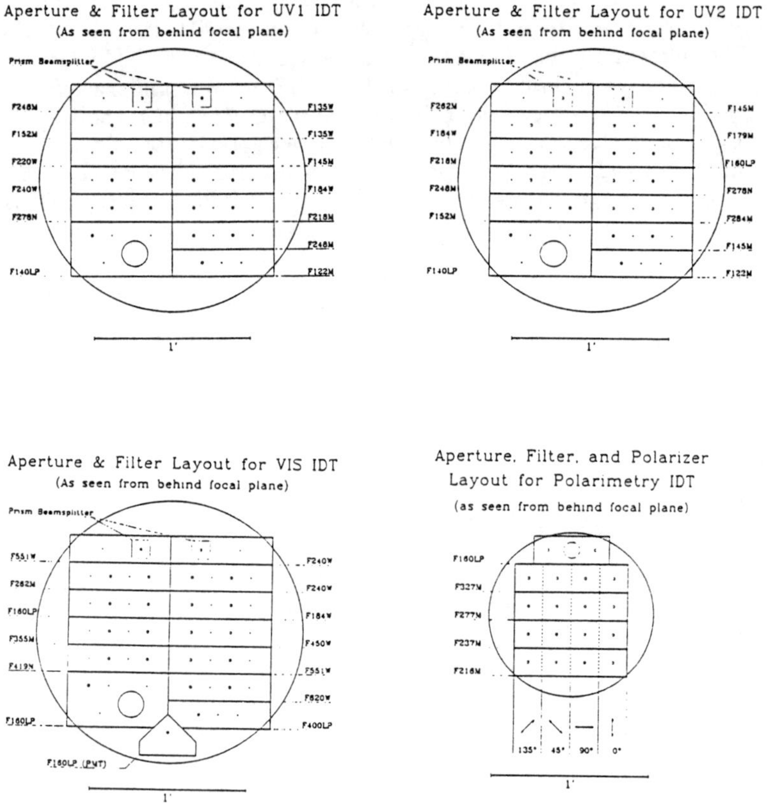

Fig. 2. Aperture plate layouts for the HSP IDTs (from the HSP Handbook).

The HSP is a simple instrument with no moving parts (Fig. 1). Each IDT has a filter/aperture assembly at the front of the instrument. The $f/24$ light bundle first passes through a filter, and then through one of the entrance apertures. The aperture plate (Fig. 2) has a 10″.0 finding aperture and a series of either 3 or 4 small 0″.4 and 1″.0 apertures for each filter. The aperture plates are at the focal plane of the HST. The light is then reimaged by off-axis elliptical mirrors onto the photocathode of the IDTs. The star is positioned in the desired filter/aperture combination by telescope motion, and the IDT read beam is commanded to the appropriate spot on the photocathode by magnetic deflection. The mechanical aperture to the multiplier section of each IDT is 180 μm in diameter, giving a read beam of 1″.0 on the sky for the VIS and UV detectors, and 0″.65 for the POL detector. The latter has no re-imaging mirror.

A variety of filters are available for each IDT, including a series of polarizers combined with filters on the detector at the focal plane of HST. Figure 3 shows the filter curves for the four IDTs and the PMT.

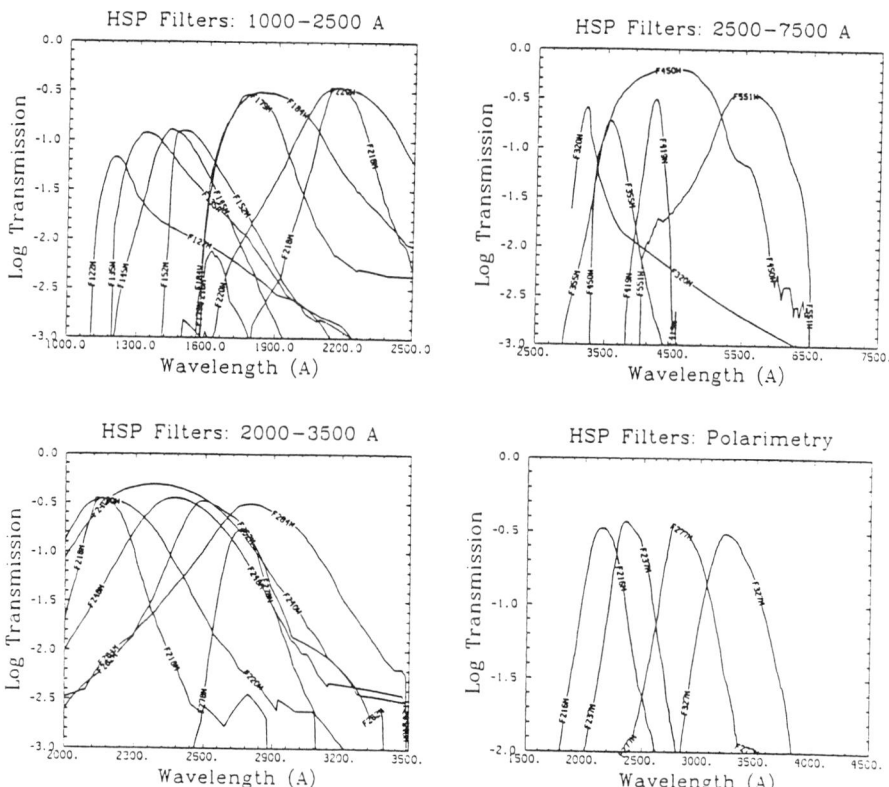

Fig. 3. HSP filter curves (from the HSP Handbook).

TESTING OF THE HSP TO DATE

The primary purpose of testing to date has been to determine various instrument parameters and HSP-specific telescope calibrations. The functional testing of the HSP has gone well, with the exception of early high-voltage testing of the PMT in which it was inadvertently operated out of specification and produced unusual data. The testing of this detector was delayed somewhat until the out of spec operation of the detector was recognized.

Another problem occurred during the first run of the aperture mapping test on the VIS detector. The bright Earth was being used as a calibration source for mapping aperture locations in detector coordinates. During this test the flux through a clear filter on the VIS detector pushed photocathode currents above operational limits such that automatic instrument safing shut down the HSP. It turned out that the bright Earth is about a factor of two brighter than predicted pre-launch, but shows flashes 5-50 times normal for short periods of time. This required a change in the operational procedures for the POL and VIS detectors. In addition, the Orion nebula had to be used to calibrate the positions of the apertures in the visual filters of the two IDTs. These changes caused a delay of several months in the test program.

The most significant difficulty with the test program was caused by the inability of HST to point reliably; miss distances were often considerably greater than 5". It was not until January 1991 that a usable calibration of the telescope's Fine Guidance System was available and used for operations and data analysis. The telescope is now routinely positioning stars to within 2" of requested positions, and acquisition of target stars is no longer a problem.

CURRENT STATUS OF THE HSP

Aperture location calibrations have been done successfully in both detector coordinates and spacecraft coordinates. The detector coordinates have the largest measurement errors, about 0".05. The spacecraft aperture locations are better known and are currently good to 0".02. Figure 4 shows contour plots of the same star in two HSP apertures. Both images show the star to be well centered in the science aperture, verifying both sets of aperture location calibrations.

The worst problem with the HSP is produced, of course, from the spherical aberration of the primary mirror in the HST. Figure 5 shows an image of a point source taken with the HST Planetary Camera. The sharp core of the image contains only about 15% of the light, and there is clearly a measurable amount of energy in the wings of the image at 0".5 radius from the image center. The 0".4 apertures of the HSP are effectively unusable because of the severe amount of light lost through them, and even photometry with the 1".0 apertures is affected. The spherical aberration degrades photometry with the HSP in two ways. First, miscentering of images in HSP apertures will lead to different throughput and appear as changes in the photometric output of the source. Second, instabilities (jitter) in the pointing of the telescope during an observation will cause fluctuations in photometry.

Figure 6 shows the effects on throughput of miscentering a star in an HSP 1".0 aperture. The data are based on fluctuations of the signal as a star was

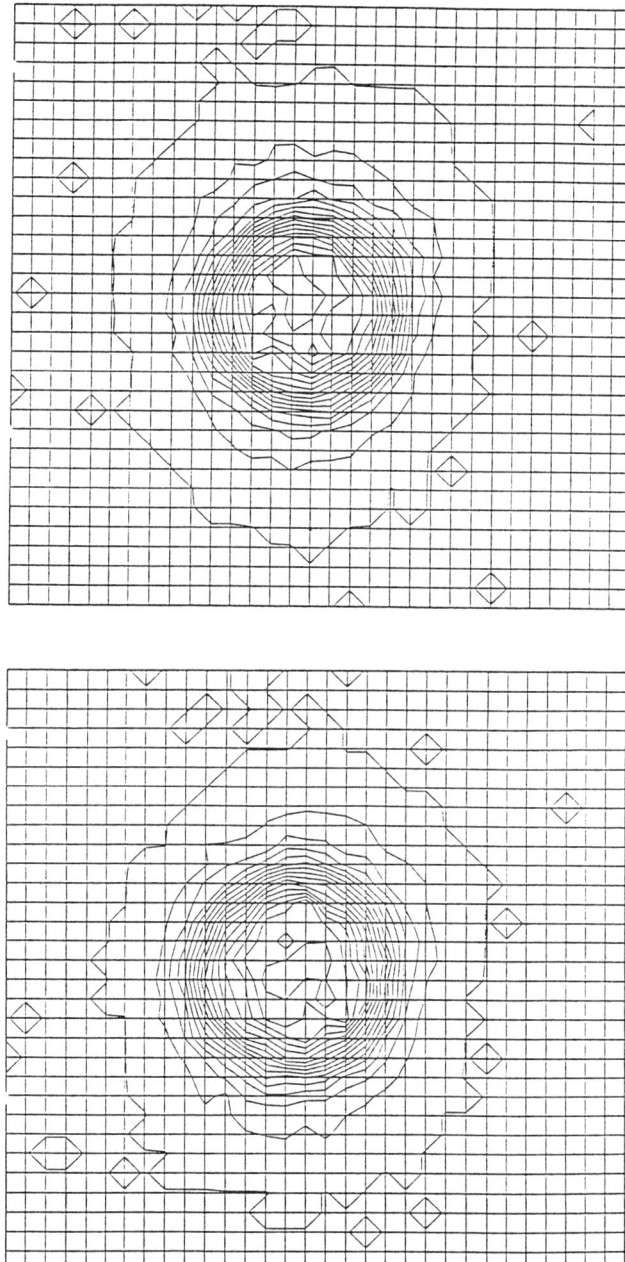

Fig. 4. Contour plots of a star in two HSP apertures.

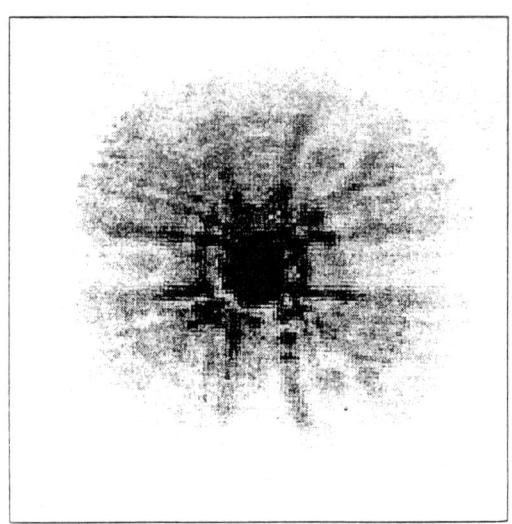

Fig. 5. Grey scale picture of a bright star with the PC (White and Allen 1990). The scale of the image is 8".3 on a side.

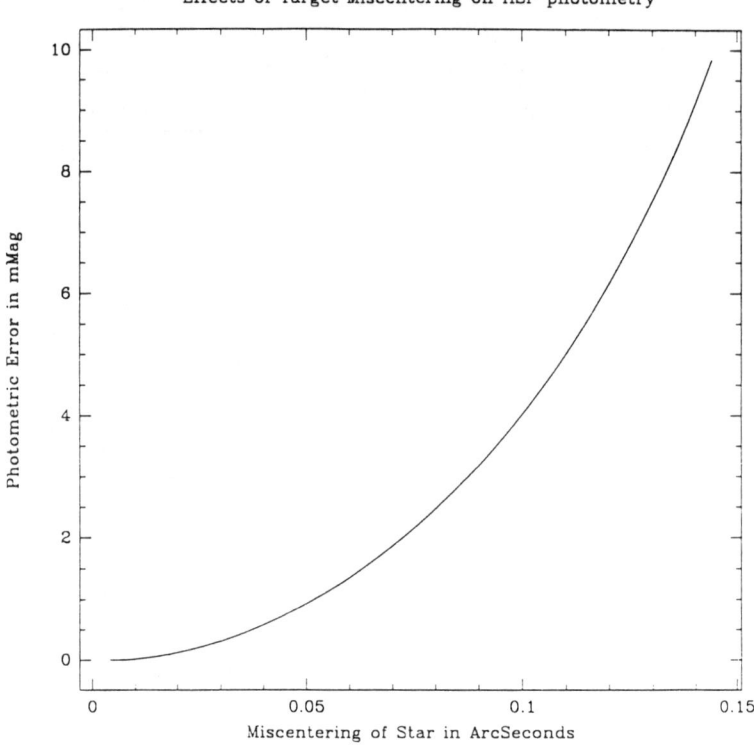

Fig. 6. Plot of photometric error versus centering error for 1" HSP aperture.

scanned though an HSP aperture for a calibration test. Figure 7 shows a power spectrum of such a data set. It exhibits a peak at 0.6 Hz, corresponding to a known periodicity in spacecraft pointing, and illustrates the effects of spacecraft jitter on photometry time series.

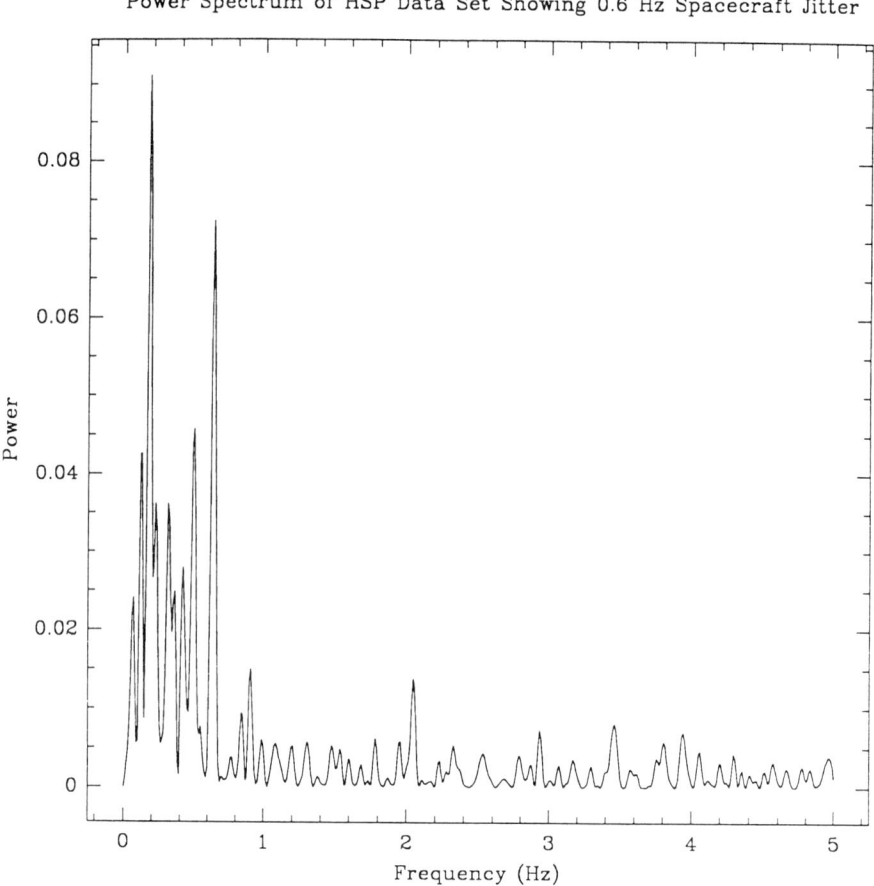

Fig. 7. Power spectrum of HSP data showing 0.6 Hz spacecraft jitter.

The HSP recently finished its photometric performance test which measured the linearity of the photometric response to different amounts of light. This test also demonstrated the photometric capabilities of the instrument. The test was run on 4 stars with well known V magnitudes (5–11) and $B-V$ colors. The test was done with 2% signal-to-noise ratio expected from the count rates of the 4 stars. The observations are shown on the next page; they are for a magnitude-count rate relation with a single color term. This test will be extended to $V = 15$ mag.

Known V	Measured V
5.111	5.119
7.247	7.267
8.060	8.035
11.070	11.067

We can see that the measurements are in good agreement with expected values to within the accuracy of the data. In fact, this test verifies that the HSP is by far the most accurate photometer on the HST. The WF/PC is capable of at best ~7% photometric accuracy, nor is the FOC capable of accurate photometry because of limitations on count rates in the instrument.

FUTURE TESTS AND PROGRAMS WITH THE HSP

Science Verification Programs

- Color transformation test
- Centering effects on photometry
- Short term photometric stability using the roAp star HD 60435
- Jitter test
- Time-resolved photometry using the Crab pulsar
- Occultations by planetary rings

GTO Programs

- Crab pulsar pulse shape in the UV
- Planetary occultations
- Search for optical pulses in X-ray pulsars
- Periodic variations in DQ Herculis stars
- Observations of ZZ Ceti stars
- Variability of high luminosity stars
- Active galactic nuclei
- Observations of cataclysmic variables
- Observations of radio pulsars
- Observations of remnant stars of supernova remnants
- Variations of B Cephei stars
- Variations of short period RR-Lyrae stars

PHOTOMETRY FROM SPACE

There are two fundamental problems with using the HSP as a platform for photometry from space. The first is that the HST is not dedicated to photometry. The HST was to have been used primarily for cosmological studies. While interesting science can be done with short bits of photometry, it is difficult to get much observing time with HST.

The other and more fundamental problem is with HST's orbit. Ideally a space-based photometer should not be constrained from observing an object by its orbit. With the 90 minute HST orbit, and the lack of a usable continuous viewing zone (CVZ) due to vehicle constraints, all HST observations have a maximum time span of ~45 minutes with ~45 minute gaps in the data set. This is deadly to stellar oscillation studies, since the aliasing in the data set from the large gaps will mask many periodicities of interest.

The ideal space-based instrument for stellar photometry would be much different from HST/HSP. The telescope need be only of modest size (~1 m). It would also not need to be diffraction limited, although good imaging characteristics would be helpful in crowded fields. The orbit of the telescope would be geosynchronous, since this would greatly simplify operations, and provide the least constraints on the observable portion of the sky. A lower orbit would be acceptable provided the spacecraft and orbit allowed for CVZs which lasted for several weeks to a month.

The most important characteristic of this ideal photometer should be that it is a dedicated payload. The need for long periods of uninterrupted observing of objects clearly implies that the photometer is the only, or at least the primary, instrument on the telescope.

In reality it appears that most of these criteria could be met except for the geosynchronous orbit at very low cost (in NASA terms). A small mirror and photometer could be launched from a platform like the Pegasus rocket into a 500–700 mile orbit. The payload could be built at very low cost using existing technology, and a simple ground system could operate the observatory because of the low data rates generated by a photometer (see Priedhorsky 1989).

One of the advantages of a low cost payload which is not found in current NASA instrument projects is its short development lifetime. Twelve years were required to build HST. Other projects have taken similar amounts of time. Small experiments could be built and flown within the lifetime of a graduate student. Some risk would have to be taken when building the payload, but the return would still be considerable since several of these payloads would equal the cost of one NASA SMEX instrument. Missions of this type are needed since the exploratory science represented by a small dedicated space-based photometer is critical to development of larger payloads. However, this type of instrument is absent from NASA programs.

ACKNOWLEDGEMENTS

This work was supported by NASA grant NAS5-29448.

REFERENCES

Priedhorsky, W. C. 1989, *Low-Cost Small Satellites for Astrophysical Missions*, NASA Workshop on High Energy Astrophysics in the 21st Century.

White, R. L. 1990, *Hubble Space Telescope High Speed Photometer Handbook*, STScI.

White, R. L., and Allen, R. J. 1990, *The Restoration of HST Images and Spectra*, STScI.

PART IV.

ROBOTIC TELESCOPES ON THE MOON

TELESCOPES FOR A LUNAR OUTPOST

JACK O. BURNS
Department of Astronomy, New Mexico State University, Las Cruces, NM 88003

ABSTRACT   This paper presents an overview of lunar-based telescopes. It begins with a review of the impressive attributes of the Moon for astronomy which include a hard vacuum, seismic stability, low gravitational field, large surface area, dark and radio-quiet skies on the far-side, cold craters at the poles, and an abundance of raw materials. These attributes, coupled with the availability of technicians at a lunar outpost, make the Moon the most attractive site for the next generation of space-based observatories. It is argued that modest, light-weight instruments such a generic 1-m optical/uv/IR telescope and a low radio frequency phased array are well suited for early deployment on the Moon. More advanced concepts such as a 16-m filled aperture telescope and an optical interferometer will make unprecedented strides in sensitivity and resolution. Although complex, such telescopes are well-suited to an evolutionary construction and deployment scheme.

INTRODUCTION

On July 20,1989, exactly 20 yrs after the first manned landing on the Moon, President Bush committed the American space program to a new direction. He called for a new Space Exploration Initiative (SEI) which consists of two related components. First, the President proposed the establishment of a permanently manned Lunar Outpost within roughly the first decade of the 21st century. The two subsequent studies of a lunar base, by the Augustine and the Stafford committees, have each called for vigorous scientific investigations on the Moon particularly in the area of astronomy. Second, the Lunar Outpost would be used as a stepping stone for an eventual manned Mars mission which might occur by the end of the second decade of the 21st century. The Moon is an attractive location for learning how to live and work on another planetary body, for discovering how to recover natural resources, and for inventing techniques for extraterrestrial scientific investigations in relative proximity to the Earth before undertaking the more dangerous and difficult task of exploring Mars.

The Moon is a particularly attractive location for siting the next generation of space-based astronomical observatories (Burns and Mendell 1988; Burns et al. 1990; Mumma and Smith, 1990). A combination of the Moon's natural attributes described below and the resident staff of technicians at the lunar outpost will permit pioneering new telescopes, ranging from low radio frequency arrays to large aperture optical telescopes, to be constructed on the Moon. For the first time, space-based observatories will be built, operated, and regularly serviced in much the same way as Earth-based telescopes. The lunar telescopes will be operated in a robotic mode with possibly direct control by astronomers at a variety of large and small institutions. Thus, the experience that we are currently gaining on the Earth in designing robotic observatories will be of key importance in operating lunar-based telescopes. Observatories on the surface of the Moon represent a logical extension of automatically controlled telescopes pioneered by numerous groups on the Earth.

The concept of lunar-based observatories has begun to be embraced by an increasing number of astronomers throughout the U.S. and abroad. For example, the recently published Bahcall committee report (Bahcall et al. 1991) contains an entire chapter devoted to telescopes on the Moon. This committee clearly recognized the considerable advantages of performing astronomy from the Moon, the opportunities for astronomy within the SEI program, and the need to begin planning now, given the long lead time for space-based missions. Among their recommendations are the following: (1) "NASA should initiate science and technology development so that facilities can be deployed as soon as possible in the lunar program. NASA should support the long-term development of technologies suitable for possible lunar observatories." A number of the advanced observatories planned for the Moon, such as a synthesis array at optical wavelengths, will require considerable further engineering and construction testbeds on both the Earth and in Earth-orbit. (2) "Site survey observations, possibly with soft landed experiments such as a small transit telescope, should be a high priority for the lunar program. The requirements for astronomical observations should be carefully considered in the selection of the site for a lunar base." Modest remotely-controlled astronomical telescopes should be a part of early unmanned explorations of the Moon which will help to determine the location of a permanent lunar outpost. (3) "NASA, along with other governmental and international agencies, should strive to have the far-side of the Moon declared a radio-quiet zone." As will be noted below, the far-side of the Moon is the only region in the near-Earth environment which is known to be free of Earth-based radio interference especially at frequencies $\lesssim 30$ MHz. I would further advocate, given the very dark night skies on the lunar far-side, the radio-quiet environment, and the unique geological features, that vast sections of the Moon's far-side should be set aside as international scientific preserves.

It is becoming increasingly clear to many that the Moon as a site for astronomical observatories offers the potential for solving many of the outstanding problems in astrophysics, from extrasolar planet detection to understanding star and galaxy formation to mapping large scale motions in the Universe (Mumma and Smith, 1990). In the following sections, I will outline the advantages of the Moon for astronomy and will consider a variety of telescopes at different parts of the electromagnetic spectrum that could be placed on the lunar surface.

WHY THE MOON?

In many ways, the Moon is a nearly ideal location from which to perform astronomical observations. It is particularly well suited for high resolution imaging using large aperture telescopes and interferometers, and for observations outside the Earth-based windows particularly at very low radio frequencies. Some particular advantages of the lunar surface as a site for the next generation of telescopes include:

Low Atmospheric Optical Depths

The average density of the Moon's atmosphere is $0.2 - 1.0 \times 10^6$ molecules cm^{-3} (night to day) which translates to extremely low optical depths at all parts of the electromagnetic spectrum (Potter and Morgan, 1988). In fact, the atmospheric density on the lunar surface is less than that in low Earth orbit. The entire mass of the atmosphere of the Moon is only 10,000 kg, or about that inside a typical basketball arena on Earth. Ninety-three percent of the lunar atmosphere is composed of Ne, H_2, and He (Hoffman *et al.* 1973). There is a similarly low ionospheric density with an estimated plasma frequency of < 90 kHz.

The lack of atmosphere also means, of course, that there will be no wind loading or weather-related problems with lunar-based telescopes. This, coupled with the low gravity, implies that much lighter-weight structures and simpler drive mechanisms would suffice for telescopes on the Moon in comparison to those on Earth.

Seismic Stability

One of the most important advantages of the Moon for astronomy is the excellent seismic stability of the surface. Average ground motions are < 1 nm. Typical subsurface seismic energy is 10^{-8} of that on the Earth (Goins *et al.* 1981). Although moonquakes were recorded during the Apollo program, they are very low level averaging 1-2 on the Richter scale and are much less frequent than on Earth (500/yr vs. 10,000/yr on Earth). The rubble which makes up the subsurface layers of the Moon is an excellent damping agent that does not permit the seismic waves to propagate to the same kind of distances as on Earth. Seismic waves are intensely scattered so the damaging effects of a moonquake are less than those of a similar magnitude quake on the Earth.

The seismic stability is an important issue when considering interferometry at submillimeter, infrared, and optical wavelengths. The baselines between elements in an interferometric array must be controlled to within a fraction of a wavelength to maintain phase coherence. In principle, this can be done in Earth-orbit for short baselines (hundreds of meters) using structures which physically connect array elements. However, for longer baseline interferometry (kilometers) at very high frequencies, maintaining the baselines for free-flying elements requires very complex and very expensive station-keeping. Such baseline stability comes for free on the surface of the Moon where baselines of 10-km or more are feasible and limited only by the curvature of the Moon (Burke, 1990; Burns *et al.* 1991). Thus, the Moon is realistically the only location where very high resolution (tens of $\mu arcsec$) imaging at optical/IR wavelengths will someday be feasible.

Low Gravitational Field

The surface gravity of the Moon is 162.2 cm s^{-2} or about one-sixth that of Earth. At first glance, any gravitational field might appear to be a disadvantage because of the gravity loading that it would produce on telescope superstructures. However, one must keep in mind the low intensity of the field relative to Earth. The lack of weather plus the low gravity will permit both very large and very "flimsy" telescope support structures on the Moon (Akgul et al. 1990; Chua et al. 1990). Thus, telescopes of 16-m diameter at optical/IR wavelengths and 1-km diameter for Arecibo-style lunar crater radio antennas become feasible. The Moon offers no practical limit to large aperture telescopes. In fact, one would not likely consider building single aperture structures anywhere much larger than those noted above since interferometers will be more cost-effective in both sensitivity and resolution beyond this point.

The Moon's gravity does have a practical advantage in terms of engineering and construction. It was demonstrated from a combination of the Space Shuttle and the Apollo programs that performing construction activities on the Moon is far simpler (although still nontrivial) in comparison to Earth orbit. The gravitational field of the Moon offers a simple yet important assist for tasks such turning a wrench where the inertial mass of the Moon comes into play. The lunar surface is an environment similar to the Earth in terms of construction activities so that adaptations of familiar Earth-based bulldozers, trenchers, etc. can be used. The pointing system of a lunar-based telescope will use a gravity-assisted mechanical system like that of Earth-based telescopes rather than the complex gyro system on HST.

Large Surface Area

The Moon has large flat areas especially within the maria which are ideally suited for long baseline, high resolution interferometers at submillimeter through optical wavelengths.

The Lunar Far-Side: A Dark and Radio-quiet Sky

At night on the far-side of the Moon, astronomers will have the darkest and coldest skies in the near-Earth environment. Backgrounds will be limited by zodical light, the density of stars near the Galactic plane, and the density of galaxies at high latitudes.

The lunar far-side is unquestionably the quietest location at radio frequencies within the inner solar system (Kaiser, 1990). In 1972, a Radio Astronomy Explorer satellite (RAE-2) carrying an enormous V-shaped antenna was placed into lunar orbit (Alexander et al. 1975). As shown by Fig. 1, a dramatic reduction in radio frequency background was observed when the Moon occulted the Earth, dropping to near that expected for the nonthermal Galactic emission. The variation between the near and far sides of the Moon becomes even more dramatic at frequencies below a few MHz. Astronomical observations on the Earth at frequencies $\lesssim 30$ MHz have become increasingly difficult because of the irregular and partially opaque (or totally opaque at $\lesssim 10$ MHz, the ionospheric plasma frequency) ionosphere which varies with solar cycle, and because of the increasing man-made interference. RAE-2 has shown that significant man-made interference breaks through the ionosphere especially above 5 MHz on the night-side making observations from Earth-orbit difficult (Erickson, 1990). In fact, LaBelle et al. (1989) claim that interference at 5 MHz observed in Earth orbit has increased by a factor of 100 over the past twenty years possibly due to new over-the-horizon radar installations. Lightning storms represent an additional source of interference above a few MHz. Below ≈ 1 MHz, the Earth's magnetosphere is a tremendous source of naturally-produced low frequency emission termed the Auroral Kilometric Radiation (AKR). The AKR, which was the most important discovery of RAE-2, is still not well understood but is likely produced by magnetic field reconnection events in the magnetotail. The only location where one can escape with assurance from these high Earth backgrounds is the lunar far-side.

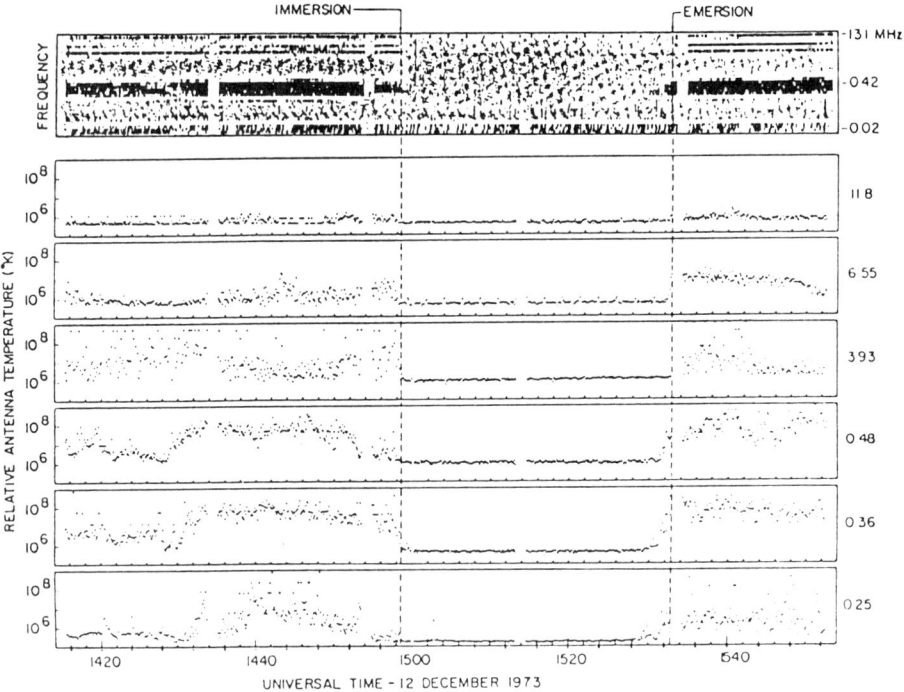

Figure 1. Radio light curves from RAE-2 showing the dramatic difference in low frequency backgrounds on the lunar near-side and far-side (Alexander et al. 1975).

Natural Cryogenic Environment in Lunar Polar Craters
There are craters near the lunar poles that are permanently shadowed and may have surface temperatures as low as 40 K (Burke, 1988). Such a cold environment would be ideal for siting a telescope which operates in the thermal infrared where both the detector and the telescope superstructure could be passively cooled (Lester, 1991).

Raw Materials
One should not forget that the Moon is rich in raw materials such as aluminum, ceramics, and an interesting high tensile strength, low thermal expansion anhydrous glass (Blacic, 1985). Such materials will be mined from the Moon

and potentially used as components for telescope construction, thus greatly reducing costs associated with transportation (e.g., Johnson and Wetzel, 1990). This will eventually become a major advantage over high Earth orbit in siting observatories.

Accessibility
Finally, the Moon will offer an important advantage over high Earth orbit in terms of accessibility to the telescopes for maintenance and upgrades. The existence of a lunar outpost will provide the manpower needed to service the observatories. We should be able to tolerate a higher failure mode, and thus a less costly telescope, because of this accessibility. We will see, for possibly the first time, a synergy between the manned and the unmanned space programs on the Moon where both are genuinely important for achieving scientific goals.

CONCERNS WITH THE LUNAR ENVIRONMENT

Although the environment on the lunar surface offers many substantial advantages over the Earth's surface or in Earth orbit, there are some significant concerns that should be considered. These include:

Cosmic Radiation
No magnetic field is currently being generated within the core of the Moon, so only a very low field is present on the lunar surface. The lunar B-field strength ranges from $3-330 \times 10^{-5}$ G, which should be compared to an average Earth field of 0.3 G (Dyal *et al.* 1974). Thus, there is little or no modulation of cosmic rays. High energy particles strike the lunar surface unimpeded. This is a considerable problem for both lunar astronauts and for sensitive solid state devices such as CCD cameras.

Fortunately, the lunar regolith (or soil) is an excellent source of shielding for most Galactic cosmic rays. This regolith is mostly powder, having been impacted by micrometeorites over billions of years (Heikien, 1975). It has been estimated that about 5 m of lunar soil would produce a reduction in the flux of cosmic rays equivalent to that in low Earth orbit (Johnson and Dietz, 1991). So, one can envision placing CCDs and other sensitive detectors below the lunar surface in a Coudé room. Alternatively, anticoincidence devices, similar to those used commonly in high energy astrophysics, would need to be developed to remove the effects of cosmic rays as a major source of enhanced background.

Micrometeoroids
The lack of atmosphere means that solar system debris particles strike the Moon's surface with speeds of 10-200 km s^{-1}. From studies of lunar rocks, Taylor (1988) calculates that 1 μm sized craters will be produced by micrometeorites at a rate of 1200 craters $m^{-2}yr^{-1}$. This flux is less than that for low Earth orbit where man-made debris is quickly becoming a major hazard. The lunar micrometeoroids will not produce significant damage to optical surfaces, but the secondary ejecta are potentially very damaging to a telescope (MSFC, 1991). Thus, precision optical surfaces may need to be shielded on the Moon by dome-like structures that will also be used as sunshades.

Thermal Changes

At the Apollo 17 site, day to night surface temperatures ranged from 384 K to 102 K (Keihm and Langseth, 1973). Just below the surface at a depth of 30 cm, the temperature is about 250 K and varies by only 2-4 K from day to night due to the low thermal conductivity. Such large surface temperature variations will place severe constraints on telescope structures which are required to maintain precise tolerances for high resolution imaging. Sunshades and/or metal-matrix composite materials with low coefficients of thermal expansion will be required for the telescope superstructure (Akgul *et al.* 1990).

These temperature variations are severe but are still not as troublesome as in low Earth orbit where spacecraft move in and out of sunlight every 45 minutes. On the Moon, there is two weeks of nearly constant temperature day and two weeks of very cold night. These long periods of thermal stability, especially during the lunar night, will offer astronomers a unique opportunity for stable, deep, and long duration exposures of celestial objects.

Pollution Near a Lunar Outpost

There are ambitious plans for mining, manufacturing, habitats, and spacecraft landings which will potentially vent large quantities of gas into the lunar atmosphere. The lunar vacuum is a fragile commodity as demonstrated by the Apollo program where the mass of the Moon's atmosphere was temporarily doubled during each mission to the Moon (Johnson, 1971). Fortunately, the lunar environment is relatively efficient in cleansing itself of such gas via thermal escape, adsorption by the regolith, and extraction of gas (photoionized by solar uv-radiation) by the solar wind. However, Vondrak (1974) has suggested that significantly higher injection rates, as might result from a vigorous lunar base, could eventually change the loss mechanism to thermal escape alone leading to the development of an atmosphere with a decay rate of hundreds of years. Our own calculations suggest that gas is more effectively dispersed near a lunar outpost (Fernini *et al.* 1990). Assuming a collisionless, isothermal atmosphere, we found that even for the most extreme scenario involving the extraction of large quanitites of He^3 (for fusion reactors on Earth) from the regolith, the density of atmospheric gases drops back to near the present ambient value beyond about 10 km from the mining operation. Thus, we propose that lunar observatories should be sited, and remotely operated, at least this distance from the base.

Dust is a more troublesome issue. The anhydrous soil is apparently highly susceptible to electrostatic and photoconductivity effects. The Apollo astronauts reported problems associated with charged dust clinging to spacesuits and interfering with the operation of the lunar rover (Neal *et al.* 1988; Johnson *et al.* 1991). In addition, Criswell (1972) noted a bright glow photographed by Surveyor 7 which he attributed to levitation of dust grains at dawn. There is some suspicion that dust "creeps" between light and dark areas due to the establishment of large electrostatic potential differences between these different regions (De & Criswell, 1977). Clearly, more theoretical calculations, laboratory experiments, and early in-situ measurements from the lunar surface are needed to determine the full extent of the dust problem and to propose solutions for lunar-based telescopes.

MODEST LUNAR-BASED TELESCOPES

It is likely that initial missions back to the Moon will be limited in scientific payload capacity. One proposal, studied at a recent workshop at the Johnson Space Center, calls for a return to the Moon in the 1990's using common or generic unmanned lunar landers built with currently existing technology and launched with the present generation of rockets. They would be limited to $\lesssim 500$ kg payload mass. Similarly, when astronauts first return to the Moon, their concern must be focused upon habitats and basic survival with only a limited capacity for scientific experimentation. Fortunately, there are a number of astronomical telescopes which are light-weight, automated and powerful when placed on the Moon that could be among the first pieces of scientific instrumentation. Table 1 lists a few suggestions for early lunar-based telescopes.

Dedicated, but generic one-meter aperture telescopes hold much promise for early deployment on the Moon (Smith, 1990; Sykes et al. 1990). They will likely have masses $\lesssim 200$ kg, will be relatively low power consumers (≈ 100 W), and will be remotely controlled from Earth using the technologies shown feasible in these conference proceedings. Although modest in aperture compared to say HST, they are powerful if dedicated to selected tasks. For example, a lunar automatic photoelectric telescope has the advantage of very high speed and the collection of ungapped data which are so important in testing models of both active stars and galaxies (Zeilik, 1988). An all-sky CCD survey in the uv and the IR using a Schmidt wide-field telescope could be the lunar analog of the tremendously useful Palomar Sky Survey.

The Lunar Transit Telescope proposed by John McGraw has the enormous advantage of simplicity. With few if any moving parts, it is less susceptible to failure on the lunar surface. Yet it is scientifically very powerful. Although its power consumption is similarly modest, a > 2-m aperture mirror would require more lander payload capacity.

Another possible early telescope for the lunar surface is a Moon-Earth Radio Interferometer (MERI, Burns 1988). We envision an experiment analogous to that performed by the JPL group using the TDRSS satellite and the NASA Deep Space Tracking Network antennas (Levy et al. 1986). They demonstrated the viability of space-based VLBI by detecting fringes on several quasars for baselines ranging from the Earth's surface to Earth orbit. Similarly, one could use either a dedicated VLBI antenna or a communications antenna coupled with current hydrogen maser clocks to form a VLBI station on the Moon. One could then attempt to detect fringes on baselines of 384,000 km (10 $\mu arcsec$ resolution at 10 GHz) or about 50 times longer than current ground-based VLBI. There may even be a possibility of imaging strong sources at these resolutions since the combination of the NRAO VLBA, the Japanese VSOP low Earth orbit VLBI satellite, the Soviet RadioAstron high Earth orbit antennas, and the lunar antenna produce quite reasonable u-v coverage (Burns and Asbell, 1987).

A further exciting candidate for an early experiment involves very low frequency radio astronomy from lunar orbit or the lunar surface. We have studied several concepts for lunar-based very low frequency telescopes that have the considerable advantages of extremely low mass ($\lesssim 20$ kg), low power (≈ 20 W), and strong scientific motivation. One proposal called LORAE (Lunar Orbit Radio Astronomy Experiment) involves placing two low frequency antennas on

Table 1. Modest Lunar-Based Telescopes

Telescope	Reference
Generic One-meter Telescopes	
- Lunar APT	1, 2, 3
- uv, IR Sky Survey	2
- Far-IR Testbed	2
Lunar Transit Telescope	4, 5
X-ray All-Sky Monitor	6
Moon-Earth VLBI	7
Very Low Frequency Interferometers	8

[1] Zeilik (1988)
[2] Sykes et al. (1990)
[3] Genet (1991)
[4,5] McGraw (1990, 1991)
[6] Peterson (1990)
[7] Burns & Asbell (1987)
[8] Burns (1991)

Table 2. Advanced Lunar-Based Observatories

Telescope	Reference
16-m Large Lunar Telescope	1, 2
Lunar Submillimeter Array	3
Lunar Far-Side Very Low Frequency Array	4
Optical/IR Imaging Interferometers	5, 6, 7

[1] Bely et al. (1989)
[2] MSFC (1991)
[3] Mahoney (1990)
[4] Burns et al. (1989)
[5] Ridgway (1990)
[6] Burke (1990)
[7] Shao (1990)

board two perpendicularly inclined lunar polar orbiters (Burns, 1991). Such orbiters would be dedicated to mapping the Moon's surface at high resolution as a precursor to a lunar base. We propose to perform aperture synthesis between the two spacecraft at 13 and 25 MHz where the resolution peaks at ≈ 2 arcsec (at 25 MHz limited by interplanetary scintillation). This resolution, > 100 times better than the best ground-based observations, would place low frequency radio astronomy on a par with centimeter-wavelength astronomy for the first time. Exceptionally good u-v coverage is obtained for 100-km altitude orbits over timescales of 1 yr for data taken only when both antennas are over the lunar far-side. Such coverages produce impressive, well-resolved imaging of 3C-like sources at low frequencies. The antenna on each spacecraft is actually a phased array of > 100 dipoles which are encoded on the surface of an inflatable structure. This unique telescope has good antenna directivity, and will accurately point without any moving parts by electronically phasing the dipole array. We have recently considered a similar inflatable structure for the lunar surface as either a single low resolution telescope for solar system observations, or as part of a low frequency array. Either LORAE or the lunar surface low frequency interferometer offers much scientific promise to investigate fundamental questions involving low energy relativistic electrons including types II and III solar flares, planetary magnetospheres, interstellar medium turbulence, and extragalactic radio sources (Kassim and Weiler, 1990).

An artist's sketch of three of the above candidate lunar-based telescopes on board an unmanned lunar lander is shown in Fig. 2.

ADVANCED LUNAR-BASED OBSERVATORIES

The real power of the Moon as a site for the next generation of space-based observatories is best seen in the advanced concepts listed in Table 2. Here, large-aperture optical/IR telescopes and high frequency interferometers feature extraordinary sensitivity and angular resolution. Most represent significant new technology developments either because of problems in deployment (e.g., far-side low frequency array) or in precision of operation (e.g., optical interferometer). However, each of these telescopes can be emplaced using an evolutionary construction strategy which is well suited to a lunar base (Burns and Mendell, 1988; Smith, 1990). Individual components or elements of these telescopes can be brought piecemeal to the Moon. Each component is individually capable of performing significant science on the Moon (e.g., one mirror segment for a 16-m aperture telescope or one element of an optical interferometer). Thus, a powerful interferometric array, for example, can be brought on line over the course of several years improving with each new element yet performing innovative observations while still in the engineering phase. The construction and operation of the NRAO VLA is an excellent example of such an evolutionary construction scheme.

A Large Lunar Telescope (LLT) with a 16-m diameter passively cooled, diffraction-limited, filled-aperture has been proposed as a possible successor to the HST (Bely *et al.* 1989, MSFC 1991). Such a telescope has the potential for some promising astronomical discoveries including detections of Earth-like planets around nearby stars and distant protogalaxies (Illingworth *et al.* 1991). The telescope primary would likely consist of $\approx$ 4-m light-weight segments with

Figure 2. Three possible early lunar-based telescopes placed on the Moon using an unmanned lunar lander. The three telescopes include a VLBI antenna, a 1-m optical/uv/IR telescope, and an inflatable low frequency phased dipole array antenna. Drawing is by Pat Rawlings of SAIC.

active wavefront sensing for diffraction-limited performance. The LLT would operate between 0.1 and 10 μm with a field of view of 2' in the optical through the IR. The resolution would be 0.01 arcsec in the visible. The telescope could be passively cooled to < 100 K and the imagers/spectrographs would be further cooled cryogenically. The visible and IR backgrounds with the LLT would be 10^{-4} and 10^{-8}, respectively, that of the Earth. Unresolved objects could be detected down to 32^m with < 3 hr integration. Thus, the major attribute of this telescope is the outstanding sensitivity, especially in the infrared.

An even more impressive leap in telescope performance, this time in resolution, is possible with lunar-based optical and IR interferometers (Burke 1990; Shao 1990, 1991; Burns et al. 1991). The particular design studied by our group consists of 42 1.5-m optical elements with a maximum baseline of 10-km. The resolution of such a telescope is $10 \mu arcsec$ at a wavelength of $0.5 \mu m$ or more than 4000 times better than that of HST. There are few occasions in astronomy where one is presented with an opportunity for such a gigantic leap in telescope performance. As noted above, the Moon is uniquely suited for such a long baseline telescope. Our design calls for the elements to be irregularly distributed along two circles; the outer circle has an overall diameter of 10-km and the inner circle of nine elements has a diameter of 0.5 km. We have shown that such a configuration produces superior instantaneous u-v coverage which is important for a lunar aperture synthesis telescope because of the Moon's slow 28 day rotation period (Burns, 1990). An optical/IR lunar interferometer is one of the most scientifically exciting instruments for the Moon. It has the potential for imaging Jupiter-like planets (with > 20 resolution elements) and resolving Earth-like planets around nearby stars. (Dynamic range is not as much an issue for an interferometer in comparison to a filled aperture telescope because an interferometer could resolve out the disk of the nearby star with a judicious choice of baselines.) An lunar optical/IR synthesis array could also map active regions on the surfaces of stars, image accretion disks near compact objects, and measure quasar proper motions to study Hubble flow anisotropies.

CONCLUSIONS

Existing terrestrial robotic telescopes, as well as the proposed South Polar observatories, are crucially important precursors for lunar-based telescopes. The fact that such complex automated telescopes are successfully operated by small staffs at individual observatories gives us hope that similar initially modest lunar-based telescopes can be built cheaply and operated by small groups back on Earth (Genet, 1991). It is not necessary to build large and costly institutes to run some of the simpler observatories on the Moon. The IUE, operated by several modestly-staffed institutions, is an excellent model for early lunar-based observatories. Such a cost effective management structure will be required if we are ever to see telescopes on the Moon given the country's on-going fiscal crisis.

The characteristics of the Moon are well suited to both small and large aperture telescopes. In fact, these attributes are so overwhelming that it is inevitable in my opinion that observatories will someday be placed on the lunar surface. This may occur as early as this decade if an unmanned generic lunar lander program begins within the next several years.

Finally, the ultimate promise of the Moon for astronomy lies in the construction of large single aperture and synthetic aperture telescopes. The scope, size, and operational wavelengths of these telescopes are such that they can only be practically built on the lunar surface. The jump in sensitivity and resolution over Earth and space-based telescopes in unprecedented. The range of scientific problems that can be addressed with these large aperture and long baseline lunar telescopes is truly impressive and mind-boggling.

ACKNOWLEDGEMENTS

There are many people who have contributed important ideas for lunar-based telescopes whom I wish to thank. Let me begin with our New Mexico Lunar Observatories working group including Stewart Johnson, Neb Duric, Koon-Meng Chua, Jeff Taylor, John Basart, Brian Dennison, Steve Castillo, Bernie McCune, Reta Beebe, and Cathleen Burns. I am also grateful for insightful discussions with Harlan Smith, John McGraw, Russ Genet, Garth Illingworth, Bernie Burke, Mike Shao, Max Nein, Carl Pilcher, Mike Bicay, Mike Kaplan, Wendell Mendell, Ray Smartt, Namir Kassim, Kurt Weiler, Dayton Jones, and Mike Mahoney. This work was supported in part by a grant from the NASA Johnson Space Center.

REFERENCES

Akgul, F., Gerstle, W., and Johnson, S. 1990, in *Engineering, Construction, & Operations in Space, Space 90*, ed. S. Johnson and J. Wetzel (New York: ASCE), p. 697.
Alexander, J., Kaiser, M., Novaco, J., Grena, F., and Weber, R. 1975, *Astr. Ap.*, **40**, 365.
Bahcall, J. *et al.* 1991, *The Decade of Discovery in Astronomy & Astrophysics* (Washington, DC: National Academy Press).
Bely, P., Burrows, C., and Illingworth, G., eds., 1989, *The Next Generation Space Telescope* (Baltimore: STScI).
Blacic, J. 1985 in *Lunar Bases & Space Activities of the 21st Century*, ed. W. Mendell (Houston: LPI), p. 487.
Burke, B. 1990, *Science*, **250**, 1365.
Burke, J. D. 1988, in *Future Astronomical Observatories on the Moon*, NASA Conference Publication 2489, ed. J. Burns and W. Mendell, p. 31.
Burns, J. 1988, in *Future Astronomical Observatories on the Moon*, ed. J. Burns and W. Mendell, NASA Conference Publication 2489, p. 97.
Burns, J. 1991, in *Radio Interferometry — Theory, Techniques, & Applications*, ed. T. Cornwell, *A.S.P. Conference Series*, in press.
Burns, J., and Asbell, J. 1987, in *Radio Astronomy From Space*, ed. K. Weiler (Green Bank: NRAO), p. 29.
Burns, J., Duric, N., Taylor, G., and Johnson, S. 1990, *Scientific American*, **262**, 42.
Burns, J., Johnson, S., and Duric, N., eds., 1991, *NASA Conference Publication*, in press.

Burns, J., and Mendell, W., eds., 1988, *Future Astronomical Observatories on the Moon*, NASA Conference Publication 2489.
Chua, K., Hicks, J., and Johnson, S. 1990, in *Engineering, Construction, & Operations in Space II, Space 90*, ed. S. Johnson and J. Wetzel (New York: ASCE), p. 697.
Criswell, D. 1972, *Proc. Lunar Sci. Conf. 3rd*, 2671.
De, B., and Criswell, D. 1977, *J. Geophys. Res.*, **82**, 999.
Dyal, P., Parkin, C., and Daly, W. 1974, *Rev. Geophys. Space Phys.*, **12**, 568.
Erickson, W. 1990, in *Low Frequency Astrophysics from Space*, ed. N. Kassim and K. Weiler (New York: Springer-Verlag), p. 59.
Fernini, I., Burns, J., Taylor, G., Sulkanen, M., Duric, N., and Johnson, S. 1990, *J. Spacecraft and Rockets*, **27**, 527.
Genet, R. 1991, these Proceedings.
Goins, N., Dainty, A., and Tuksoz, M. 1981, *J. Geophys. Res.*, **86**, 378.
Heikien, G. 1975, *Rev. Geophys. Space Phys.*, **13**, 567.
Hoffman, J., Hodges, R., and Johnson, F. 1973, *Proc. Lunar Sci. Conf. 4th*, 2865.
Illingworth, G., et al. 1991, in *Working Papers: Astronomy & Astrophysics Panel Reports* (Washington, DC: National Academy Press), p. IV-1.
Johnson, C., and Dietz, K. 1991, in *OE/Aerospace Sensing, SPIE Proceedings*, Vol. **1494**, in press.
Johnson, F. 1971, *Rev. Geophys. Space Phys.*, **9**, 813.
Johnson, S., Chua, K., Burns, J., and Slane, F. 1991, in *OE/Aerospace Sensing, SPIE Proceedings*, Vol. **1494**, in press.
Johnson, S., and Wetzel, J., eds., 1990, *Engineering, Construction, & Operations in Space II, Space 90* (New York: ASCE).
Kaiser, M. 1990, in *Low Frequency Astrophysics from Space*, ed. N. Kassim and K. Weiler (New York: Springer-Verlag), p. 3.
Kassim, N., and Weiler, K., eds., 1990, *Low Frequency Astrophysics from Space* (New York: Springer-Verlag).
Keihm, S., and Langseth, M. 1973, *Proc. Lunar Sci. Conf. 4th*, 2503.
LaBelle, J., Treuman, R., Boehm, M., and Gewecke, K. 1989, *Radio Science*, Nov. 1989, 730.
Lester, D. 1991, these Proceedings.
Levy, G., et al. 1986, *Science*, **234**, 187.
Mahoney, M. 1990, in *Astrophysics from the Moon*, ed. M. Mumma and H. Smith (New York: AIP), p. 508.
Marshall Space Flight Center (MSFC) Study 1991, *Large Lunar Telescope*.
McGraw, J. 1990, in *Astrophysics from the Moon*, ed. M. Mumma and H. Smith (New York: AIP), p. 433.
McGraw, J. 1991, these Proceedings.
Mumma, M., and Smith, H., eds., 1991, *Astrophysics from the Moon* (New York: AIP).
Neal, V., Shields, N., Carr, G., Pogue, W., Schmitt, H., and Schulze, A. 1988, *Extravehicular Activity at a Lunar Base, Phase II Final Report*, NASA NAS9-17779, pp. 41, 62, 70.
Peterson, L. 1990, in *Astrophysics from the Moon*, ed. M. Mumma and H. Smith (New York: AIP), p. 345.
Potter, A., and Morgan, T. 1988, *Science*, **241**, 675.

Ridgway, S. 1990, in *Astrophysics from the Moon*, ed. M. Mumma and H. Smith (New York: AIP), p. 495.

Shao, M. 1990, in *Astrophysics from the Moon*, ed. M. Mumma and H. Smith (New York: AIP), p. 486.

Shao, M. 1991, these Proceedings.

Smith, H. 1990, in *Astrophysics from the Moon*, ed. M. Mumma and H. Smith (New York: AIP), p. 273.

Sykes, M., *et al.* 1990, in *Astrophysics from the Moon*, ed. M. Mumma and H. Smith (New York: AIP), p. 328.

Taylor, G. 1988, in *Future Astronomical Observatories on the Moon*, NASA Conference Publication 2489, ed. J. Burns and W. Mendell, p. 21.

Vondrak, R. 1974, *Nature*, **248**, 657.

Zeilik, M. 1988, in *Engineering, Construction, & Operations in Space I, Space 88*, ed. S. Johnson and J. Wetzel (New York: ASCE), 1095.

MULTI-USE LUNAR TELESCOPES

RUSSELL M. GENET and DAVID R. GENET
AutoScope Corporation, and the Fairborn Observatory, 3435 East Edgewood Ave., Mesa, AZ 85204

DAVID L. TALENT
Lockheed Engineering and Sciences Company, Solar System Exploration Department, 2400 NASA Rd. 1, C-23, Houston, TX 77058

MARK DRUMMOND
Sterling Software, NASA-Ames Research Center, Moffett Field, CA 94035

BUTLER P. HINE
NASA-Ames Research Center, MS 244-4, Moffett Field, CA 94035

LOUIS J. BOYD
Fairborn Observatory, 3435 East Edgewood Ave., Mesa, AZ 85204

MARK TRUEBLOOD
Winer Mobile Observatory, P.O. Box 42556, Tucson, AZ 85733

PHILOSOPHICAL AND POLITICAL UNDERPINNINGS

A fundamental characteristic of human nature is to explore. In ancient times, an intrepid individual was able to fulfill his or her need for exploration in a fairly inexpensive and direct manner. The oceans were a vast open frontier — new continents lay beyond, waiting to be discovered, explored, and developed. As the centuries have passed, these earthly frontiers, except for the ocean depths, have now been explored quite completely. The greatest frontier left to us lies above. However, exploring this new frontier requires more than the far-sighted vision and the daring of a single individual. The frontier is space, of course, and its exploration must be achieved by mankind as a whole. Any nation that intends to take the lead in this exploration must bear in mind that it is the collective body of its citizenry that forms the bulwark of that exploration. Not only will they pay for such exploration, but they must be made an integral part of this great endeavor.

In the United States of America, NASA is the clearly identifiable agency that carries the standard for the dream of exploration. In these days, the early 1990s, as we once again look towards the Moon, it is important to help the public maintain focus on the relevance of our exploration goals. One of these goals is found in the oldest of the sciences — astronomy. The average man can still buy a telescope and look to the heavens and, in fact, people do so by the hundreds of thousands. No science has such a large following as astronomy. Astronomy is also the science that many youngsters first get excited about as they try to understand their place in space and time.

Therefore, as we contemplate returning to the Moon, one of the most important things we can do, from the standpoint of science, society, and the politics of this endeavor, is to place astronomical telescopes on the Moon that are inexpensive and accessible to both research astronomers and members of the population at large — particularly young students. This vanguard of our aggressive return to the Moon would provide early and outstanding scientific returns.

We believe it is possible to create actual public involvement in science. This can be done with small robotic telescopes in the earliest phases of our return to the Moon. Inherent in the design of these telescopes would be simplicity at every possible stage. The optical technology would be straightforward and historically proven. The telescopes themselves, and there would be many more than just one, would each be dedicated to a single, well defined, and achievable mission; i.e., each would have a dedicated instrument.

One of the great advantages of small, deployable, lunar telescopes is that they can go from a concept to working systems on the lunar surface in much less than a decade. A number of such telescopes could be made very accessible to individual researchers in the astronomical community. The format for this use could be made highly streamlined with a minimum of overhead. Individual observers, with approved programs, would be able to access the lunar telescope that met their requirements directly from their home institution via computer networks such as Internet, or through a simple modem and telephone connection. Unlike telescopes operating in low-Earth orbit, the operation of multi-use lunar telescopes should not require an extensive infrastructure. It should be no more difficult for future astronomers to remotely operate lunar-based telescopes than it is for present-day astronomers to operate existing Earth-based robotic telescopes. Some research programs might require extensive, on-line interaction between the astronomer and the lunar telescope (i.e., real-time operation); other programs might only need to uplink a "pointing list" of targets.

In calling these lunar telescopes "multi-use" we actually have two aspects in mind. First, the same basic telescope (and supporting electronics, etc.) will be used with a wide variety of instrument payloads — one instrument with each payload. Thus, it will not be necessary to design, develop, and test a number of different telescopes. Second, we are proposing multi-use in the sense that the telescopes would be used for both research and education.

Since, under the multi-use lunar telescope concept, each telescope will have a dedicated function (photometry, imagery, etc.), the prospective user groups for each telescope should be fairly well-defined. Telescope time could be allocated by a Telescope Allocation Committee made up of a rotating subset of telescope users. A Principal Astronomer (PA), who would also be an astronomer/user, could serve as the primary point of contact between NASA and the user community. But, paramount at the outset, should be the minimization of bureaucratic overhead.

One great advantage of the multi-use lunar telescope concept is its ability to accommodate observing programs that require extended intervals of observing time — not always possible from low-Earth orbit. Long-term programs could and should be encouraged. A large number of observers may be accommodated by using automated scheduling and prioritization with uplink of the pointing sequence being made daily.

The multi-use telescopes should be viewed not only as being a collection of telescopes with multiple dedicated purposes, but also as serving multiple user types. We have already mentioned the most obvious group of users — professional astronomers employing the telescopes to push out the frontiers of knowledge in their particular branch of astronomy, using capabilities which do not exist on the ground. But the application of these systems may be considerably extended beyond this with significant overall benefits.

The United States will maintain its leadership in science and technology only if there is a new generation of well-trained scientific minds prepared to take the place of their elders. If we place telescopes on the Moon and do a hundred worthwhile scientific projects in the next few decades, but fail to educate the next generation of young minds to carry them forward, then we will have been remiss. In this regard the multi-use lunar telescope concept may be readily employed in several ways.

First, graduate students may be allowed, under the guidance of their major professors, to conduct research using these systems. It would be perfectly sensible for an astronomy graduate student to attack some interesting problem using a combination of ground-based and lunar-based telescopes. This, however, is an obvious use of such telescopes — to educate the next generation of astronomers. But what of other gifted students — perhaps those who show some promise in the sciences, but are only at the undergraduate level? It would seem to make sense to set aside some fraction of time on lunar telescopes, or even to provide for the establishment of a dedicated system for undergraduate education; this resource would be available to students from every state of the Union. Projects could be developed and proposed, and regional and state-wide conventions and symposia could be held for the students to present their proposals. Judges, made up of a group of astronomical professionals, could select projects from every state. The students would then be allowed to carry out their research under the guidance of a professional astronomer. Not only would the student experience a great educational gain, but the prestige and honor of being selected could be highly motivational.

In addition to use by graduate and undergraduate students, a limited amount of guided access could be provided for high school students — perhaps even in conjunction with the undergraduate program. Not only would this be of value to the students involved, it would clearly demonstrate a national commitment to science education. Not every state currently has a heavy involvement in the space program; however, there isn't a congressional district in the country that doesn't have a high school, college, or university in it. The average citizen-voter may find it a lot easier to be sympathetic to the goals of the space program if the high-school kid who mows his grass were selected to make observations of a distant galaxy using a telescope on the Moon in her senior year.

The multi-use lunar telescope concept, if kept simple, would be inexpensive enough for concerns smaller than the federal government to take on the expense of establishing one on their own. Several states might get together and fund a telescope for their universities — even an individual state. For any state or consortium willing to fund the building of a telescope, NASA could provide the launch service.

In summary, the multi-use lunar telescope concept:
- is a good way to do inexpensive, high-return astronomy from the Moon within a decade;
- is a good way to learn how to do astronomy in the lunar environment on a scale such that our mistakes will not cost too much;
- is a good way to put astronomical research from the Moon directly in the hands of the astronomical community with rapid return of results;
- is a great tool for education;
- is a potentially great motivator of today's students; and
- is a way of involving the whole nation in the space program in a fashion that helps them feel that it is their program.

MULTI-USE LUNAR TELESCOPES

The objective of multi-use lunar telescopes is to reduce the initial and operational costs of space telescopes to the point where a fair number of telescopes (a dozen or so) would be affordable. The basic approach is to have a common telescope, control system, and power and communications subsystems that can be used with a wide variety of instrument payloads, such as imaging CCD cameras, photometers, spectrographs, etc. By having such a multi-use (and multi-user) telescope (a common practice for Earth-based telescopes), development costs can be shared across many telescopes, and the telescopes can be produced in economical batches.

As mentioned earlier, users would themselves directly operate the telescopes from their home institutions on Earth. This is already being done with Earth-based fully automatic telescopes. For instance, the Fairborn Observatory and the Smithsonian Institution operate seven fully automatic telescopes at an automated observatory on Mt. Hopkins in southern Arizona. The approximately four dozen institutions that use these telescopes operate them via modem or computer network. Only two persons, working part time, are required to care for the equipment and handle the miscellaneous paperwork. One of the users on each telescope is designated the principal astronomer (PA). The PA is responsible for resolving any scheduling conflicts that the automated scheduling cannot handle, for assuring the smooth flow of observational requests and results, and for monitoring the quality of the results. The same general approach could be taken to the operation of telescopes on the Moon.

The Moon provides a solid and highly stable platform from which to operate the telescopes, obviating the need for an expensive space stabilization system. These multi-use lunar telescopes could be placed on the Moon either by a soft lander, such as the proposed Common Lunar Lander (Artemis), or taken as part of the cargo manifest to a manned lunar outpost. If taken to the Moon by a soft lander, we have assumed that power and communications would have to be a part of the multi-use lunar telescope, while if taken to a manned lunar outpost, power and communications could be provided from the outpost. This would, presumably, lower the cost per telescope.

The Common Lunar Lander, Artemis, is a concept developed by Stephen Bailey, Alan Binder, and others at the Johnson Space Center. A conference on Artemis was held in July 1991, and a summary is available from Stephen Bailey.

We have assumed that the most constraining size and weight requirements would be the case where a single telescope and attendant power and communications capabilities along with a lunar lander were carried by a Delta II launch vehicle. This vehicle provides a payload capability (after allowing for the lunar lander) that in size is about 90 inches in diameter and 137 inches in height (assuming the telescope were pointed straight up at launch), and in weight is about 200 kg (440 lbs). As will be suggested, we believe that a 1-m class aperture telescope may be accommodated within these constraints, and a 0.8-m (32-inch) telescope almost certainly could be.

SCIENTIFIC REQUIREMENT

From a scientific perspective, what is proposed is a 1 meter or smaller multi-use lunar telescope, along with attendant control and, in some cases, power and communications, that would accept a wide range of astronomical instruments. With the appropriate instrumentation modules, we envision such telescopes being used for the following purposes:

(1) General purpose imaging and area photometry in the ultraviolet (UV), optical, and near-infrared (IR) regions of the spectrum. This could be done with the same CCD camera on the same telescope, although separate cameras might be considered for widely different spectral regions. With diffraction-limited optics and absence of an atmosphere, such a system would compare very favorably with the current Hubble Space Telescope (HST), would be much more accessible for use (as there would be more of them) and, at some lunar locations, would provide continuous coverage for months on end if desired.

(2) High precision aperture photometry in the UV, optical, and near-IR regions (again, perhaps with the same instrument module on the same telescope or, more likely, with different detectors for different spectral regions). Such high-precision photometry could reveal fine details (at the sub milli-magnitude level) of microvariability in stars, active galactic nuclei, quasars, etc. High-precision photometry could also be used to detect Earth-sized planets transiting other stars.

(3) UV spectroscopy similar to that of the International Ultraviolet Explorer (IUE), but with a much larger aperture and modern instrumentation. This would not only significantly increase the limiting magnitude (and hence access an order of magnitude more objects), but it would improve temporal resolution and, with its greater capacity, serve an even larger number of users.

(4) Narrow field-of-view imaging of planets, the Sun, and other solar system objects. The secondary mirror for this system would be different from that in the other systems to provide a much longer effective focal length. These objects could be frequently monitored at very high resolution .

While many other applications are possible (such as polarimetry, etc.), we mention just one of them as a matter of interest and variety. This is IR imaging and area photometry, including the far-IR. In this region of the spectrum it is not only necessary to cool the detector and filters, but also the entire telescope.

If one of the multi-use lunar telescopes were soft landed in a lunar crater near one of the Moon's poles, it has been suggested that the entire telescope would cool to about 40 K, making it suitable for many IR applications without the need for the sort of helium-filled jacket that was required for the Infrared Astronomical Satellite (*IRAS*, which eventually ran out of the helium coolant).

MULTIPLE INSTRUMENTS

Single telescopes can, of course, be equipped with multiple instruments. The *HST*, for instance, is equipped with five different instruments. If one has only a single telescope, such as the *HST*, then it makes sense to equip it with many instruments. However, if one has a number of multi-use telescopes in operation, it would be more efficient and considerably lower in cost to equip each of them with only one instrument, thereby making each telescope dedicated to a single (but widely utilized) form of astronomical observation.

A case can be made, however, for equipping each telescope with a lightweight CCD camera that could be used not only for general imaging and area photometry in the UV, optical, and near-IR, but also for orientation of the telescope, identification of objects in crowded fields, etc. This would have the added advantage that should the main instrument (such as a photometer or spectrograph) fail, then the telescope could still be operated in an imaging or area photometry mode (both always in great demand). We believe this case to be fairly compelling, so each telescope would have, as part of the standard configuration, a CCD camera with appropriate filters, etc. In the cases where the main instrument were also a CCD camera, we presume that this camera would have a different, probably wider, field of view.

LUNAR LOCATIONS

A primary location for operation would be a manned lunar outpost or a location very near such an outpost. Even in the case of soft-landed telescopes prior to manned re-occupation of the Moon, strong arguments can be made for placing telescopes at or very near the eventual outpost location.

Major Earth-based telescopes are typically utilized for 50 years (or perhaps even longer) before they are considered obsolete. Achieving such a long life requires, of course, that failed modules be replaced, but more importantly that instruments be upgraded as technology advances and astronomical research requirements change.

The main portions of the telescope (optics, mount, mechanical drives, etc.) might last longer than a human lifetime (although the effects of the lunar environment may make such components have a different life on the Moon than we are used to here on Earth). By making control electronics, computers, and instruments modular appendages to the main telescope, they can be replaced, leaving the main telescope intact. Considering the cost of even "low cost" lunar telescopes, utilizing them over long life times has much to commend it.

As human occupancy on the Moon matures, one might expect at least occasional human presence at more than one location on the Moon, and it would make sense to concentrate telescopes at these locations. As astronomy itself will, in the long run, be an important reason for human presence on the Moon (a lunar "growth industry"), it would also be appropriate to consider which locations are most favorable from a purely astronomical viewpoint.

If we assume that telescopes are initially placed at the first lunar outpost location, and further assume that this is near the Moon's equator and near the midline as the Moon faces the Earth, then it might make sense to eventually establish two outpost sub-stations also near the lunar equator but on the back side of the Moon, each about 120° in longitude from the near-side outpost. This would provide continuous coverage of objects as they pass from a telescope at one outpost to another as the Moon rotates. In making many types of astronomical observations, there are considerable advantages to be gained if objects can be observed more frequently than the once per month period allowed by the lunar cycle. Many astronomical changes of interest occur on time periods of hours or days, and the ability to sample observations at these intervals adds much to the science that can be accomplished. Some types of observations, such as "stellar seismology," require continuous observations that go unbroken for many weeks or even months.

One of the disadvantages of a minimal multi-telescope network for obtaining continuous or frequently sampled observations is that should any one of the telescopes on the network fail, then a sizeable gap can be introduced that essentially ruins the continuous nature of the results. This problem can be overcome, however, by having overlapping or redundant coverage. Furthermore, one must maintain calibration between the participating telescopes (instruments) in the network.

Another approach to obtaining continuous or frequently sampled observations is to locate a telescope at or near one of the Moon's poles. In these locations, the sky just wheels around in essentially a great circle. The price of this advantage is, of course, loss of complete sky coverage — one can now only look at about half of the celestial sphere, but one can look at it essentially all the time. A special case is a location very close (essentially at) the poles (and in a continuously shadowed crater), and this is an advantageous location where one requires a "cold" telescope for IR observations. For other situations, however, a location down from the poles roughly 10–15° may be advantageous.

By moving down a few degrees from the poles, one can be assured of having sunlight (for power purposes) during half of the lunar cycle. The Moon is tilted from the orbital plane about the Sun by just a few degrees (unlike the Earth which has a large 23.5° tilt). To be assured of sunlight, one has to move far enough south to not have the Sun blocked by local mountains on the horizon, etc. While providing full power when it is up, the Sun would always be close to the horizon during the day, making observations away from the Sun possible in almost the entire available sky (and minimizing problems of scattered light in the telescope, to be discussed later).

If one is located on the near side of the Moon and desires direct communications with Earth, then one has to go further away from the poles than just a few degrees, or the Earth will, during the 18 year Earth-Moon cycle, dip below the horizon. By going about 15° away from the poles on the midline of the Moon facing the Earth, one can be assured of both continuous power during the lunar day, and continuous communications with the Earth. If one is off the midline, then one might have to go further yet from the poles.

It might be noted that while the Earth will always be visible, it will also always be fairly low in the sky, not too far above the horizon, and thus observations away from the Earth will be possible for most of the sky at all times. With the Earth always low in the sky, scattered light problems from earthshine might also be reduced.

Another advantage of moving down somewhat from the poles is that a strip of sky at the celestial equator would be visible by telescopes near both the northern and southern lunar poles. This simplifies calibration of instruments at these two locations, as they can both use the same set of equatorial calibration objects.

The arguments given above for near polar locations can also be applied to locations on the far side of the Moon. The advantage of the far side is that one does not have to contend with earthshine reflecting through baffles and onto detectors. Thus, one might be able to observe fainter objects than would otherwise be possible. The disadvantage is that direct communications with the Earth would not be possible, necessitating relayed communications via a lunar satellite or a satellite parked at a Lagrange point.

It might be noted that it is not known yet whether radiation-induced noise in detectors or earthshine bouncing off multiple baffles will be more limiting when it comes to long integrations trying to go to the faintest possible limits. There have been discussions of using stacked CCD arrays (for instance) as coincidence detectors to reduce the effects of unwanted radiation during long exposures.

OPTICS AND OPTICAL ASSEMBLY

We considered HexTek gas-fusion optical blanks (manufactured by a firm in Tucson) as an example of what could be done, weight-wise, for a 1-m telescope. With current manufacturing techniques, HexTek can produce a 1-m mirror blank that weighs 90 kg. With some (but not extensive) efforts, they could produce a mirror blank that weighs 70 kg. HexTek uses a Pyrex-type material.

While very low-expansion materials could be used for the optics, the temperature range over which they have low "expansion" is somewhat limited. The late Harlan Smith pointed out that there might be some merit to "living with" the day/night thermal cycle and simply refocusing after each transition. In maintaining the philosophy of building many systems, keeping optical prices modest may have some merit, and Pyrex-type materials are certainly inexpensive. Harlan also suggested that fused silica is known to keep its optical shape over a wide temperature range, and therefore should be given consideration. An alterative to living with the wide temperature swings in the optics is to work at reducing the swing via environmental control — perhaps with appropriate "shutters," etc. Whatever approach is taken, there should be no sacrifice in optical performance. Diffraction-limited performance would be expected at all wavelengths (except perhaps the UV).

The approach being taken to focusing in variable temperature environments by the Lunar Transit Telescope (McGraw 1992) is to have an assembly between the primary mirror and instrument at prime focus that is designed to have, overall, essentially a zero expansion coefficient. This approach is worth considering. Another, more traditional approach would be to have a focus mechanism, such as a linear stepper motor, move a secondary mirror (or move the instrument payload if at prime focus). Some limited tilt adjustment of the secondary might also be worth considering. After going to all the trouble and expense to place a telescope on the Moon, we must be assured that optical performance will be superb. If this requires controllable adjustments, so be it.

DUST COVERS, BAFFLES, AND TUBE ROTATION

For soft-landed telescopes, the payload shroud will presumably be ejected in Earth orbit, and this will leave the telescope out in the open when landing occurs. Dust kicked up during the powered landing (or by other nearby landings) must not be allowed to contaminate the optics. Thus, the optics need to be enclosed by a tube with a dust cover. If there were never to be any later nearby landings, then the dust cover could be opened or ejected a single time. However, if later landings or nearby human activities were expected that could contaminate the optics with dust, then it might be appropriate to be able to command the dust cover shut. If so, some redundancy would be called for here, as it would be a shame to have an otherwise functional telescope with a dust cover that couldn't be opened.

It might be briefly noted that there is some possible advantage to leaving the telescope on a lunar lander from the viewpoint of getting the telescope somewhat above the surface of the Moon. There have been reports of dust levitation near the surface during the day-night transition periods.

Interior baffling to reduce stray light must be given very careful consideration, as this will probably limit how faint the system can observe during the lunar day. It might also be noted that optics with a high surface polish produce less stray light. Furthermore, avoiding dust on the optics (as above) reduces stray light.

Another strategy for reducing stray light is the use of a Sun shade (and perhaps an Earth shade). The end of the optical tube, extending out beyond the secondary mirror (or prime focus instruments), can be cut at an angle, and the entire optical tube, optics, instruments, etc., rotated as a unit to place the high side of the cut towards the Sun (or Earth). This then places the entire opening from the tube to the sky in the shade. Rotation of the optical assembly can serve other purposes as well, such as positioning an instrument to some preferred orientation.

In a non-equatorial mount (which is likely to be the case), the tube can be slowly rotated to counter field rotation in long exposures. Failure to do this can, for long exposures, smear the image via field rotation. As previously discussed in the context of optics, one should not do anything to degrade optical performance, and this includes the effects of field rotation. The ability to take very long exposures (to the limit imposed by stray light or background radiation on the detector) is vital for space telescopes. One of the many potential advantages of a steerable (controllable) telescope over a transit telescope is that the objects being investigated can be kept exactly positioned on the detector for long periods of time, allowing fainter objects to be observed than might be possible with non-tracking systems.

There are other, more exotic approaches that might be considered to reducing stray light from the Sun or Earth. For systems located near (but not at) the Moon's poles, the Sun and Earth would always be near the horizon on one side of the telescope (the side towards the Moon's equator). A large "pop up" shade could cover this portion of the sky at all times except when observations of this small fraction of the sky were desired. To reduce scattered light from the ground around the telescope, it has been suggested (perhaps with slight humor) that carbon black could be sprayed over the Moon's surface for a few hundred meters all around the telescope.

TYPE OF MOUNT

We have primarily considered three types of mounts. These are (a) equatorial, (b) alt-alt, and (c) alt-az. Each of these will be discussed below.

The primary advantage of an equatorial mount (when properly physically aligned) is that the field does not rotate during long exposures. The main disadvantage of an equatorial mount is that its primary axis must be aligned to be parallel to the Moon's axis of rotation. A secondary disadvantage is that the "tilt" of an equatorial mount would depend on the Moon's latitude where it was placed. This can mean, for instance, that a mount near the Moon's equator would be very different from one near the poles.

For soft-landed systems, the requirement for equatorial alignment would suggest that the soft lander would have to orient itself on landing to within a few degrees of some prescribed orientation. Furthermore, after landing, observations would have to be made, and the results from these then used to physically adjust the altitude and azimuth of the mount, albeit over some limited range and, presumably, just a single time (or perhaps a few times early on to allow for some fine tuning). Once positioned, such adjustments would no longer be required. At the Lunar Outpost (or other manned or occasionally manned sites) the astronauts could be called on to make telescope alignment adjustments but, other things being equal, it would be best not to add to their busy work schedule out on the lunar surface. While not totally eliminating equatorial mounts from consideration, we consider such requirements to be a severe handicap.

It might be noted, before leaving the subject of equatorial mounts, that at the equator, an alt-alt mount (if oriented north-south and leveled) is an equatorial mount. Similarly, at the poles (if leveled) an alt-az mount is an equatorial mount.

An alt-alt mount (assumed not equatorial) suffers from field rotation, so would probably need a third axis, perhaps as discussed above in terms of rotating the entire optical/instrument assembly. On the other hand, it can be placed in any orientation, and need not be completely level. Thus, it could just be soft-landed with random orientation. Observations could be made to figure out the orientation, and the pointing equations could subsequently use this information to properly point the telescope thereafter.

The alt-alt mount, unless it is heavily counterbalanced, has two "blind spots" (similar to "gimbal lock") at two opposite locations on the horizon. On the other hand, as one rarely wants to look at the horizon, these are good locations for blind spots.

An alt-alt mount, however, is bulkier than the alt-az mount to be discussed below. Furthermore, this bulkiness is in a horizontal direction, and thus is not a good form factor for sitting on top of a lunar lander in a shroud on a launch vehicle. This is a serious handicap that, while not totally eliminating such a mount from consideration, makes it seem somewhat unattractive to us.

Finally, there is the alt-az mount. As with all non-equatorial mounts, it suffers from field rotation. It has a single blind spot near the zenith (straight overhead). This is a true disadvantage, but (except at or very near the Moon's poles) one can usually catch objects that transit this blind spot either before or after the blind spot (as the Moon rotates). Typically the blind spot is just a degree or so in diameter, a very small percentage of the total sky available, although the size of the blind spot is dependent on the available slew rates, the Moon's rotation rate, and the location of the telescope.

The alt-az mount may be placed in any orientation, needs only to be very roughly level (off 10° or so should not hurt), can be placed at any latitude, and nicely matches the form factor available for launch. At this point in time we favor the alt-az mount type.

PLANNING AND SCHEDULING

A lunar telescope will be an extremely valuable resource, and requests for time on the telescope will greatly overwhelm the number of observing slots that are available. Telescope observing time will be a precious commodity that must be carefully apportioned among members of the scientific community. The management infrastructure that supports the day-to-day operation of a set of lunar telescopes must be able to allocate telescope time in an efficient, flexible, and safe manner.

Efficiency is important since we want to make maximum use of the telescope resource. Of course, efficiency is a goal both for the use of the telescope and the process by which the resource allocation is actually carried out. Efficiency in the first sense affects how much of the telescope's time is spent making useful observations; efficiency in the second sense affects how much of the scientists' time is spent wading through the bureaucracy that surrounds the telescope. We need a system for allocating telescope observing time that is efficient in both senses.

Flexibility of scheduling is also extremely important. Different users will have different scientific goals and strategies for achieving those goals. A telescope scheduling system must cater to a wide variety of observational requests. For instance, a scientist might simply require brightness data on a particular clean eclipsing binary at a precise moment. Such an observing goal is easily translated into a request for telescope observing time, and can be directly and competitively scheduled with other such requests. However, another scientist might want to find flare stars, and might decide that the best way to do this is to iteratively scan a set of stars, comparing brightness values with preset norms, detecting deviations, and locking onto and observing any star exhibiting a significant deviation. Such a scientific goal is not easily translated into a specific observation request at a specific moment in time. The scientist in this case must be allowed to write a procedural observing strategy for the telescope, and the procedure must be run on the telescope for a given interval of time. The point here is simply that different scientific goals engender different sorts of telescope requests, and a telescope scheduling system must be flexible enough to accommodate a wide variety of request types.

Safety is also an important consideration in telescope scheduling. Safety relates to the way that a particular telescope is used; essentially, a schedule must operate the telescope without breaking it. For example, it is clearly unsafe to point sensitive telescope instruments at extremely bright objects, and it is also typically unsafe to drive mechanical linkages through their limits of motion. All safety constraints for a given telescope must be articulated in advance, and all observational schedules must be formulated so as to respect these constraints. This is particularly important when local telescope control is surrendered to an astronomer's procedural observing strategy (as in the flare star example).

Initial deployment of telescopes on the Moon is obviously a rather expensive venture, so we seek to minimize the recurrent operational costs by using automation whenever appropriate. Current technology provides for a reasonable level of automated scheduling (e.g., Johnston 1989; Liu 1988), and there is a significant amount of on-going research dedicated to extending this functionality in various ways. In particular, there are now some systems that provide a combination of scheduling and planning (Currie and Tate 1991; Drummond and Bresina 1990). The key idea here is that a scheduler can only sequence a given set of observations, while a combined planner-scheduler can reason about alternative observations that might be made in order to satisfy a given scientific goal. There is little doubt that planning and scheduling the operations of lunar-based telescopes will offer significant new challenges for planning and scheduling systems, but it seems clear that existing levels of automation promise to address a significant part of the problem.

Telescope planning and scheduling automation can reside in computers remotely on Earth, or local to the telescope on the Moon itself. As we discuss below, the telescope will have limited power, so a large computer at the telescope is probably out of the question. This Earth-Moon automation split imposes certain interesting limits on the amount and nature of the computation that can be done at the telescope. For instance, it might not be possible to install extremely sophisticated scheduling software at the telescope itself. Instead, it might be necessary to do all planning and scheduling using Earth-based computers, and to use the local telescope computer only as an "execution engine" to carry out individual observations and monitor results. Errors of execution requiring significant replanning might necessitate further computation on Earth. Of course, the details of the Earth-Moon automation split depend on the sort of computational horsepower available at the telescope, and this in turn depends on the amount of electrical power available. In theory, we would like to place as much of the automation as possible at the telescope, to allow maximally flexible modes of telescope operation and control. In practice, however, the amount of automation we can actually place at the telescope will depend on the available electrical power.

Perhaps the key point to be made with respect to scheduling and command is that the approach we are advocating is much simpler than most space telescope approaches (such as that of *HST*), and is similar to the operation of current automatic telescopes on Earth (at Mt. Hopkins and elsewhere). In this approach, each telescope has a single, dedicated instrument. One astronomer, the Principal Astronomer (PA), is responsible for the overall scheduling and use of the telescope, and is also responsible for initial data reduction, monitoring quality, etc. Since all the astronomers using a given telescope are all using the same (single) instrument, their observations will tend to be similar, and thus scheduling, quality control, and initial data reduction all are greatly simplified compared to multiple-instrument telescopes. This approach has worked out very well on Earth, and PAs have been able to handle the scheduling, quality control, and initial data reduction for an entire telescope as a part-time job.

POWER STRATEGY

Lunar telescopes should operate day and night (and especially at night!). We have rejected any notion of having telescopes that would operate only during the lunar day when relatively large amounts of power would be available from solar cells. The exception to this, of course, are any telescopes devoted primarily to observing the Sun itself.

A potential advantage to lunar telescopes will be their ability to observe very faint objects (even with modest aperture telescopes). To sacrifice this advantage by operating only in the daytime when, almost certainly, stray light will hurt the faint limit, would be sad indeed. Design consequences follow from this decision, however.

If power were provided by a radioisotope thermal generator (RTG), then power could be the same day and night. This might be a necessity for a system located in a permanently shadowed crater near the pole, as discussed earlier.

An alternative approach is solar cells and batteries. Batteries are heavy, but we are willing to accept this penalty in order to achieve our other goals, even if it results in a somewhat smaller aperture telescope. The use of batteries (and their weight penalty) places a very high premium on achieving low power consumption, especially at night.

To some extent, power consumption can be reduced at night by scheduling faint objects for night time observation (when they are best made anyway), as this reduces telescope movement, communications, and computations. Movement lengths can also be reduced via appropriate scheduling (i.e., not running back and forth across the sky).

Low-power environmental control can be optimized for night time efficiency. For instance, electronics could be packaged in insulated containers such that naturally consumed power will keep them at the proper temperature. During the day, when much greater power is available, active cooling might be considered.

How low-power consumption might be achieved during the night remains to be determined, but it is our hope that, on the average, it can be brought down to just a few watts. We have set, as a goal, a total system weight (including telescope, control, solar cells, batteries, communications, etc., but not including any lunar lander) of less than 200 kg. We are assuming, at this point, that almost 45 kg would be devoted to an RTG or to solar cells and batteries, suggesting the importance we attach to night time operation.

COMMUNICATIONS

We are assuming that communications will consist of three somewhat separate areas. These are the command uplink, the science and engineering downlink, and the Earth network.

The command uplink would be a low bandwidth, low baud rate channel for sending commands to the telescope (objects to be observed, etc.). As the telescope itself will be quite intelligent, these commands can be at a fairly high level and thus be brief. We assume that it will only be necessary to uplink commands on infrequent, perhaps prescheduled intervals.

The downlink would send both scientific observations and engineering (housekeeping) data. Again, we assume that the downlink would operate only on occasion, again perhaps at prescheduled times.

As controllable telescopes are capable of very precise pointing, consideration should be given, for the main downlink, to using these capabilities to reduce power consumption. Specifically, a highly directional, high frequency (perhaps Ka band) antenna might be fastened to the telescope itself. When it was time for a prescheduled communication to take place, the telescope would stop regular observations, and not only point to the Earth, but point to a very specific place on Earth.

An extension of this idea is the realization that, at optical frequencies, a telescope is, of course, a very highly directional antenna in its own right. At a prescheduled time, the telescope would stop its normal observations, and point at and track a highly specific location on Earth (where a larger optical telescope would also momentarily stop its normal astronomical observations). A simple detector selector would move to communications (transmit), and a very low-power optical diode laser would dump the information to Earth. The detector selector might then go to the communications receive port and new instructions might be uploaded from Earth.

JPL is considering something similar for very deep space communications (Jupiter and well beyond), where the spacecraft would be equipped with about an 8-inch telescope. If this works, imagine how much easier it would be with a 1-m class telescope on the very nearby Moon!

Whatever type of downlink is used, there is a tradeoff between the amount of data stored and how frequently data needs to be sent to Earth (and to some extent how many Earth stations are required). For instance, at one extreme, one might store up data for a full Earth day and downlink only once a day. This would allow the use of a single Earth station, but would require storage of large amounts of data, at least in the case of imaging systems. At the other extreme, one might insist on not storing any data at all, and downloading data as, for instance, it was clocked off a CCD chip. This latter extreme, however, would require continuous communications.

LUNAR LANDER REQUIREMENTS

We envision a 0.8-m to 1.0-m telescope system, complete with power, communications, etc., weighing a bit less than 200 kg. For most locations, the landing accuracy requirements would be very modest — perhaps measured in miles.

The landing orientation, at least in some configurations we are considering, could be entirely arbitrary. Furthermore, the system would need to be only approximately level — perhaps within 10° or so.

Landing G forces would, hopefully, not greatly exceed launch G forces. (If they did, this would increase system weight somewhat, or result in a slightly smaller aperture.)

We would prefer that the system be permanently mounted on the lander and stay on the lander after landing.

EARTH-BASED PRECURSORS

We strongly recommend that a fair number of remotely located, Earth-based precursors be utilized in "routine" astronomical observations by a wide cross section of researchers and students for at least a couple of years before somewhat similar systems are placed on the Moon. This will assure that all the bugs have been worked out of the systems, that the user interfaces are made as friendly as possible, that direct operation by users without intermediaries is practical, and that the systems are highly reliable.

The remote location or locations chosen for the Earth-based precursors could be better than any current location occupied by telescopes here on Earth. Harlan Smith, before his untimely death, pointed out that there is a mountaintop in northern Chile (elevation 20,000 feet) where it almost never rains (about once every 10 years), where it is unusually cloud free (perhaps the most cloud-free location on Earth), and where there is a good road to within 400 feet of the summit (the highest road on Earth). This location looks like the Moon. Astronomical observations made from this inhospitable and remote site might be of the highest possible quality anywhere on Earth.

There have also been discussions of an 18,000 foot elevation plateau not far from the Earth's South Pole. Long, continuous observations from this location would be possible, and it certainly qualifies as being one of the most remote and inhospitable places on Earth, perhaps more inhospitable than the Moon itself.

By operating a number of precursor lunar telescopes at very remote and inhospitable locations on Earth, considerable confidence could be developed in the operation of such systems while, at the same time, making higher-quality observations than are possible from any current location on Earth.

REFERENCES

Currie, K., and Tate, A. 1991, "O-Plan: The Open Planning Architecture." *Artificial Intelligence* (North Holland), **52**, 1.

Drummond, M., and Bresina, J. 1990, *Fifth IEEE International Symposium on Intelligent Control* (Philadelphia: IEEE Computer Society Press).

Liu, B. 1988, "Scheduling Via Reinforcement." *Artificial Intelligence in Engineering*, **3**, 2.

Johnston, M. 1989, "Reasoning With Scheduling Constraints and Preferences." *SPIE Technical Report 1982-2* (Baltimore: Space Telescope Science Institute).

McGraw, J. T. 1992, *These Proceedings*.

THE LUNAR TRANSIT TELESCOPE: A MOON-BASED STRIP SEARCH OF THE UNIVERSE

JOHN T. McGRAW
Steward Observatory, University of Arizona, Tucson, Arizona, 85721

ABSTRACT The Lunar Transit Telescope (LTT), a unique, non-moving astronomical survey telescope, is described. The autonomous operation of the soft-landed telescope is discussed and scientific programs enabled by LTT data are listed. The moon is discussed as a viable spacecraft for this mission. In particular, perturbations to the moon's stable rotation are shown to negligibly impact LTT image quality. The stable bearing properties of the lunar regolith are shown suitable for LTT support, as well. Meteoroid impacts on the telescope optics are shown to be a minor problem. A cosmic ray mitigation system which ensures data integrity is described. An opportunity for international collaboration in the reception of LTT data is proposed.

INTRODUCTION

We are now into the fifth decade of space physics and astrophysics accomplished by launching rockets carrying spacecraft or probes. The most common scenario for space-based science involves mounting one or more instruments on a spacecraft and placing the spacecraft into an appropriate orbit. Every spacecraft has certain characteristics, such as rotational or active stabilization, thermal control and communications capabilities, with parameters most often determined by the requirements of the mission. Very often it is the spacecraft which is the principal limiting mission attribute with respect to cost, capability or lifetime. There exists, however, for a wide range of scientific purposes, an ideal but recently neglected natural spacecraft - the moon. In this paper we shall discuss the use of the moon as the spacecraft which carries a unique astronomical telescope which, once emplaced, remains motionless, yet observes an immense volume of the Universe.

The telescope is the Lunar Transit Telescope (LTT) (McGraw 1990, McGraw and Benedict 1990), which utilizes the moon for many purposes. The moon provides:
- a seismically stable telescope mount
- a stable, well-determined rotation rate
- shielding from cosmic rays and micrometeorites over half the sky
- capability for continuous communication directly to earth

- large area for deployment of efficient thermal shields, etc.
- access to the entire electromagnetic spectrum
- high resolution direct imaging unaffected by an atmosphere
- high earth orbit
- cryogenic operation
- long mission lifetime.

These attributes in a single man-made, earth-orbiting spacecraft would be prohibitively expensive, thus arguments can be made that, for appropriate missions, the moon represents a very cost-effective vehicle. It simply needs to be instrumented! Figure 1 shows artist Nancy McDonough's concept of LTT as an early, soft-landed mission to the moon.

Clearly, there are negative aspects to using the moon as a base for observations (*cf.* Burns 1991, Vaniman, *et al.* 1991 [VHROM], Taylor 1990). Some problems to be overcome include the effects of lunar dust, possibly electromagnetically levitated, the relatively high cosmic ray flux, micrometeorite impacts and large diurnal temperature variations. These are, however, surmountable problems, some suggested solutions to which we discuss in this paper.

The description of the moon as a spacecraft is, of course, context sensitive - to a planetary geologist the moon is a rocky body which hold secrets about planetary systems and should be studied *in situ*, possibly utilizing humans on the surface. In this paper we assume a scenario for utilizing the moon at the earliest possible time to accomplish scientific programs which cannot be done in any other way. Early use virtually certainly implies no people. It also implies preference for the simplest early experiments which simultaneously justify their own existence and provide data on operational and environmental problems associated with utilizing the moon for manned and unmanned scientific and exploration programs. The LTT fulfills these criteria for an early lunar mission in that it is very simple, requiring no motions once emplaced, but returns astronomical data of extremely high value for use by the entire astronomical community. LTT complements HST and other of the Great Observatories and other astrophysical missions. It utilizes current and emerging technology and, therefore, could be developed and emplaced on a rapid timescale.

THE LUNAR TRANSIT TELESCOPE

LTT is unique in that, except for initial alignment mechanisms, it has no moving parts. The basic principle upon which LTT operates is schematically illustrated in Figure 2. The telescope utilizes frame transfer charge-coupled devices (CCDs) in the focal plane to produce an image of the sky using the "time-delay and integrate" technique. For single element detectors, this is also known as the "push broom" detection scheme. A mosaic of CCDs in the LTT focal plane is accurately aligned with columns, that is, the "vertical" shift direction, running from east to west. The vertical clocking rate is made to match the apparent sidereal rate in the focal plane. In a frame transfer CCD,

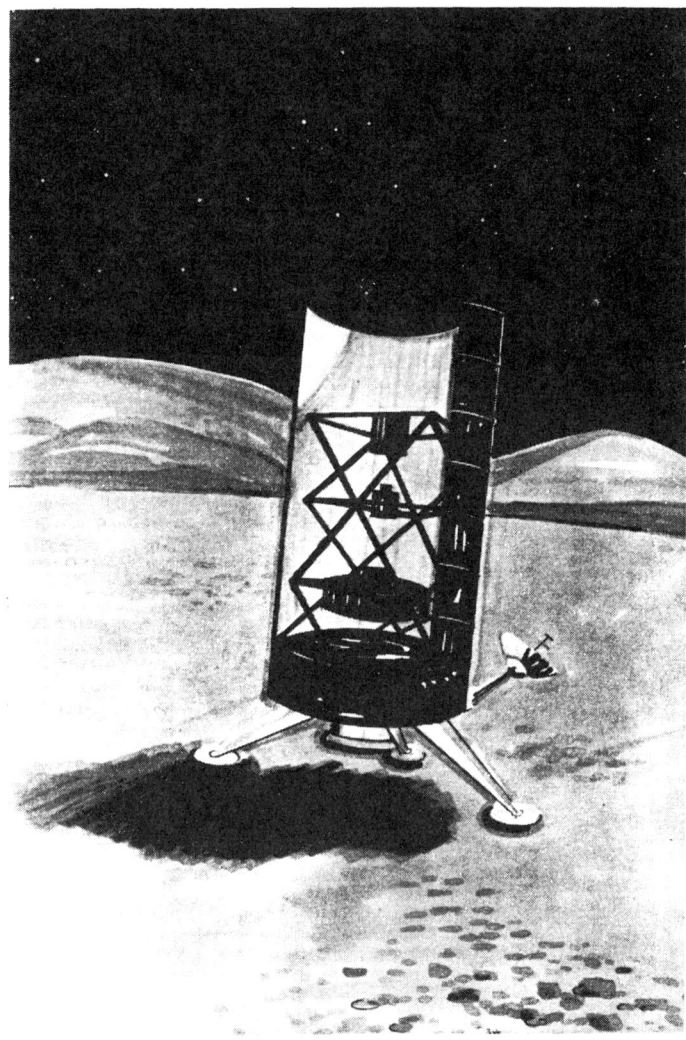

FIGURE 1. Artist's concept of the Lunar Transit Telescope (LTT) as a soft-landed mission. The two meter, f/2.2 telescope produces a high resolution, large area survey of the Universe in bandpasses ranging from 0.1 to 2.5 microns. Data are transmitted directly from the LTT site near the lunar limb to Earth, where they are processed to produce a "strip search of the Universe" which addresses many fundamental astronomical problems.

as charge is clocked into the horizontal shift register from the last row of pixels, all other rows are simultaneously moved one pixel in the same direction, as well. This is very much like the operation of an old-fashioned "bucket brigade." The result is that the charge image being created in each CCD remains precisely under the optical image formed by the telescope for the duration of the transit time across the CCD. Thus, the column length of the CCD determines the integration time for each object and the row length determines the north - south angular extent of the surveyed area. As charge reaches the horizontal shift register at the apparent western edge of the CCD, it is rapidly read out, digitized and stored in memory. The entire horizontal readout sequence is carried out in less than one vertical clock period. The wavelength and bandpass observed by each CCD is determined by a fixed filter in front of it. In this manner, each CCD produces a "seamless" image of a strip of the sky which is as long (east - west) as clocking continues and as wide (north - south) as the angular extent of the CCD.

As an example of this technique, assuming existing CCDS, for a 2048 X 2048 CCD with pixels which subtend 0.1 arcsecond on the sky, the vertical clock period, determined by the lunar rotation rate, is about 0.2 second. Assuming a readout time per pixel of 10 microseconds, the 2048 pixel horizontal shift register can be completely read out in about 0.02 second - much shorter than the vertical shift period. The integration time on the sky is 2048 pixels X 0.2 sec/pixel, or about 6.6 minutes. As data are digitized, they can be transmitted directly to earth where the image from the entire mosaic can be digitally "stitched" together.

SCIENTIFIC PROGRAMS OF THE LUNAR TRANSIT TELESCOPE

The most exciting aspect of LTT is the breadth and depth of the scientific programs addressable by its data. This telescope really accomplishes a "strip search of the Universe" not only in terms of the geometry of the survey, but also in terms of rigor and completeness. Table 1 gives parameters for a "strawman" two meter LTT.

With this definition, LTT will survey about 2% of the sky - but to limiting magnitudes which ensure that we obtain a "fair" sample of the Universe for an incredibly large number of programs.

The scientific programs of LTT can be set in context by a simple overview:

• The LTT will enable an imaging survey of the Universe with higher angular resolution and broader wavelength coverage over a larger fraction of the sky than has ever been attempted or can be attempted until this telescope is placed on the moon.

• LTT enables a deep, unbiased, statistically significant complete survey of virtually every type of object. It is an ideal instrument for statistically describing the content, structure, texture and evolution of the Universe.

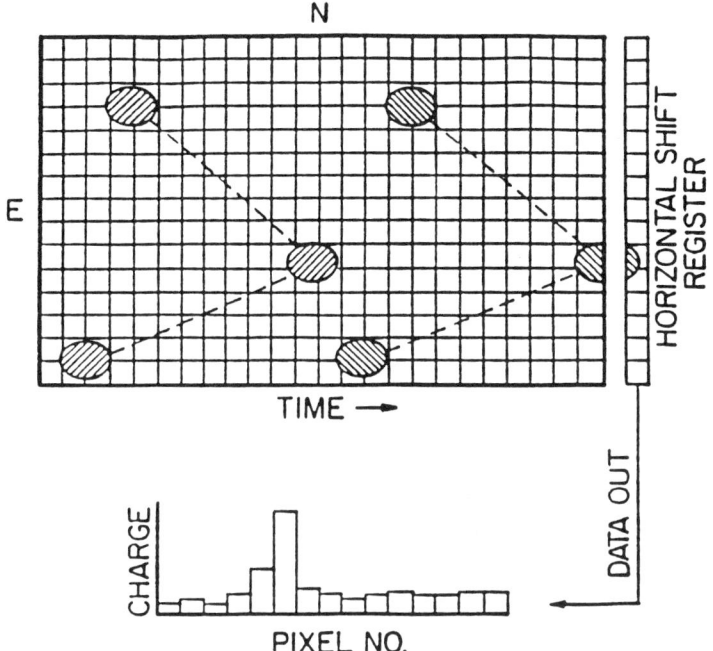

FIGURE 2. The LTT utilizes as the "drive" of the telescope the readout system of a frame transfer CCD. In this readout system, a clock sequence in the "vertical" (time) direction of the CCD results in the last (western) row of charge being transferred into the horizontal shift register (HSR). At the same time, every other row is shifted towards the HSR, ie. to the west. The shift rate corresponds to the apparent sidereal rate. As data "fall off" the west edge of the CCD they are rapidly shifted out, digitized and stored. A star field is shown at two times as it transits a CCD in the LTT focal plane.

- LTT surveys literally hundreds of millions of objects, allowing selection of specific targets for followup with other space and ground based telescopes. We learn about the content and physical processes in our Universe by studying extreme cases - LTT observes so many objects so completely that it is certain to discover extreme examples in virtually any physical domain.

- Because LTT surveys a large volume of the Universe to faint limiting magnitude over more than four octaves of the electromagnetic spectrum, it is a virtual certainty that the most significant contribution of this telescope cannot be predicted - it will be a serendipitous discovery.

TABLE 1

THE LUNAR TRANSIT TELESCOPE

Detectors:

Si CCDs (Ultraviolet and Optical)
HgCdTe with CCD Readout (Infrared)
0.1 Arcsecond Pixels
Diffraction Limited Beyond 2 Microns
CCDs are Mosaicked in the Focal Plane

Field of View:

All Reflective Optics (Paul-Baker Three-Mirror System)
2 Degree (Wide, ie. N-S) Field of View
575 - 720 Sq. Degrees Surveyed Each Lunar Day
Precession of Lunar Orbit Widens Observed Strip to $5°$
More Than 75% of Strip Observed Each Lunation

Bandpasses:

5 Bandpasses (R = 0.1 - 0.25), One Bandpass Per Octave

Estimated Limiting Magnitudes
  (S/N = 10, single 360 s integration):

V = 27	Estimates scaled from HST WF/PC and
H = 25	NICMOS calculated performance and
K = 24.5	CTI measured performance. Assumed quantum efficiency of 60%.

Average Time on Target (18.6 Year Mission):

Average Annual Integration Time = 3500 seconds

To provide an instrumental perspective, LTT and the familiar Hubble Space Telescope (HST) are very similar in size. If the HST achieves an operating efficiency allowing ten WFPC survey images to be observed each day (two fields in each of five bandpasses), the annual sky coverage will approach 1.3 square degrees with no overlap. The LTT will inventory more than five hundred times this area per year with equivalent or better resolution and sensitivity. Multiple observations will allow addition of data to produce deeper, higher signal-to-noise measurements, as well as the ability to detect and measure variability and motions.

We briefly describe current scientific programs which can be addressed by LTT data.

Extragalactic Astronomy

•   Imaging/Morphology of Distant Galaxies - At isophotal R magnitudes greater than 25, we may observe more than 10,000 galaxies per square degree and the sky rapidly becomes confusion limited in galaxies. For galaxies brighter than this, surface photometry and morphology can be accomplished for field and cluster galaxies.

•   Galaxy Evolution - Multiple bandpass imaging of galaxies to the confusion limit will allow investigation of galaxy evolution at $z > 1$.

•   The Distance Scale - Discover and measure the light curves of Cepheids to about 40 Mpc. Image the brightest blue and red supergiants in galaxies to about 40 Mpc for use as secondary distance indicators. Calibrate using Cepheids.

•   Search for Gravitationally Lensed Objects and the Lensing Material - Approximately 6 - 7 lensed quasars have thus far been discovered with separations in the arcsecond range. The frequency of occurrence and separation of lensed objects give estimates of the mass density of the universe. Repeated (monthly/annual) observations can yield information on individual lines of sight if variability is detected and correlated between the images.

•   A Search for Cosmic Strings - Cosmic strings have been postulated as "seeds" for the observed structure of the universe. A search for clusters of identical pairs of galaxies lensed by a continuous string would demonstrate their existence or limit their space density.

•   Evolution of Clusters of Galaxies - The LTT will observe many clusters of galaxies to relatively large redshifts ($z > 3$). With the LTT one can investigate the color and morphology of cluster members, as well as cluster spatial correlations as a function of redshift. It will also be possible to examine biasing by comparison of cluster and field galaxies.

•   Distribution of Extragalactic Background Light - Determine the color and distribution of extragalactic background light and use these measurements to limit luminosity evolution and extend galaxy clustering to higher redshifts. Resolve Olber's Paradox.

•   Active Galactic Nuclei - The majority of the energy distribution of AGNs will be sampled in each image. Variability as a function of wavelength can be investigated with time resolution of one lunar day.

•   Quasars - This survey will isolate by color and variability a sample of quasars for an investigation of space density, evolution, obscuration and environment.

- Geometric (Loh and Spillar, 1986) Test of q_o - With color information and an image format, the area and depth of the LTT galaxy survey will allow geometric tests of q_o to be made. If redshifts can be obtained by introducing a grating or grism, the test becomes rigorous.

Solar System Astronomy

- Kuiper Belt Comet Search - The Kuiper Belt is a disk of comets reaching from the orbit of Neptune to about 50 AU. This disk has been postulated as the source of short-period (Jupiter family) comets.

- Studies of Asteroid Belt Populations and Structures - The LTT would sample one kilometer and smaller asteroids throughout the asteroid belt and compare them statistically with near-earth and other main belt asteroids. A 100 m diameter asteroid will have a magnitude of approximately 27 at a distance of 3 AU.

- Investigations of Individual Asteroids and Comets - Distant, faint, slow-moving asteroids and comets will occasionally recur in the LTT field of view. These objects can be well studied over a broad spectral range.

- Composition and Distribution of Zodiacal Dust - Reflected sunlight and thermal emission in the infrared can be used to investigate zodiacal dust, a major source of the natural background in the optical and infrared.

Galactic Astronomy

- Search for the Faintest Stars and Brown Dwarfs - The faintest stars and brown dwarfs will be visible at V in a volume greater than 250 cubic kiloparsecs, ie. in a sphere of radius four kiloparsecs.

- Measure Motions of Stars/Kinematics of the Galaxy - Repeated observation of the field will allow derivation of proper motions at the sub-pixel level for stars of sufficiently high S/N e.g., V < 27. At this limit, single measurement errors are approximately 1 - 2 milliarcsecond. A parallax program for stars at distances of hundreds of parsecs can be initiated.

- Galactic Structure - Multicolor Star Counts - The surveyed strip will include a wide range in galactic latitude. Multicolor star counts to faint limits (V > 28) will define the scale heights of virtually every stellar population and allow an investigation of the galactic potential.

- The IRAS Infrared Cirrus - Measure the Optical and Infrared Colors of the ubiquitous IRAS IR cirrus, which is detectable in visible light. Measure the colors and smallest spatial scales of the IR cirrus and correlate to extinction derived from foreground/background stars.

Relativity

- <u>The Eddington Experiment Revisited</u> - At the limb of the sun light rays are deflected by about 1.7 arcsec, decreasing inversely as the impact parameter to approximately 4 milliarcsec normal to the Earth - Sun line. Utilize LTT astrometric measurements of thousands of stars over an 18.6 year baseline to map the sun's point-source gravitational potential.

Spectrophotometry of the Surveyed Strip

- A diffraction grating or grism can be introduced into the beam of the LTT to produce low resolution "slitless" spectra of virtually every surveyed object to a limiting magnitude two to five magnitudes brighter than the imaging limit (resolution dependent). This implies that galaxy and quasar redshifts can be measured for a very large number of objects. This is viewed as a logical upgrade to the LTT following an imaging survey, or as the function of a second Lunar Transit Telescope.

SPACECRAFT LUNA

We present order of magnitude arguments that the moon, in many respects, makes an excellent spacecraft on which to mount astronomical telescopes.

The Moon as a Telescope Mount

Clearly, given the operational principle of LTT, the most critical attribute of the LTT spacecraft is its rotation and the orientation in space of the rotation axis. The lunar sidereal rotation period is 27.322 ephemeris days, which, because the moon is in a synchronous orbit, is almost exactly the orbital period about the earth, as well. The lunar orbit is inclined to the ecliptic by 5.145 degrees and the lunar equator is inclined to the plane of its orbit by 6.683 degrees. Thus, the lunar equator is inclined to the ecliptic by 1.542 degrees. This geometry is shown in Figure 3. The lunar orbit precesses in a retrograde direction in the plane of the ecliptic with a period of 18.61 years. This is the dominant effect modifying the direction in space of the lunar rotation axis. There exist small physical librations of 0.02 degrees in longitude with a period of one year and 0.04 degrees in latitude with a period of six years (Allen 1973). Only the longitudinal libration will affect LTT and the form of the variation is sufficiently well known that small changes to the CCD vertical clocking rates will accommodate it. There also exist optical librations, principally due to the ellipticity of the lunar orbit and the 6.7 degree tilt of the rotation axis to the moon's orbital plane. These do not affect LTT operation as they are apparent rather than physical effects.

Any single point in the sky can be included in the LTT strip simply by tilting the telescope (in the selenographic meridian) during emplacement to point at it. For example, a point at or near the north or south Galactic pole or the Galactic center might be scientifically legitimate targets for inclusion in

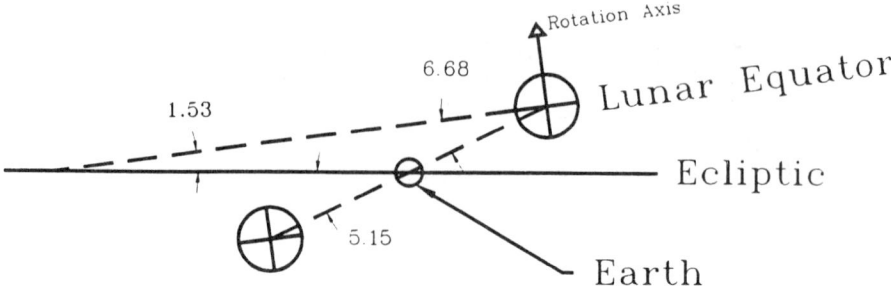

FIGURE 3. Geometry for the moon's rotation axis. The moon's orbit is inclined to the ecliptic by 5.145 degrees. The lunar equator is inclined to its orbit by 6.683 degrees. The lunar equator is, therefore, inclined to the ecliptic by 1.542 degrees. The lunar orbit precesses in the retrograde direction in the plane of the ecliptic with an 18.6 year period. All angles in this figure are labelled in degrees, and plotted using a scale factor of five for clarity.

the LTT survey. Two possible areas surveyed by LTT are schematically illustrated in equatorial coordinates in Figure 4. Each curve is generated by plotting a single point at the right ascension and declination observed by LTT at terrestrial midnight for an 18.6 year period and the strip is thus broadened by the precession of the lunar pole.

The pointing precision of the LTT affects both image quality and astrometry. The ultimate reference coordinate system for LTT will derive from the sky, possibly from positions of quasars. Because pointing and "tracking" stability are key to the performance of LTT, we briefly estimate the capability afforded by the moon.

At the shortest wavelength LTT is meant to operate (0.1 micron), the Rayleigh criterion for a two meter telescope yields a resolution of 0.013 arcsecond. As a stringent "telescope mount" stability limit, we require of the moon rotational, differential regolith settling and seismic stability of a fraction of this angle during timescales of the order the integration time on a CCD. Because the focal plane is a mosaic of relatively small format CCDs, we adopt as the required time for stability 1.8 hours, the transit time on a nominal one degree field of view, that is, a time much longer than the transit time of an individual CCD. The resulting stability criterion is motion less than about 2 $\times 10^{-6}$ arcsec/second. This stability criterion is met by the lunar rotation rate *even if no correction is made for the known physical librations.* Any other physical motions which modulate the lunar rotation we assume to be smaller than the librations. In fact, we shall modulate the CCD vertical clocking rates to

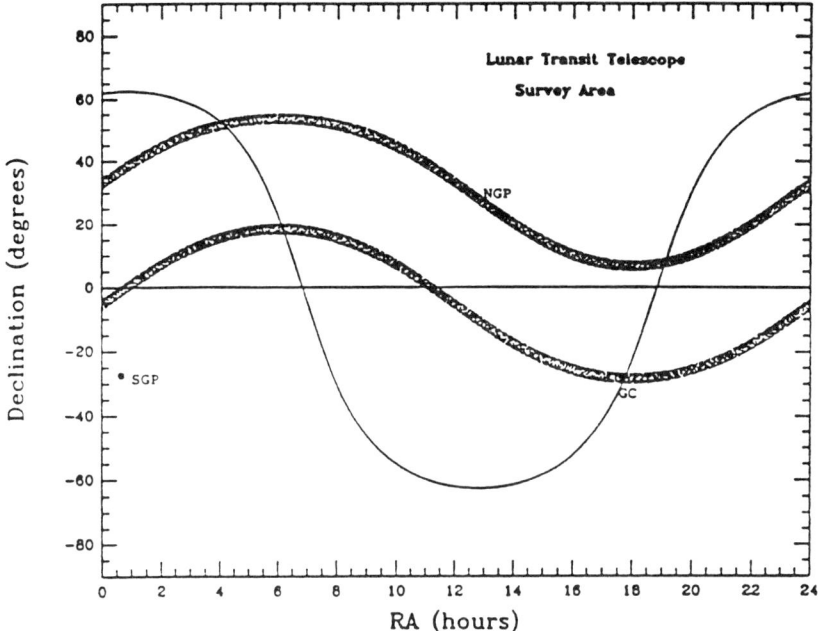

FIGURE 4. The LTT survey area in the equatorial coordinate system. LTT will survey a strip of the sky, the position of which depends upon the elevation at which the telescope is pointed. The three labelled points correspond to the north and south Galactic poles and the Galactic center. The curved solid line indicates the Galactic plane. Two possible survey areas are shown in this figure. The upper area includes the north Galactic pole, while the lower area includes the Galactic center. Multiple LTTs could be used to survey different regions of the sky, or a single LTT could simply be tilted to a new elevation to survey another area.

accommodate librations and any other known physical motions.

The LTT can be sized by considering launch vehicles currently capable of boosting a payload to the moon. The largest of these is a Titan IV with a Centaur upper stage. This vehicle is capable of putting a net payload (lander plus telescope) of about 4000 kg on the moon. If we elect to "float" the LTT (and its lander) on the lunar regolith, and if we assume three half-meter diameter landing pads, the total surface contact area is 0.6 m^2. The maximum surface strain is then 11.2 kPa, almost exactly the same footpad as the Lunar Module (VHROM). This strain is more than two orders of magnitude less than the ultimate bearing capacity of the lunar regolith (Carrier, Olhoeft and Mendell 1991). The initial settlement of the payload into the regolith is estimated to be about 1.0 ± 0.4 cm. After emplacement and initial settling, we assume as an upper limit additional elastic differential settling of 1 cm over a 5 m spacecraft base. The Maxwell relaxation time for lunar regolith can be roughly estimated by taking the ratio of the viscosity to the shear modulus for the *earth's* crust, which yields a timescale estimate of 3 - 30 years. This results in an angular rate of change of about 7.1×10^{-7} arcsec/sec over the mission lifetime of 18.6 years, or about the same variation as the precession induced change. This change is again negligible in terms of impact to the imaging quality of the telescope, though bulk and differential regolith settling must be considered more rigorously, as all available physical data are from the lunar equatorial regions visited by Surveyor and Apollo landers. While the compaction properties of the lunar regolith are favorable for this type of soft-landed experiment, the time rate of change of compaction is not well known and supplementary or complementary landing pads incorporating penetrating spikes or foundations might be necessary to ultimately anchor a telescope.

Finally, the total seismic energy of the moon is about 10^8 time less than the earth. This includes intrinsic and meteoroid-induced activity (Taylor 1990). Both the amplitude of individual moonquakes and their frequency of occurrence is greatly reduced from that of earth, making the moon an ideal site from this standpoint. We again make an order of magnitude estimate, scaling from the displacements typically left by terrestrial earthquakes - say one meter. Scaling by the relative <u>amplitude</u> of lunar to terrestrial quakes, the resulting lunar displacement is of the order 0.1 mm, resulting in a potential angular displacement in the focal plane of about four arcseconds. This would be an episodic displacement which is fortunately extremely rare, but could be digitally "repaired" after the fact if it did occur.

In summary, our conservative estimates indicate that from the standpoint of stability and precise knowledge of its motions, the moon represents an admirable spacecraft on which to mount the LTT and other astronomical telescopes! The point of this discussion is that no spacecraft which we might launch into any orbit can provide the precise, long duration pointing and tracking obtainable from the moon. The moon is an excellent telescope mount! For moving telescopes, this perhaps indicates a technical challenge to design devices which move slowly and precisely enough in the lunar environment to capitalize on the moon's stability. For the LTT mission of surveying a large area of the sky with high angular resolution, this stability is critical, but the necessary precise motions are all electronic.

Shielding on the Moon

Operation of a telescope on the moon is, of course, a mixed blessing. The lack of atmosphere, which makes available the entire electromagnetic spectrum, also allows micrometeoroids with velocities in excess of 10 km/s to impact the surface. The absence of a lunar magnetic field implies no shielding from cosmic rays. In addition, solar radiation directly illuminates the surface, resulting in extreme thermal fluctuations. The mass of the moon directly shields half of the sky from these effects. This generally represents a significant improvement from the high earth orbit (HEO) environment. Moreover, the lunar surface presents a stable platform for deploying additional shielding. We address shielding requirements, in turn.

Lunar meteoroids derive from asteroidal and cometary materials. Lunar Orbiter instruments indicated a "typical" rate of 0.16 impacts/m²/day (VHROM). The meteoroid size spectrum indicates that the vast majority of impacts will be of sub-millimeter particles with masses much less than one gram. The LTT structure can be protected from significant damage by but a few millimeters of composite material shielding. In particular, the baffling and sun shielding necessary for operation of LTT can probably be made to do double duty as meteoroid shielding. By the nature of the experiment, however, the primary mirror is directly accessible to meteoroids coming in along the optical axis. Neglecting shielding by the telescope secondary assembly, for a two meter f/2.2 telescope, the optics requires an angular opening to the sky of about 25 degrees, or a solid angle of about 0.2 steradians, or equivalently about 1.5% of the sky. We should thus expect approximately 3 significant, direct micrometeoroid impacts per year on the primary mirror, assuming isotropic arrivals. Given the uncertainties, this rate is consistent with the estimate given by Taylor (1990) of a meteoroid flux sufficient to produce microcraters 10 microns in diameter of about 300 events/m²/year, or about 14 events per year on the LTT primary. This corresponds to direct destruction of about 2×10^{-8} m² of the primary over the mission lifetime. Clearly, the effect of impact ejecta will affect several times this area, but this is still a negligible fraction of the mirror's surface. More important will be the wavelength dependent diffraction effects of surface cratering and degradation from migrating lunar dust.

The risk of cataclysmic damage from meteoroid impact does exist, though it is extremely small. The risk of a lunar astronaut being hit by a centimeter sized meteoroid was estimated to be about one chance in 10^6 or 10^8 per year of surface time (VHROM). The risk to LTT will be similar, as the critical surface area is only a few times the surface area of an astronaut. These are clearly better odds than are afforded terrestrial telescopes which are subject to damage (repairable, to be sure) by wind, rain or dust, as most "experienced" observers will attest.

Minimization of thermal variation in the optics and structure of LTT requires shielding from direct illumination by earth and sun. Shielding the Earth becomes rather simple by siting LTT near a limb of the lunar nearside. From this site the Earth always remains low on the horizon where a fixed shield excludes its reflected light from the field of view. A site at an intermediate selenocentric latitude allows easier shielding from direct solar illumination. Figure 5 schematically shows a conceptual shield which never allows the LTT structure to "see" the sun. In fact, by making the telescope and focal plane radiate efficiently and by making the interior of the shield as smooth as possible and its emissivity very low, it is possible to passively cool the telescope to cryogenic temperatures even during the lunar day. The large opening angle to cold space aids efficient radiation. Multiple shields might be used to shift the radiative load. All of this shielding is supported by the lunar surface, which itself is shielded in this concept. This is necessary because the moon is basically at the temperature for thermal equilibrium at one AU from the sun, with a sub-surface temperature of about 250 K (VRHOM, Taylor 1991). In addition to shielding sunlight, earthlight and the moon's thermal radiation, these shields might help mitigate the effects of micrometeoroids and lunar surface dust.

Preliminary estimates of the efficiency of this cooling system indicate that LTT could rather easily be kept at 150 K - 170 K, even during the lunar day. While ultraviolet and optical CCDs operate quite happily at these temperature, this is too warm for the IR CCDs, which will saturate on the thermal background radiated by the telescope itself. For operation in the near infrared, the detector package must be cooled to 65 K or colder, and the telescope itself should be as cold as possible, with a low emissivity light path. During the lunar night, the surface temperature rapidly drops to about 100 K, however (VRHOM, Taylor 1990). Small amounts of focal plane cooling would, therefore, allow operation in the infrared during the lunar night. Thus, LTT could operate in the optical and ultraviolet during the entire lunar day, with operation in the near-infrared after lunar nightfall, at which time the telescope becomes cryogenic. Clearly, the passive cooling achievable by appropriately shielding the LTT must be investigated in detail to define the telescope's operation. The presence of the lunar surface upon which to place shielding allows options for innovation in deployment.

The last effect we consider here is the degradation of information from the telescope focal plane caused by cosmic rays. In this discussion

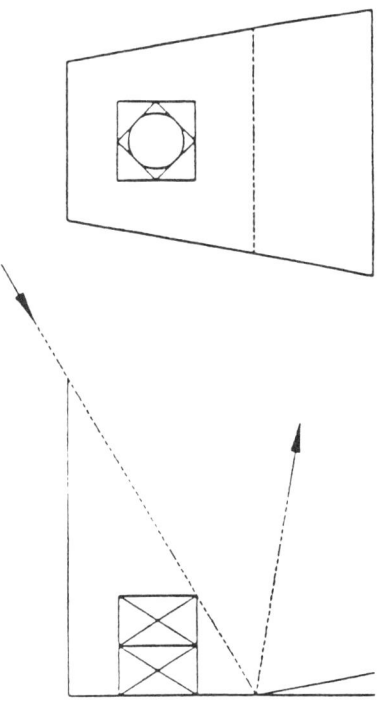

FIGURE 5. Schematic diagram of a sun shield for LTT. Above: top view. Below: side view. The interior of the shield is flat and highly reflective. The grazing ray for the highest sun altitude is shown as a dashed line which is reflected by the upward-angled horizon shield. The geometry ensures that rays which enter the shield are reflected out without intersecting the telescope. The large opening angle allows efficient radiation of the telescope to space. Multiple shields can be used to shift the thermal load.

"cosmic rays" includes all ionizing particles, including the galactic cosmic rays, solar flare and solar wind particles, with the flux of the latter two obviously time variable. The particles are principally protons and electrons, with rare occurrences of heavier nuclei. The solar wind component of the cosmic ray flux represents the largest flux, $\sim 10^8$ protons/cm^2/sec, with the lowest energies per nucleon, ~1 keV, and penetration depths, typically microns (VHROM). These particles are easily stopped by the shielding discussed above. Though they will impact the telescope optics directly, they will have virtually no effect on the CCD mosaic and associated electronics. The solar cosmic rays have energies of typically 1 - 100 Mev/nucleon, fluxes of about 100 particles/cm^2/sec and penetration depths of centimeters. At these energies, it is impractical to fully shield even the focal plane of a lunar based telescope. At still higher energies (1 - 10 GeV/nucleon) are the Galactic cosmic rays, with penetration depths of centimeters to meters, but with much lower flux of 1 - 4 particles/cm^2/sec.

Secondary particles generated locally by high energy interactions can further degrade the performance of detectors and electronics. Thus, unless detectors and electronics can be shielded by meters of material to stop both the primary particles and their secondaries, inappropriate shielding could, in fact, degrade the performance of the instrument. Clearly, the largest negative impact on LTT data quality will be caused by the solar protons, which have both high flux and energy. Because it is impossible for an early, unmanned lunar mission to provide meters of shielding, we must re-think the problem.

The first effect of ionizing radiation is to deposit charge in the CCD detectors, thus compromising the integrity of the data. It is, in fact, the data integrity which we wish to preserve, thus we investigate cosmic ray mitigating techniques in the data domain.

Because CCDs are very thin (~10 microns) compared to their width and height (~ centimeters), most cosmic ray events penetrate nearly normal to the imaging surface. This results in cosmic rays most often impacting single pixels (or but a few neighboring pixels). One mitigating approach to maintaining data integrity is thus to make pixels as small as possible so that a smaller percentage of pixels per image is affected. The full-well charge capacity of each pixel should be as large as possible to avoid charge "bleeding" into adjacent pixels from energetic events. These are, of course, exactly the properties one requires for a high resolution, fast imaging telescope anyway!

A second promising technique we are developing is to produce a "vertical mosaic" of CCDs which acts as an anti-coincidence detector for cosmic ray events. A vertical mosaic is shown schematically in Figure 6. The basic theory is that cosmic rays will completely penetrate a CCD, or even a stack of CCDs, leaving behind an ionized trail. Light is, of course, absorbed in the first CCD, which produces the image. Reading out the CCDs in parallel provides knowledge of where cosmic rays affected information in the image and some measure of the magnitude of the effect. One can thus "correct" by interpolation for most cosmic ray events, choose to ignore others or at least assess the quality of the image data. Similar anti-coincidence techniques might be applicable to other electronics, such as processors and memory.

The lunar surface is totally unprotected, which is its great virtue in terms of observing the electromagnetic spectrum (or particles) from this site. Operation of LTT (or any instrument, for that matter) in the lunar environment thus requires shielding from particles and radiation deleterious to the desired observation. In the case of LTT, shielding from solar illumination and the necessary structure of the telescope do "double duty" by shielding the large flux of solar wind particles, as well. Mitigation techniques such as vertical mosaics of CCDs can be used to ensure data integrity from the focal plane array. The effectiveness of these techniques can legitimately be assessed by ground-based experimentation.

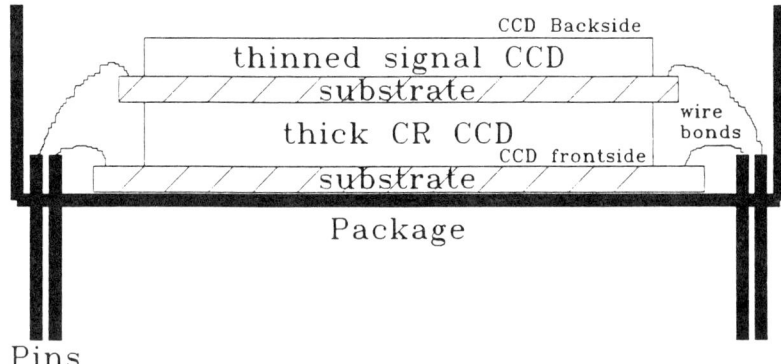

FIGURE 6. A vertical mosaic of CCDs for cosmic ray mitigation. The "signal" and "cosmic ray" (CR) CCDs are read in parallel. Anti-coincidence techniques are used to determine where cosmic rays affected the image, and how badly. Images may be "repaired" by interpolating over affected pixels, or pixels may be set to "don't care" values and ignored in further processing. At the very least, the vertical mosaic will allow direct determination of data quality and image integrity.

ASTROPHYSICS FROM THE MOON

Once in place LTT will send a continuous deluge of data to earth. An estimate of the performance of the strawman telescope indicates this rate to be about 2 Mbytes/s (170 Gbytes/day) (McGraw 1990). The command and control uplink rate will be very small, far less than 1 Kbit/s.

Receiving the downlinked LTT data is a major challenge, but one which offers potential for international collaboration. In particular, a worldwide network of receiving antennae could be organized to continuously record data from LTT. The basic technique is familiar to radio astronomers who utilize VLBA and VLBI techniques for continent- and hemisphere-wide aperture synthesis. The proposal is to collaborate with universities and agencies in other countries to fund duplication of and support for receiving stations based on approximately 35 m dishes. Because there do not currently exist a sufficient number of dishes to ensure continuous coverage, additional dishes would have to be produced for appropriately distributed sites. Existing designs which might be duplicated are specified to operate at higher frequencies than the K band assumed to be used for LTT data transmission, thus in terms of surface figure would be more than adequate to the task. In return for commitment to receive LTT data, the data would be returned to participants in the shortest possible time for complementary collaborations. The scenario is for two or more sites always to receive LTT data. They will transmit raw data to a central site where the instrumental signature will be removed, a first-order data base created and the pixel data archived. These data will be made available to collaborators as rapidly as possible - perhaps on a one day timescale. When not receiving LTT data the receiving dishes at the distributed sites could be incorporated into a VLBI network. This could be very advantageous to radio astronomers as the optimum distribution of receivers for LTT purposes would extend from mid- to high latitudes in both hemispheres, thus expanding the phase plane coverage of the network. It should be noted that it is distinctly possible that other soft-landed missions could utilize this network, as well. A schematic LTT data reception network is illustrated in Figure 7.

SUMMARY

We have described the basic operation of the Lunar Transit Telescope and listed but some of the scientific programs which would be addressable by its data. We have investigated the moon as a spacecraft and found that it has attributes, such as stability and shielding capability, which make it an attractive site on which to base scientific experiments. In fact, in some cases duplicating in an artificial satellite the positive attributes provided by the moon would be prohibited by technology or cost. The large area, high resolution LTT survey is made possible by the inertial and rotational stability of the moon, its lack of appreciable atmosphere, as well as its shielding and communications capability. There is no fundamental reason why the moon cannot be used as an observatory site. Remaining problems have engineering solutions. As an example, cosmic ray mitigation by a technique such as vertical mosaicking appears promising and should be developed as a standard technique for space observatories. Of course, the rather remote site for LTT requires autonomous operation in some guise, perhaps incorporating an earth-based network of receiving stations to receive the continuous stream of image data. Certainly, the value of these data for a wide variety of astronomical research topics coupled

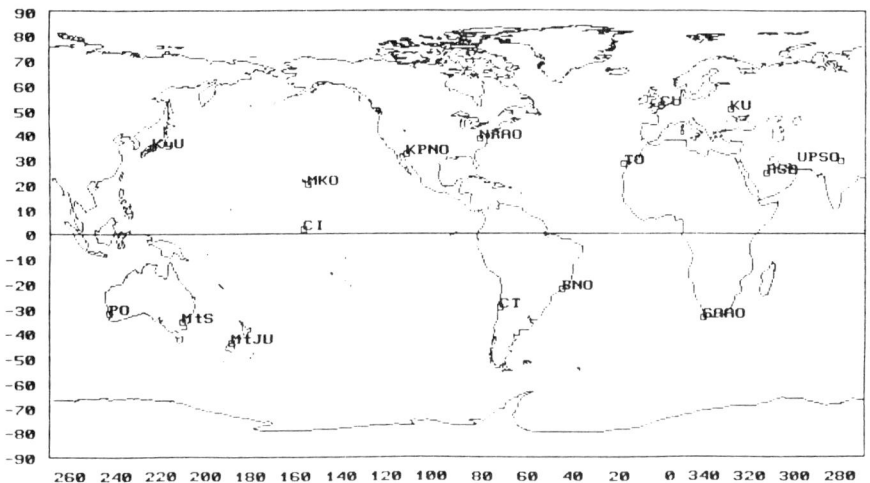

FIGURE 7. Schematic depiction of a global LTT data reception network. Radio dishes approximately 35 m in diameter could be optimally deployed by collaborating universities and agencies. When not used for LTT (or other lunar experiment) data, these dishes could be used in VLBI synthesis observations.

with the relatively low cost of the mission, which utilizes existing and emerging technologies, creates a very high benefit-to-cost ratio for the project. LTT appears to be an excellent mission to spearhead our overdue return to the moon for legitimate reasons.

Thanks to Lisa LaFlame and Anne Lierman for their help in investigating the feasibility of the LTT mission and to Dave Harvey and Mike Lesser for some of the figures. UA undergraduate Lexi Moustakas was the origin of the idea for the LTT relativity experiment. As usual, G. F. Benedict at McDonald Observatory thought up some really good LTT science projects and managed to explain the real motions of the moon. Finally, thanks to Harlan Smith, who has long pointed out the importance of the moon as a scientific site - and more importantly, that an individual can influence the human adventure. This work is supported by NASA grant NAG8-884 for definition of the LTT.

REFERENCES

Allen, C. W. 1973, *Astrophysical Quantities*, (Athlone: London), p. 147.

Burns, J. 1991, in ASP Conf. Proc., ed. Filippenko, A. V., article in this volume.

Loh, E. D. and Spillar, E. J. 1986, *Ap. J. (Letters)*, **307**, L1.

McGraw, J. T. 1990, in AIP Conf. Proc. **207**, *Astrophysics from the Moon*, eds. Mumma, M. J. and Smith, H. J., p. 433.

McGraw, J. T. and Benedict, G. F. 1990, in AIP Conf. Proc. **207**, *Astrophysics from the Moon*, eds. Mumma, M. J. and Smith, H. J., p. 464.

Taylor, G. J. 1990, in *Astrotech 21 Space Infrastructure Handbook*, JPL D-7430, ed. Fordyce, J., p. 10-1.

Vaniman, D., Reedy, R., Heiken, G., Olhoeft, G. and Mendell, W. 1991, in *Lunar Sourcebook*, eds. Heiken, G., Vaniman, D. and French, B., (Cambridge: New York), p. 27. [VRHOM]

A LUNAR POLAR INFRARED OBSERVATORY

DAN LESTER
Department of Astronomy and McDonald Observatory, University of Texas, Austin, TX 78712

ABSTRACT The prospects for an infrared telescope on a permanently shadowed crater floor at the lunar pole are discussed, and the state of our knowledge about conditions in lunar polar craters is reviewed. The unique thermal environment in such a crater allows the cooling of large telescopes without expendable cryogens. This advantage, as well as other more generic conveniences of a lunar site, suggest that the long-term future for infrared astronomy may be strongly coupled to lunar development.

INTRODUCTION

In the infrared part of the spectrum, a cold telescope in the absence of an atmosphere is well known to be a potent prescription for discovery. The new richness in the infrared sky that was revealed by the IRAS satellite, the important cosmological revelations that resulted from COBE, the enormous promise of the forthcoming ISO and IRTS missions, and the proposed SIRTF observatory which will detect sources that are thousands of times fainter than those observable from ground-based telescopes, underscores this assertion. In this paper, I explore the advantages of infrared astronomy with passively cooled telescopes at the lunar poles. With the caveat that these regions have not been adequately surveyed, it appears that these advantages are extraordinary, and that the long-term future for infrared astronomy may be strongly coupled to lunar development.

BIG COLD TELESCOPES ARE NICE TO HAVE, BUT HARD TO GET

In a background-limited part of the spectrum, such as the thermal infrared, with a diffraction-limited beamsize λ/D, it can be shown that the length of time required to detect a point source goes as D^{-4}, compared with the D^{-2} dependence of a detector-limited experiment. Thus, while large telescopes are always better, the payoff from larger aperture size is especially spectacular in the IR. Our atmosphere prevents us from cryogenically cooling telescopes on Earth. What prevents us from putting enormous, cold telescopes into space? Mainly the lifting hardware that is available. Upper-stage shroud sizes on existing boosters place strict limits on the aperture sizes that can be accommodated, and the price of cryogen tanks that surround and cool the telescope is a severe constriction on the fraction of the shroud diameter that can actually be used for optics, as well as the more obvious penalty of limited cold lifetime. The Titan IV is the largest, most powerful expendable rocket available (by the middle of this decade) to the U.S., and this booster is just capable of placing a 1-m cryogenically cooled telescope with a five-year lifetime (SIRTF) into high Earth orbit (HEO). The ESA Ariane 5 will offer similar capabilities. While low Earth orbit (LEO) offers opportunities for assembly of larger apertures, the heat load from the Earth (both reflected and thermalized sunlight) and the large zones of avoidance of a short period orbit make this option unattractive to a long-lived, cold telescope project.

BIG COLD TELESCOPES DON'T NEED EXPENDABLE CRYOGENS

In a purely radiation dominated thermal environment, in which the heat sources (Sun, Earth) are localized in a very cold sink (dark sky), a spacecraft that is properly shielded from the Sun will get cold all by itself. This technique of passive cooling, familiar to owners of solar homes, is a potentially powerful method for capitalizing on the "free lunch" that is offered by the extraordinarily simple thermodynamics of space. This technique has been used successfully to cool several Earth-observing satellites in LEO to temperatures of order 120 K (see also Tulkoff 1990). While this is not cold enough to compete with cryogenically cooled telescopes for the purpose of doing most infrared astronomy, more careful thermal management, as well as higher, more optimal orbits, offer this potential. For example, Edison, a proposed passively cooled telescope that can be as large as 3 m, can be launched into HEO and attain temperatures < 50 K, making it a powerful facility for work out to at least 100 μm, and more specifically, enabling zodiacal background-limited performance in the mid-infrared (Thronson et al. 1990; Thompson 1990).

The ultimate opportunity for passive cooling, however, will be realized when we return to the Moon. Its solid surface and low gravity are advantageous for any telescope, as is the lack of atmosphere, slow rotation rate, and low seismic activity. At the end of the long lunar night, mid-latitude surface temperatures get down to approximately 100 K; somewhat lower telescope temperatures could be reached by proper radiative decoupling from the surface. Such temperatures are low enough to remove all telescope background in the near-infrared, but still will limit work in the mid-infrared and far-infrared. These limitations are largely avoided, however, at the observing sites that await us at the lunar poles.

In summary, there is a very stringent limit on the size of telescopes that can be launched in one piece. This limit is severely reduced when expendable cryogens are attached. In any case, the largest, most sensitive infrared telescopes will be those that are passively cooled.

LUNAR POLAR CRATERS: OPTIMAL SITES FOR BIG COLD TELESCOPES?

While the orbit of the Moon is inclined by 5.1° to the ecliptic, with a line of nodes that precesses around the ecliptic with an 18-year period, it is not commonly appreciated that the rotational pole of the Moon is only 1.5° from the ecliptic pole, an alignment that was reached relatively quickly in the evolution of the Earth-Moon system by the same tidal process that locked the rotation. Thus, as seen from a site at the lunar poles, the Sun stays within 1.5°, and the Earth within 6.6°, of the horizon (Fig. 1). As a result, the bottom of a crater at a lunar pole can exist in perpetual solar and terrestrial shade, while nearby peaks may be high enough to be in perpetual sunlight. Thus, we see that landforms on the lunar pole make it a thermodynamically extraordinary place. This realization is not new, and the reader is referred to discussion of the advantages of a base at the lunar poles in the papers by Burke (1985, 1988). More general discussion about the potential of lunar observatories can be found in the reviews by Smith (1990) and Burns (1990, 1992). Some of the concerns relevant to a lunar IR telescope are addressed by Johnson and Wetzel (1990).

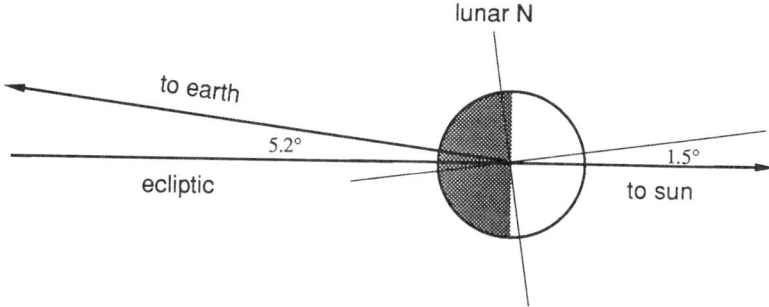

Fig. 1. The lunar orbital geometry is compared with that of the rotation axis of the Moon. The rotation axis of the Moon is only 1.5° away from the axis of the ecliptic.

What do we know about the conditions in lunar polar craters? Unfortunately for mission planning, virtually nothing. While the poles were imaged by the Lunar Orbiter missions, the lack of illumination that is so important to the potential for putting cold telescopes there is the very characteristic that prevents us from seeing inside of them! Figures 2a and 2b show representative Lunar Orbiter images of the polar regions. These pictures are included simply to illustrate the dearth of observational information that we have about these regions. A more accurate pictorial representation of the bottom of a lunar polar crater would be entirely black.

While our direct observational information about lunar polar craters is next to nonexistent, there are several inferences that can be made with some confidence. First, these craters must be very cold, and this fact is the raison d'être for putting an infrared telescope there. A small amount of energy is

Fig. 2a. Lunar Orbiter IV photo of the north polar region of the Moon. The north pole is at the upper left.

Fig. 2b. Lunar Orbiter IV photo of the south polar region of the Moon. The south pole is at the lower left.

deposited inside the crater by the thin sliver of sunlit rim. Conduction of heat on the outside of a crater wall into the inside is rendered negligible by the extremely low thermal conductivity of lunar soil. As a result of the permanent shade, and the lack of an atmosphere, it is probably safe to say that the bottom of a crater on the lunar pole is among the coldest places in the entire solar system.

For regions that are optimally situated, Arnold (1979) has shown that the surface temperature can be close to the lower limit given by the lunar heat flow through the regolith. This heat flow is set up by radioactive decay in the lunar interior, and was measured directly at several Apollo sites as ~ 20 mW m^{-2} (a figure about half that of the Earth). This heat flow is identical to that produced by a blackbody at ~ 25 K. At such temperatures, the surface will be an efficient cold trap for all gases except hydrogen and helium. Arnold (1979) has argued that the deposition rate of water ice, in particular, from comet debris, is larger than the sublimation rate from residual heat sources over large areas at the poles. Over the $> 10^8$ years that have elapsed since the poles were last "cooked" by polar wandering, substantial deposits of ice will have been built up. Such ice deposits would profoundly aid lunar base development, both in terms of human needs for water, as well as a convenient mineable source of hydrogen for fuel.

Deposited ices of lower sublimation temperature gases could, however, seriously complicate the deployment of a telescope in one of these lunar polar craters. Expeditions into such craters by warm astronauts with warm equipment would cause rapid outgassing of the trap. It should be noted that a 300K astronaut at a distance of about fifty feet contributes as much energy to a unit area of lunar surface as does the Moon underneath it! It should also be noted that the possibility of large deposits of ices in lunar polar craters is a matter of some controversy. Lanzerotti, Brown, and Johnson (1981) contest the conclusion of Arnold (1979). They suggest that spallation by cosmic rays can remove the ice faster than it can form, and that such craters, while still very cold, will probably remain mostly dry. Arnold (1987) counters that, in regions well shielded from the solar wind, such spallation is not likely to be important, and that trapped ices are eventually protected by overlying layers of meteoritic debris, and debris from landslides on the inner crater walls.

The question of whether there are large deposits of ice at the lunar poles is one of the outstanding issues of lunar astronomy. Haines and Mezger (1986) have proposed that gamma ray spectroscopy from lunar orbit can answer this question. Bombardment of surface materials by energetic Galactic cosmic rays produces high-energy neutrons. These neutrons are thermalized by scattering, and can then be captured via H(n, 2.122 MeV γ)D. A search for these 2.122 MeV photons from the lunar pole would then test for the existence of large quantities of water ice there, since the hydrogen content of the lunar soil is otherwise very small (especially in regions shielded from the H-rich solar wind). Other possible remote sensing experiments that would address this question might be near-infrared reflection spectroscopy using low-level starlight or scattered sunlight illumination, and multiband SAR imaging.

The shielding of a lunar polar crater from direct sunlight results in a radiation environment that is conducive to a long-lived observatory. Solar UV is the primary source of reflectivity degradation of thermal shielding on spacecraft. Such thermal shielding is still important on the floor or a permanently shadowed crater. It will be argued below that careful thermal shielding will allow a lunar polar telescope to equilibrate at temperatures well below that of the ambient ground temperature. The walls of a lunar polar crater also provide complete shielding against energetic particles that originate in solar flares. Such particles contribute to general instrument degradation and system noise. While the lunar polar crater telescope is still exposed to infrequent, but highly damaging GeV Galactic cosmic rays, the net flux is lower than in HEO simply because the instrument is shielded from at least half of the celestial sphere.

A UNIQUE TELESCOPE WITH UNIQUE QUALITIES (AND PROBLEMS)

Taking advantage of the "free lunch" provided by passive cooling comes at a steep price, which is that required to get to the Moon in the first place. In this latter respect, we adopt the philosophy that a return to the Moon is a matter of time, politics, and our long-range plan for space exploration. It will eventually happen. The price of the lunar base infrastructure that will support the observatory will be the largest part of the budget, and its cost will be absorbed in other ways. No predictions about the timescale are offered here, except to note that long-range NASA plans include permanent lunar base development in the early part of the next century.

Quite unlike the implementation of telescopes that are cooled by expendable cryogens, in which the lifetime clock starts ticking as soon as the dewar is topped off for the last time before launch, and the science program is a race against this clock, science with a passively cooled telescope can take a more relaxed, and perhaps thoughtful, pace. Indeed, for a structurally robust telescope, the thermal time constant may be weeks to months, and optimal performance of the observatory at the longest wavelengths will demand some patience. The Edison project can be used as an illustrative model in this context. Edison, a high orbit, passively cooled 3-m class infrared telescope (Thronson et al. 1990), is a very instructive model for a cold lunar polar telescope, and can be regarded as the logical precursor for it. Much of the thermal engineering development that a lunar polar infrared telescope will require will be needed for Edison as well. Assuming 3 μm diffraction-limited optics, and a consequent cold mass of $\sim$ 3000 kg, a cooling time to $\sim$ 50 K is predicted to be less than a year (T. Hawarden, private communication), and this is for a telescope that is insulated from sunlight simply by several layers of superinsulation. Using this as an example, and noting that the thermal environment of a lunar polar crater is much more optimal, we conclude that thermal inertia is not an important obstacle to complete realization of passive cooling in a lunar polar crater.

As discussed above, development of a cold crater that has trapped large quantities of low sublimation temperature gases (Ar, N, O) will release clouds of vapor that may interfere with the installation of the telescope. As long as the telescope is always warmer than the ground around it, there will be no danger that surface outgassing, whether because of development efforts, or from natural processes, will produce condensation or frost on the telescope optics. In fact, however, careful thermal management of the telescope will allow it to be colder than the ground. In this case, the optics will have to be protected, perhaps just by line-of-sight baffles, since any neutral gas molecules released on the Moon will follow basically ballistic trajectories. It should be noted that the trajectories of ions are likely to be more complicated. Ionization of atoms in the rarified lunar atmosphere will cause them to follow the field in the solar wind, and they can be implanted in the regolith at some distance from their source. In any case, support activities outside of the crater will have to be done with considerable caution, since the cold telescope will be extremely susceptible to contamination. Rocket exhaust, and venting and leakage of breathing gases, must be prevented from reaching the crater once the telescope is cold. As noted above, for the best shielded craters, the very appearance of a warm object on the crater rim (such as an astronaut, a lunar vehicle, etc.) can act as a substantial thermal load to the surface inside, driving off lightly bound gases.

Pre-dawn streamers in the lunar sky seen by Apollo 17 astronauts are considered to be evidence for a light haze of lunar dust (cf. Cadogan 1981). While this dust is little more likely to stick to a cold telescope than to a warm one, such contamination of an observatory devoted to thermal infrared operation could seriously degrade its performance. There is some indication that this haze is levitated electrostatically by photoionization of soil components, so we might expect this to be less of a problem in a permanently shadowed region.

Clearly, power and communication cables will have to be run to the outside of the crater, where direct link to Earth, and warm power units (presumably solar arrays) can be located. Thus, the facility cannot be self contained. For most lunar polar craters, even in the absence of the annoying affects of trapped gases, the low surface temperature will make direct intervention by astronauts rather unlikely. The importance of operating this telescope as a robotic observatory is thus evident. Local support electronics will have to function well when they are cold and, ideally, dissipate little heat, in order to maintain the lowest possible equilibrium temperature. This would be an excellent opportunity to use superconducting circuits.

A lunar polar infrared observatory would, unlike a free-flyer space mission, have very specific (but not time dependent) areas of the celestial sphere that are excluded from its view. Observations of the celestial sphere in the opposite ecliptic hemisphere would clearly require a second telescope. The success with which a lunar polar crater telescope avoids direct sunlight is precisely the reason that the ecliptic region is inaccessible to it, an inescapable circumstance that would make such an observatory (at either pole) of limited value to solar system astronomers. For observations of the faintest and most distant extrasolar system sources, however, this region of exclusion, if there must be one, is ideal. IRAS measurements have shown that at wavelengths from $\sim$ 10 to 100 μm, emission from zodiacal dust defines the noise floor in this wavelength range for a large, cold telescope equipped with current generation detectors. This emission is strongly concentrated in the ecliptic plane and so, in the mid-infrared, the sky is darkest near the ecliptic pole, conveniently at the zenith at the lunar polar observatory site.

CONCEPTS FOR BIG COLD TELESCOPES IN A LUNAR POLAR CRATER

This paper considers lunar polar craters as potential sites for passively cooled telescopes. The structural characteristics of such telescopes will only be broadly discussed, as they depend on the science to be accomplished and the level of development effort that can be undertaken. The first lunar polar infrared telescope will probably be a survey instrument similar to that proposed by McGraw (1992). This instrument has a minimum of moving parts. With a fixed orientation, it surveys a strip of sky as the slow lunar rotation scans the sky across the telescope. For a lunar polar site, small strips around the lunar poles can be surveyed with very long integration times. A more ambitious telescope would be one that was steerable, though considerable care would have to be taken to prevent steering motors from contributing a significant heat load to the telescope.

A cartoon sketch of a possible lunar polar infrared telescope is shown in Figure 3. The size has been intentionally left ambiguous. In the absence of any thermal engineering, such a telescope would equilibrate at a temperature close to that of the crater floor. In fact, by paying attention to optimal baffling strategies, one should be able to do much better than this. Taking care to shield the telescope from the sliver of sunlight that illuminates the crater rim removes an important potential heat load. Such a shield would be reflective on both upper and lower sides, radiatively coupling a blackened inner telescope structure to cold sky, and decoupling the entire telescope from the crater floor. With a far-infrared shield emissivity of < 10%, further temperature reduction by a factor of two could probably be realized. Such temperatures of 10–20 K are in the range of optimal temperatures for present generation doped-silicon impurity band conduction detectors that work out to ~ 40 μm. Thus, such detectors could operate essentially "bare" at the focal plane of the telescope with no further cooling. The Astrotech 21 Sensor Technology panel has specifically recommended that efforts be made to identify and develop infrared sensors that operate at the higher temperatures (than cryogenically cooled) that would be characteristic of passively cooled instruments.

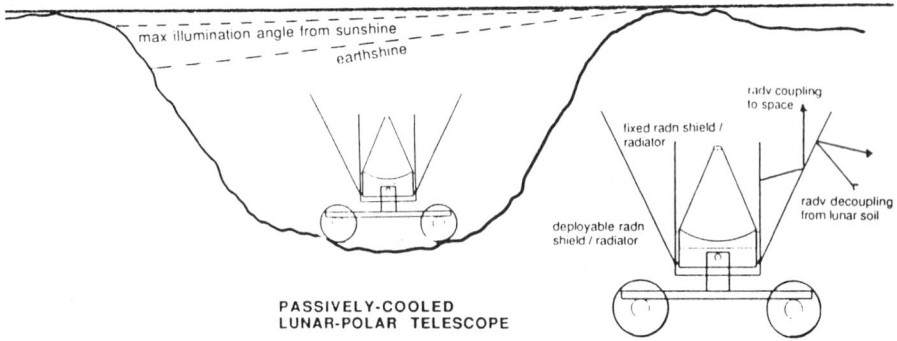

Fig. 3. A highly simplified sketch of a possible lunar polar infrared telescope. The telescope is based on the floor of a lunar polar crater, and the telescope takes advantage of the extraordinarily cold temperatures that must characterize this region.

In the cartoon, the telescope is shown on wheels, which would be used to deploy the observatory, over the crater rim, onto the crater floor. Self-deployed in this way, human intervention into the frigid depths of the lunar polar crater is obviated.

Other, more ambitious telescope concepts should certainly be mentioned. The shape of a lunar crater naturally suggests its use for an Arecibo-like installation. While thermal shielding of such a telescope probably can't be done as optimally as for a telescope that is small compared to the size of the crater, and deployment must be considered enormously complicated, the huge apertures that can be accommodated are tantalizing. Construction of such a telescope that is diffraction limited at the near-infrared and mid-infrared wavelengths that most strongly justify telescope temperatures of 20–40 K is, however, not within reach of current technology. Far-infrared installations can, of course, benefit from such temperatures, compared to warm telescopes, but not as spectacularly as those for shorter wavelength work.

ACKNOWLEDGEMENTS

I would like to thank J. Burke for information on lunar polar bases. The Edison project (particularly H. Thronson and T. Hawarden) has been a source of stimulus on the possibilities for passive cooling of large telescopes. Finally, I appreciate discussions with the late Harlan Smith, who first kindled my interest in lunar polar telescopes, and whose excitement about the enormous potential and inevitability of lunar observatories inspired this work.

REFERENCES

Arnold, J. R. 1979, *J. Geophysical Res.*, **84**, 5659.
Arnold, J. R. 1987, *Proc. Lunar Plan. Sci. Conf. XVIII*.
Burke, J. D. 1985, in *Lunar Bases and Space Activities of the 21st Century*, ed. W. Mendell, NASA/Lunar and Planetary Institute, p. 77.
Burke, J. D. 1988, in *Future Astronomical Observatories on the Moon*, ed. J. Burns and W. Mendell, NASA Conference Publication 2489, p. 31.
Burns, J. 1990, in *The Next Generation Space Telescope*, ed. P. Bely, C. Burrows, and G. Illingworth (Baltimore: STScI), p. 341.
Burns, J. 1992, *These Proceedings*.
Cadogan, P. 1981, in *The Moon — Our Sister Planet* (Cambridge: Cambridge Univ. Press).
Haines, E. L., and Mezger, A. E. 1984, *Nucl. Inst. and Meth.*, **226**, 509.
Johnson, S., and Wetzel, J. 1990, in *Observatories in Earth Orbit and Beyond*, ed. Y. Kondo (Dordrecht: Kluwer), p. 348.
Lanzerotti, L. J., Brown, W. L., and Johnson, R. E. 1981, *J. Geophysical Res.*, **86**, 3949.
McGraw, J. T. *These Proceedings*.
Smith, H. A. 1990, in *Observatories in Earth Orbit and Beyond*, ed. Y. Kondo (Dordrecht: Kluwer), p. 365.

Thronson, H. A., Jr., Hawarden, T. G., Mountain, C. M., Davies, J. K., Lee, T. J., and Longair, M. 1990, in *Observatories in Earth Orbit and Beyond*, ed. Y. Kondo (Dordrecht: Kluwer), p. 501.

Thompson, R. I. 1990, in *The Next Generation Space Telescope*, ed. P. Bely, C. Burrows, and G. Illingworth (Baltimore: STScI), p. 310.

Tulkoff, P. J. 1990, in *The Next Generation Space Telescope*, ed. P. Bely, C. Burrows, and G. Illingworth (Baltimore: STScI), p. 279.

TELEROBOTICALLY DEPLOYED LUNAR FARSIDE VLF OBSERVATORY

P. B. LANDECKER, M. A. CAYLOR, D. U. CHOI, R. J. DREAN,
C. R. EDELSOHN, J. G. GURLEY, F. A. HAGEN, G. W. SU,
M. L. TILLMAN, and C. R. WASSGREN
Space and Communications Group, Hughes Aircraft Company,
P. O. Box 92919, Los Angeles, CA 90009

ABSTRACT A low cost aperture synthetic array ALLFA (Astronomical Lunar Low Frequency Array) is described, capable of measuring very low frequency radio waves in the range 100 kHz - 30 MHz from astronomical objects. Frequencies below 10 MHz cannot effectively be observed from Earth (including from Earth orbiting satellites) due to absorption from the Earth's ionosphere and interference from both manmade and natural sources. The farside of the Moon is shielded from this interference, thus enabling sensitive observations of weak astronomical sources (as low as 10 Jansky for a 1 day integration by ALLFA) as well as providing simple stable long baselines required by aperture synthesis. The scientific objectives of ALLFA are described, and include measurements of various solar system, galactic and extragalactic objects and phenomena. The strawman array design consists of 40 telerobotically deployed sensor elements and a centrally located station. The rationale for the frequency range selected, the sensor element design and placement is described, as well as communications, power, battery and duty cycle trades performed. The largest array dimension is 25 km, corresponding to an angular resolution of about 20 arc min at 1 MHz. The anticipated observatory sky coverage is given. The lifetime of the facility is 12 years, covering slightly more than one solar cycle; factors relating to this lifetime are also described.

INTRODUCTION

The Moon has several natural advantages for observing very low frequency electromagnetic radiation. Compared to the Earth, the Moon has very little ionosphere to attenuate incoming signals. In addition, it provides a very stable platform for aperture synthesis applications, and permits long integration times. A further key advantage of the lunar farside is that observations are completely shielded from the very significant natural (auroral kilometric radiation and lightning) and manmade (shortwave broadcast and ignition) radio noise from the Earth (Erickson 1988, 1990). Note that a key discovery by the Radio

Astronomy Explorer (RAE) satellites is that the Earth is a very intense emitter of radiation in the range 0.1-10 MHz, masking weak radio astronomy sources. RAE-2 demonstrated the clear advantage of lunar shielding of terrestrial sources for VLF radio astronomy, as shown in Figure 1. The lunar farside is therefore a unique, low interference site for such measurements (Smith 1990). Finally, low gravity and the absence of weather simplify the engineering design.

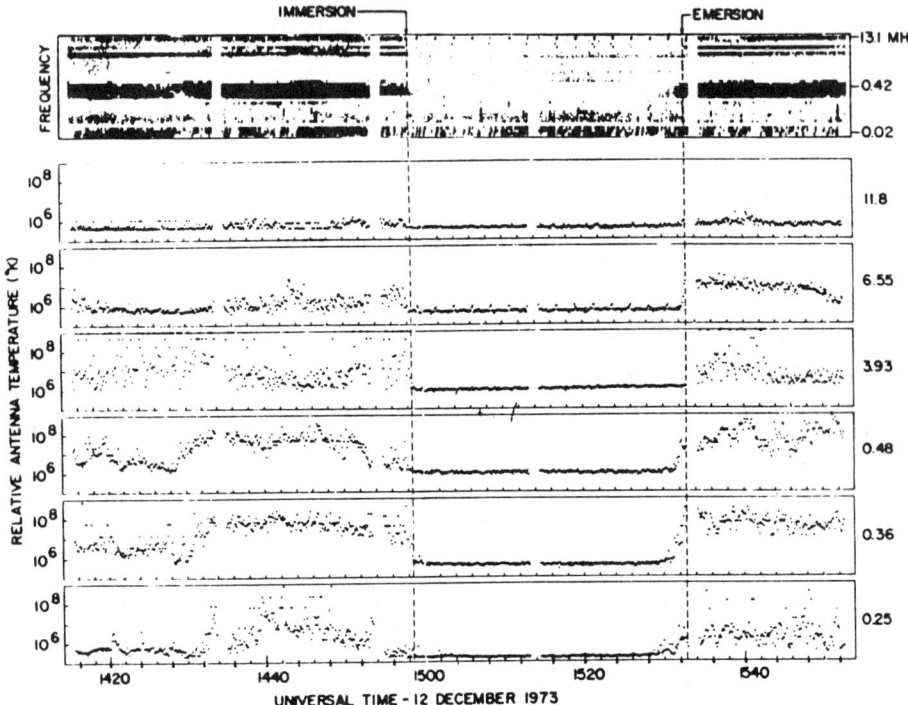

Fig. 1. RAE-2 data from lunar orbit as observed by Alexander et al. 1975. Note the sharp reduction in signal levels when the Moon's limb occults the Earth.

There has been significant previous interest in the use of the lunar farside for VLF astronomy (Basart and Burns 1990, Smith 1990, Burns and Asbell 1986, Douglas and Smith 1985). ALLFA (Astronomical Lunar Low Frequency Array) extends the previous expressions of scientific interest and develops a specific concept demonstrating the engineering feasibility of farside VLF astronomy. A review of the scientific objectives, a general description of the ALLFA concept, a rationale for the frequency range selected and a detailed detector design are given in this paper.

ALLFA consists of 40 sensor elements arranged in an ellipse around a central element (CE) and located on the lunar farside just south of the equator, with the maximum baseline of 25 km in the north-south direction. A satellite orbiting the L2 libration point is then used to relay the astronomical data to Earth for interpretation and analysis by astronomers. The overall system description is illustrated in Figures 2 and 3.

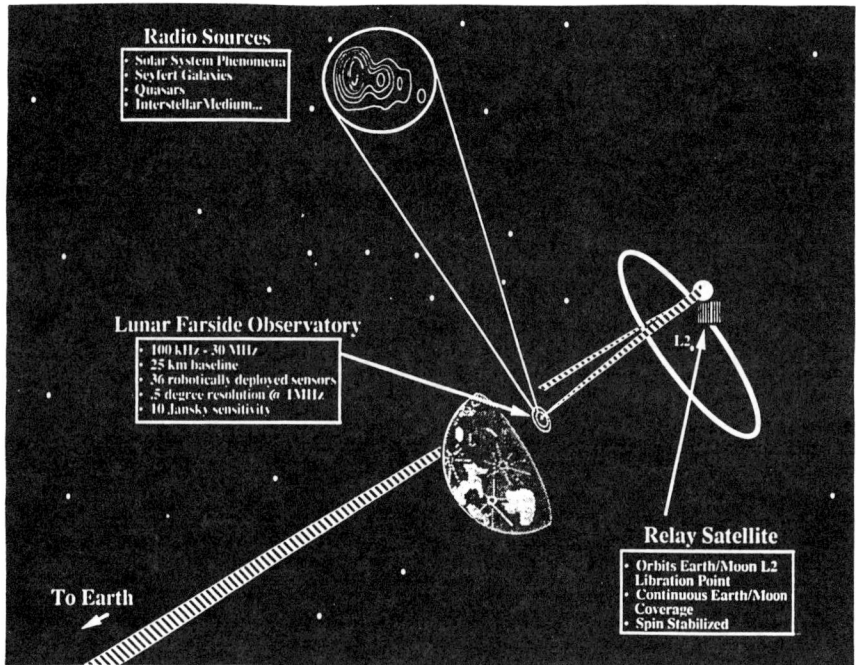

Fig. 2. ALLFA Concept. Radio astronomy data recorded on the lunar farside are relayed to the Earth.

Fig. 3. Schematic illustration of ALLFA. A telerobotic lunar rover was used to place the sensor elements around the central element. After deployment, the rover is available for other scientific uses.

The diameter of ALLFA was chosen to permit the angular resolution achieved by aperture synthesis to be comparable to the intergalactic medium scattering which for most sources is 20 arc min at 1 MHz. An elliptical configuration was selected to optimally view sources at various declinations. Pseudo-randomized element spacing provides nearly uniform coverage in the u-v plane.

Various additional aspects of the ALLFA engineering design are given in a companion paper (Drean et al. 1991). These two papers show the scientific justification, engineering feasibility and low technical risk associated with all aspects of this proposed project.

SCIENTIFIC OBJECTIVES

A VLF array such as ALLFA is uniquely capable of studying astrophysical plasmas under a wide range of conditions. At higher frequencies, low energy galactic cosmic rays and the environments of various radio sources can be investigated. At lower frequencies, the radiation observed is highly affected by the intervening interstellar medium, and also (at the very lowest frequencies) the interplanetary medium. These effects can be used to study the distributions of matter and fields in these media. Finally, it is possible at the lower frequencies to observe coherent solar radiation, turbulence of the solar wind, and emissions from the magnetospheres (aurora) or atmospheres (lightning) of certain planets.

The numerous scientific goals and objectives of ALLFA are listed in Figure 4 and described in much greater detail by Dennison et al. (1986). In addition to the measurements listed, covering the range from the lunar ionosphere to the most distant extragalactic objects, unexpected results and exciting discoveries in this relatively unexplored spectral region are likely.

- **SOLAR SYSTEM**
    - Observe energetic solar coronal electrons
    - Observe planetary magnetospheres
        - Jupiter, Saturn, Uranus
    - Observe planetary lightning
        - Venus, Io
    - Observe interplanetary medium
        - Solar wind turbulence
    - Observe lunar ionosphere
- **GALACTIC**
    - Study neutron star magnetosphere plasma instabilities
    - Measure CR electron galactic distribution
    - Generate galactic background maps
    - Study individual discrete sources
    - Interstellar scattering/refraction/turbulence
    - Determine spatial structure of pulsars
    - Observe H_{II} regions/interstellar absorption
    - Determine extent of galactic halo
    - Measure low energy electron distribution
    - Study supernova remnant loop features
    - Find spectral turnover of nonthermal sources
- **EXTRAGALACTIC ($\geq$ 1 MHz)**
    - Observe Seyfert galaxies
    - Study coherent radiation emission (quasars)
    - Measure intergalactic medium
    - Study individual discrete sources
        - Measure source spectral indices
    - Search for fossil radio halos
        - Determine radiative lifetimes
    - Determine galactic cosmic ray distribution
- **UNEXPECTED**
    - See unusual phenomena at new frequencies
    - Detect new processes and object classes
    - Serendipitous discoveries likely
    - Search for steep radio source populations

Fig. 4. ALLFA Scientific Objectives.

As shown in Figure 5, there are many new and important astrophysical observations that will be made by ALLFA. Measurements of the various electromagnetic wave properties are used to develop an understanding of the emission mechanisms and their modification by the intervening medium.

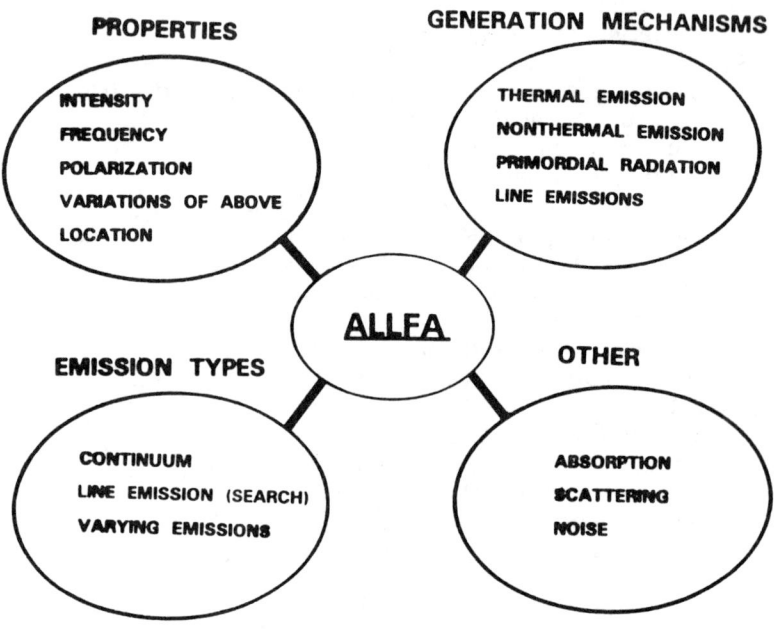

Fig. 5. Astrophysical Observations. Many new and important discoveries will be made by a VLF array such as ALLFA.

FREQUENCY SELECTION

The core frequency range is 1-10 MHz. The range will be extended down to 100 kHz since planetary objects can still be studied at these frequencies, provided it can be done at marginal extra cost. The limiting factors which determine this requirement are the interplanetary and lunar ionosphere plasma frequencies. The plasma frequency of the interplanetary medium is 20 kHz. This is therefore the effective low frequency limit for interplanetary observations with ALLFA. However, the nighttime plasma frequency of the lunar ionosphere may be about 90 kHz corresponding to an electron density global average of 100 cm^3. From the relatively uncertain data currently available on the lunar ionosphere, a conservative low frequency requirement of the observatory is 100 kHz. This value may change in the future based on better lunar ionosphere data, or engineering difficulties associated with the wideband frequency range.

The upper frequency limit of 30 MHz was selected to permit some overlap between the radio astronomy data recorded from the surface of the Earth (typically down to 30 MHz and occasionally down to about 10 MHz at a few radio-quiet locations with smaller ionosphere attenuation) and ALLFA.

SENSOR ELEMENT DESIGN

The function of the sensor element is to gather both N/S and E/W polarized low frequency data from astronomical sources and transfer the data to the central element (CE) for processing. Each sensor element includes orthogonal data collection antennas. They are placed in such a way so as to minimize coupling between the axes.

Each sensor element is autonomous, being equipped with its own power system and front-end electronics. A 0.3 m diameter center-fed Cassegrain dish is used for transmitting data to and receiving commands from the CE. Each element is designed to be compactly stowed for transport and easily deployed by a telerobotic rover (see Figure 6 and Drean et al. 1991).

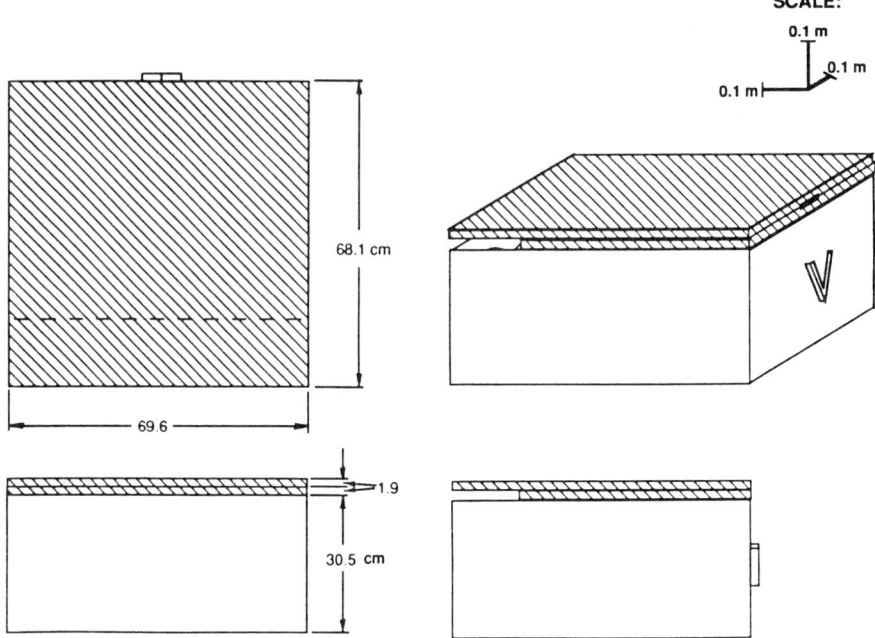

Fig. 6. Sensor element stowed configuration. This is a compact and rugged design. The folding solar panel covers the stowed antenna.

A self-contained power source is used since the large distances necessary for high spatial resolution measurements would require massive amounts of cable to be deployed. NaS batteries located under the solar panel on the sensor element body were chosen to power each sensor element. Since the batteries operate at a temperature of about 325°C, special consideration is required for their integration. Power from the solar panels is used to initially melt the Na and S which then, by the process of charging and discharging, continue to generate enough heat to keep the components molten. Since the lunar regolith is a poor thermal conductor, the Moon cannot be used to dissipate excess heat. Instead, thermal louvers are used in combination with thermal insulation for temperature control.

Solar panels are used for operating the element and charging the NaS batteries during the lunar day; the batteries are used for lunar night operations. The total area of the solar panels (0.84 m^2) is split into two sections; one which folds open and one which is on the top of the element body. Non-tracking solar panels were chosen over tracking ones in order to simplify the design and increase the reliability of the system over its lifetime by eliminating problems due to dust collecting in movable joints. Dust contamination on the solar panel surfaces should be kept to a minimum with the use of a conductive oxide coating which helps decrease the attraction of statically charged dust particles. Furthermore, the sensor element is designed to deploy the solar panels after the rover has left the vicinity thus reducing the chances of dust being thrown onto the panels.

The deployed sensor element layout is given in Figure 7. The sensor element is ready to gather data once the communications dish and solar panels deploy and the link to the CE is established. The communications dish is located on a bent mast on the southern side of the element in order to keep from casting a shadow onto the solar panels. The dish has two degrees of freedom which allow it to locate the CE.

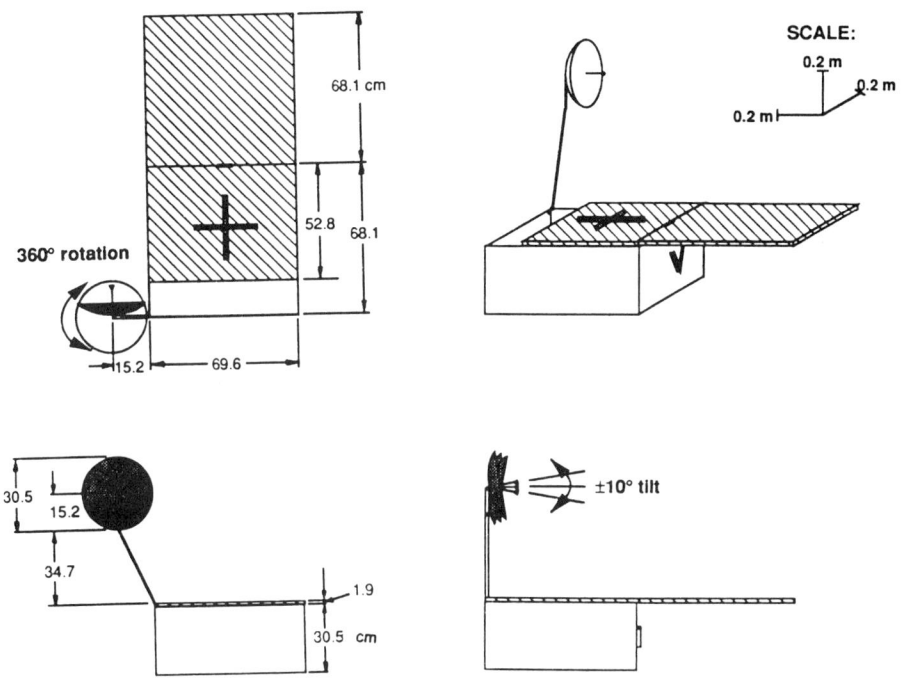

Fig. 7. Sensor element configuration. Deployment is a simple procedure. Once the sensor element is placed on the lunar surface in the correct orientation, the solar panel will unfold and the dish antenna will deploy and establish a link with the CE.

A sensor element functional block diagram is given in Figure 8. Data from each antenna will be I/Q sampled with a sampling clock provided by the CE. The bandwidth will be commandable from 100 kHz to 1 MHz in steps of 100 kHz, with an option for a bandwidth as low as 10 kHz. The band center frequency will be commandable from approximately 500 kHz to 30 MHz in steps of 1 MHz. The mission data will be time division multiple accessed with the engineering telemetry and sent via the RF link to the CE.

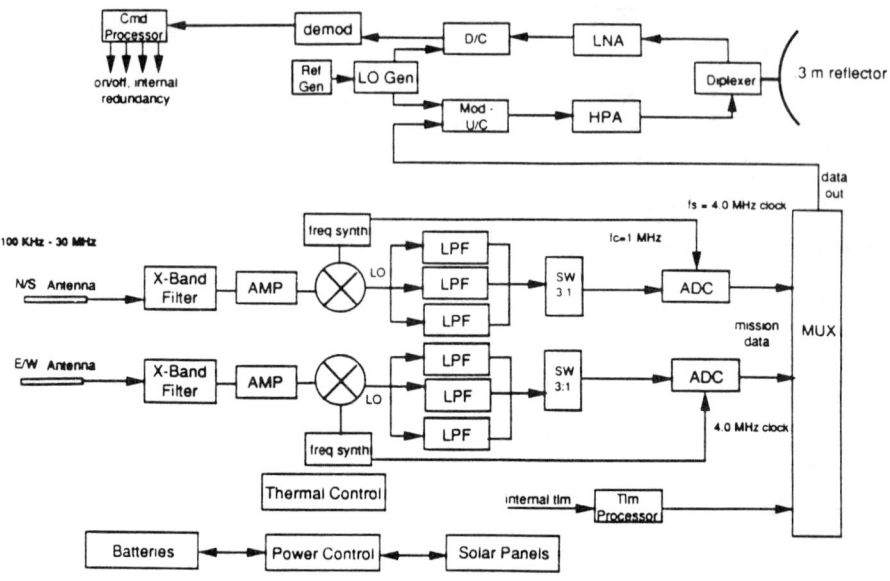

Fig. 8. Sensor Element Functional Block Diagram. The central element collects the individual sensor element data and provides the data, telemetry, and command links to/from the Earth (via the relay satellite).

SENSITIVITY

Typical astronomical source spectra are given in Figure 9. Note that an ALLFA minimum detectable source threshold of 10^{-25} W m^{-2} Hz^{-1} (10 Jy) for a 24 hour integration permits observation of many key sources. Many source strengths in our solar system are large compared to the ALLFA minimum detectable sensitivity level, so variations will be measured with high signal-to-noise.

Observatory lifetime was chosen to be 12 years, long enough that data from weak celestial sources as well as time variations from stronger sources can be recorded. The key limitation is fuel in the relay satellite. The 12 year span was selected to include at least one solar cycle.

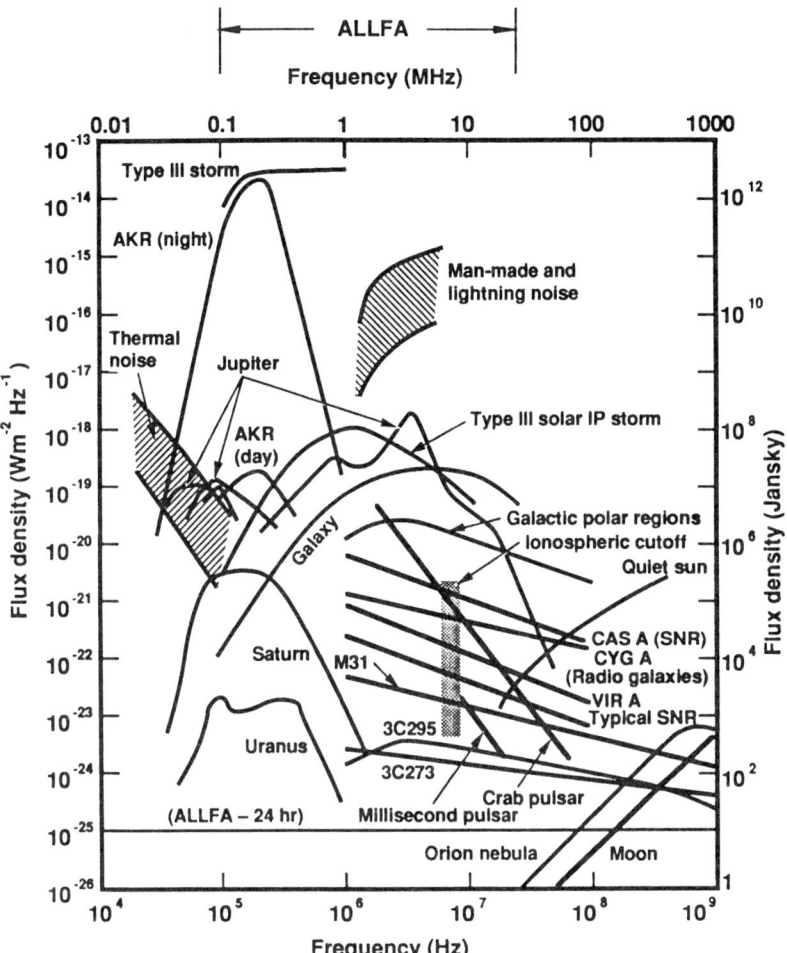

Fig. 9. Low Frequency Radio Sources. The ALLFA frequency range and sensitivity levels are shown.

LUNAR SITE SELECTION

Lunar crater basins provide large relatively smooth regions for deployment and operation of telescope array elements. Ideally, the best coverage possible of the celestial sphere would be obtained if there were observatories in both the southern and northern hemispheres of the lunar farside. However, it is unlikely that two observatories will be developed initially. Therefore, an equatorial location was selected giving almost 100% coverage of the lunar sky with one observatory. Celestial sphere coverage as a function of assumed visibility above the local horizon for an assumed deployment in Chaplygin Crater (with lunar coordinates 150 E and 5 S) is given in Figure 10.

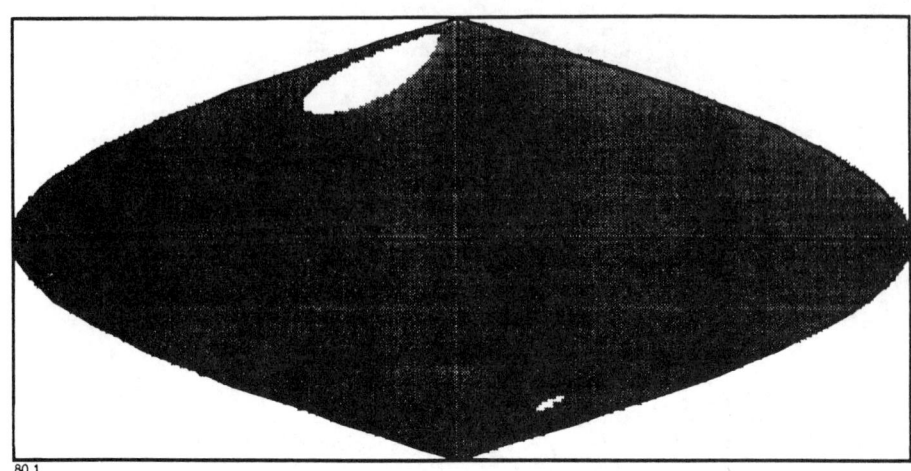

- 10° Elevation angle based on sun shadows in Lunar Orbiter photograph
- 98% of sky covered during 12 year mission

——— Earth's Equatorial Plane

| Declination

Fig. 10. Celestial sphere coverage from Chaplygin Crater.

REFERENCES

Alexander, J. K., Novako, J. C., Grena, F. R. and
  Weber, R. R. 1975, *Astron. and Astrophys.*, 40, 365.
Basart, J. P. and Burns, J. O. 1990, *Proc. Low Frequency Astrophysics from Space*, ed. Kassim, N. E. and Weiler, K. W., Crystal City, VA, Springer-Verlag, Berlin, p. 52.
Burns, J. O., and Asbell, J. 1986, *Proc. Workshop Radio Astronomy from Space*, Greenbank, WV, p. 29.
Burns, J. O., Duric, N., Johnson, S. and Taylor, G. J., ed. 1989, *Lunar Far-Side Very Low Frequency Array*, NASA CP-3039, 88 pages.
Dennison, B. K., Weiler, K. W., Johnston, K. J., Simon, R. S. and Spencer, J. H. 1986, NRL Memorandum Report 5905, 35 pages.
Douglas, J. N., and Smith, H. J. 1985, *Lunar Bases and Space Activities of the 21st Century*, ed. W. W. Mendell, LPI, I, p. 301.
Drean, R. J., Caylor, M. A., Choi, D. U., Edelsohn, C. R., Gurley, J. G., Hagen, F. A., Landecker, P. B., Su, G. W., Tillman, M. L. and Wassgren, C. R. 1991, these proceedings.
Erickson, W. C. 1988, NASA/JPL Publication 88-30, 20 pages.
Erickson, W. C. 1990, *Proc. Low Frequency Astrophysics from Space*, ed. Kassim, N. E. and Weiler, K. W., Crystal City, VA, Springer-Verlag, Berlin, p. 59.
Smith, H. J. 1990, *Proc. Low Frequency Astrophysics from Space*, ed. Kassim, N. E. and Weiler, K. W., Crystal City, VA, Springer-Verlag, Berlin, p. 29.

ENGINEERING DESIGN OF AN UNMANNED LUNAR RADIO OBSERVATORY

R. J. DREAN, M. A. CAYLOR, D. U. CHOI, C. R. EDELSOHN,
J. G. GURLEY, F. A. HAGEN, P. B. LANDECKER, G. W. SU,
M. L. TILLMAN, and C. R. WASSGREN
Space and Communications Group, Hughes Aircraft Company,
P. O. Box 92919, Los Angeles, CA 90009

ABSTRACT   Key technology challenges and requirements for unmanned lunar exploration missions are derived based on a candidate mission to telerobotically place a radio observatory (Astronomical Lunar Low Frequency Array or ALLFA) on the farside of the Moon. Highlights of the detailed ALLFA engineering design are presented, and include all phases of the project from launch to data retrieval. The entire payload complement is launched on a single Titan IV. The array is delivered to the lunar surface by means of a lunar landing vehicle which provides for autonomous hazard avoidance during the terminal landing phase. A six-wheeled telerobotic rover provides the required surface mobility without need for autonomous navigation. A six-DOF positioning arm on the rover supports all observatory deployment needs. The entire deployment process requires less than one lunar day. A simple spinning relay satellite in an L2 halo orbit accommodates communications relay requirements. Other applications of this technology include cargo carrying, communications, and surface operations for other unmanned precursor missions and development and operation of the SEI manned lunar base. It is shown that all required technology is well within reach with development required over the next 5 years to perfect NaS batteries, build throttleable cryogenic engines and some advancement in telerobotics. Commencement of lunar operations prior to the end of this decade is feasible.

INTRODUCTION

The Astronomical Lunar Low Frequency Array (ALLFA) is a multi-element radio observatory delivered by a single Titan IV launch to the Moon, and placed robotically on its the far side. The observatory operates at frequencies below 30 MHz which cannot be effectively observed from Earth (including from Earth orbiting satellites) due to absorption from the Earth's ionosphere and interference from radio noise of both human and natural origin. The farside of the Moon is shielded from this interference, thus enabling sensitive observations across this frequency range.

Robotic emplacement presents many engineering challenges whose solutions have applications to future space exploration missions. Particularly relevant to the Space Exploration Initiative (SEI) are the telerobotic rover, capable of deploying the observatory across a large expanse with limited autonomy, a lunar lander, capable of soft landing on the surface of the Moon with minimal a-priori knowledge of the surface terrain, and a relay satellite, in orbit about the lunar libration point L2 to provide continuous communications between the Earth and the lunar farside.

A discussion of ALLFA scientific objectives and element design is given by Landecker et al. (1991).

INSTRUMENT DESIGN

The ALLFA instrument consists of an elliptical array of independent sensor elements, supported by a rover vehicle for deployment, a central data collection element, and a relay satellite in a halo orbit about the Earth-Moon L2 libration point. Pseudo-random spacing of the sensor elements along the ellipse provides nearly uniform coverage in the spatial frequency (u, v) plane. The 25 km north-south dimension of the array exceeds the 17 km east-west dimension, in order to compensate for image foreshortening when observing high declination radio sources. An artist's depiction of the ALLFA concept is shown in Figure 1.

Fig. 1. ALLFA Concept with the sensor elements, the rover which deployed them, and the central element base. This layout is not to scale.

Each sensor element receives energy from astronomical sources in selected frequency bands, at two orthogonal polarizations. The low frequency radio signals are collected by antennas and analog receivers, separated into real and complex components, and digitized to 3 levels. The digital data streams are transmitted via the central element and the relay satellite to the Earth-based control station. Timing and synchronization signals, and commands to control the operating schedule and frequency selection, are transmitted by the reverse path, from the ground control station through the relay satellite and central element to the various sensor elements.

An individual microwave link, using a 0.5 watt transmitter and a 0.3 meter diameter steerable dish, connects each sensor element to the central element. The central element has a bicone antenna located on top of a 29.6 meter deployable boom, to ensure line-of-sight reception in spite of possible curvature of the lunar surface. The central element creates a composite data stream for relay to Earth via the bent-pipe communications satellite. A dedicated ground station correlates the data between various pairs of sensor elements, and reconstructs an image using aperture synthesis techniques.

With 40 elements providing good u-v plane coverage, ALLFA will have excellent snapshot imaging capability, using conventional two-dimensional image reconstruction. Rotational synthesis, using three-dimensional inverse Fourier transforms, will provide images covering a major part of the celestial sphere. The angular resolution of the resulting image is equivalent to that of a large dish antenna whose diameter is twice that of the array.

Primary power for lunar operations is provided by separate solar electric power sources for each element, with sodium sulphur batteries for night operation. The instrument can operate throughout the lunar day, whenever the ground station is visible, using power directly from the solar arrays. Night operation, which gives better sensitivity because solar noise is absent and the lunar ionosphere is thinner, is limited to a 40% duty cycle because of the limited battery capacity imposed by weight constraints of a single Titan IV.

Key instrument parameters are given in Figure 2. The ALLFA instrument is designed to cover a range of frequencies from 0.1 to 30 MHz. The upper portion of this band, 10 to 30 MHz, overlaps the spectral region observable from Earth, providing an opportunity for calibration against Earth-based instruments. The region from 0.1 to 10 MHz is unexplored, except for early space vehicles having much less sensitivity and angular resolution than the proposed array.

Frequency range	0.1 to 30 MHz
Site, Lunar longitude	$150°$ East
Lunar latitude	$5°$ South
Longest baseline,	NS 25 km
	EW 17 km
Number of sensor elements	40
Minimum lifetime	12 yr
Bandwidth	Up to 1 MHz
Polarization	2 ortho linear
A/D conv. per sensor pol. axis	phases: 2 (I&Q)
	levels: 3 (-,0,+)
Minimum integration time	10 msec
Sensitivity	10 Jy, 1 day int
Available center frequencies	500 kHz - 30 MHz
Duty cycle	night 40%
	day 40%
Min. viewing elevation angle	$10°$
Sky coverage	98%

Fig. 2. Key instrument parameters.

ENGINEERING DESIGN

A single Lunar Landing Vehicle (LLV) carries all the sensor elements, the central element electronics, and the rover. The relay satellite attaches beneath the LLV enabling a single launch of an upgraded Titan IV/ Centaur to accomplish the ALLFA mission. The integration of the ALLFA components into the Titan IV payload fairing is given in Figure 3.

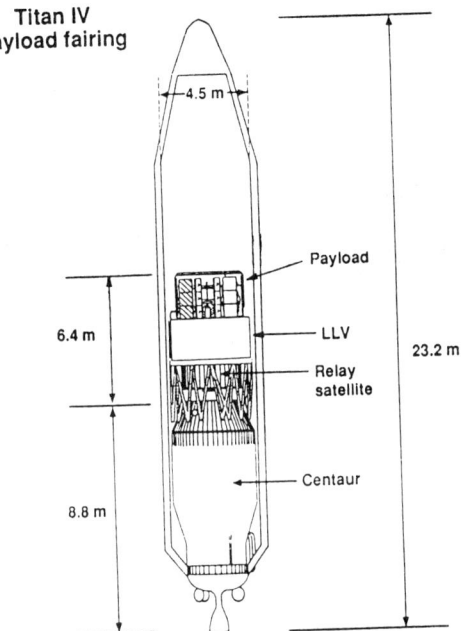

Fig. 3. Launch vehicle integration.

The mission begins with the launch of the Titan IV/Centaur combination into low Earth orbit (LEO). At this point, the LLV, payload, relay satellite and upper stage coast for up to 90 minutes (one orbit) to the correct true anomaly. Next, the Centaur upper stage provides the required increase in velocity for insertion into the trans-lunar trajectory, and spins up the LLV and relay satellite. Both the upper stage and the relay satellite then separate from the LLV, and the LLV performs a reorientation to allow its solar panels to be aligned perpendicular to the Sun's rays. The transfer time to the Moon is approximately 140 hours (almost 6 days). The mission design is shown in Figures 4, 5, and 6.

When the LLV reaches the Moon, a burn is performed for insertion into a lunar parking orbit. The LLV remains in this orbit for about 72 hours while the relay satellite continues to the L2 libration point. Once the relay satellite has been injected into the L2 halo orbit, and a communication link has been established, the LLV propulsion system provides deboost and terminal burns to land in a selected crater in the southern hemisphere of the Moon.

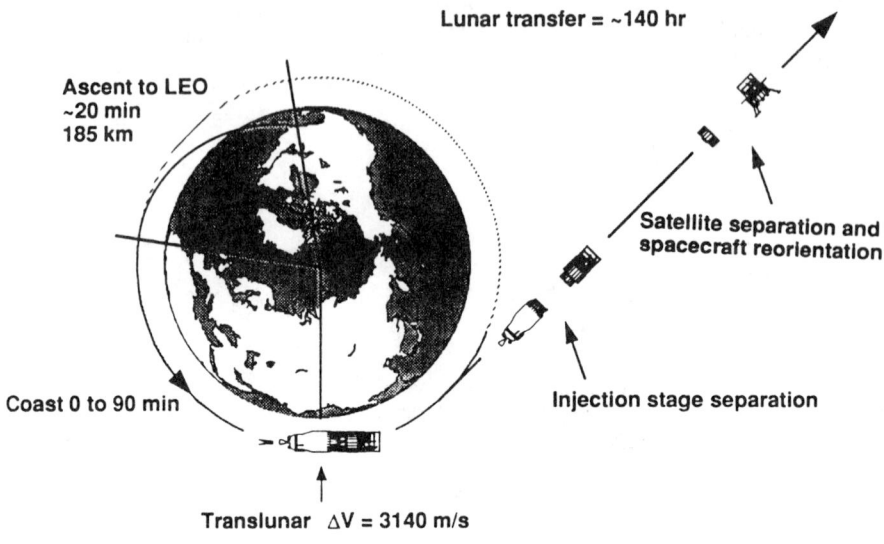

Fig. 4. Mission Design: Earth Phase.

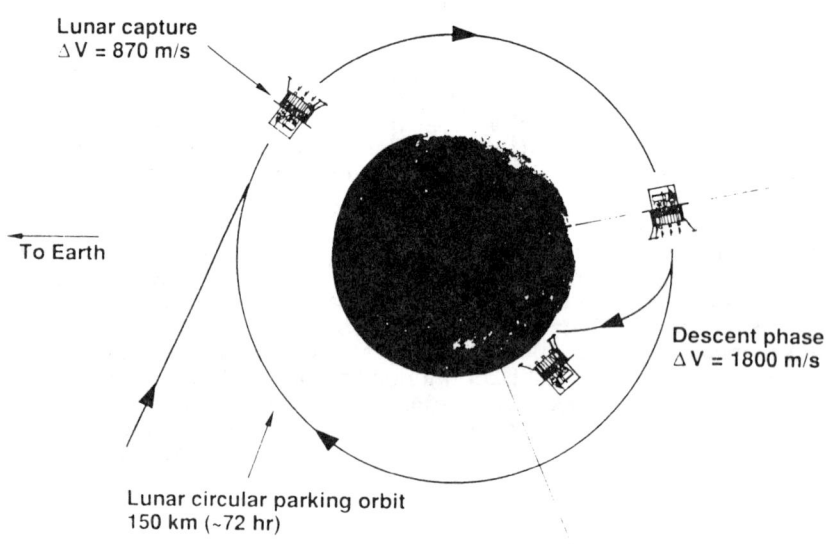

Fig. 5. Mission Design: Lunar Phase for LLV. The LLV is captured into lunar parking orbit while the relay satellite continues to the L2 libration point.

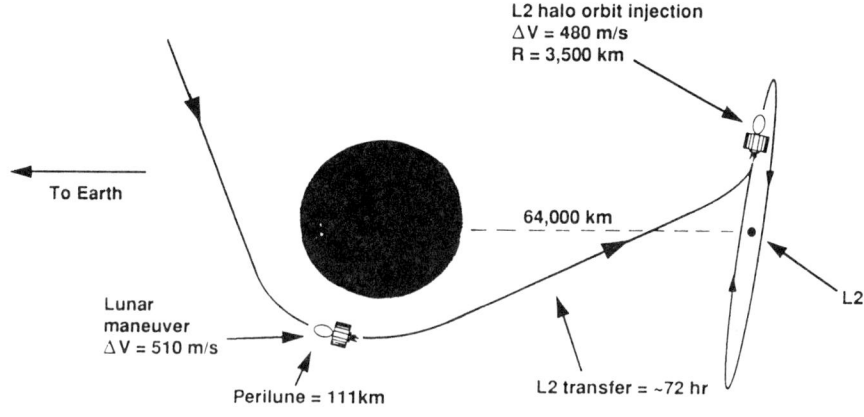

Fig. 6. Mission Design: Lunar Phase for Relay Satellite. The relay satellite is injected into a 14.7 day halo orbit about the L2 libration point.

The integration of the payload on the LLV is shown in Figure 7. This side view of the LLV shows the integration of the payload and stacking arrangement. The sensor elements are stacked in eight groups, each three elements high. One remaining element is placed between the rover wheels and has enough clearance so as to not interfere with the rover. The location of the sensor elements on the perimeter of the LLV shelf is such that the rover arm can easily reach each element while the rover is on the lunar surface. The LLV will require a terminal hazard avoidance system in order to safely land on the surface of the Moon.

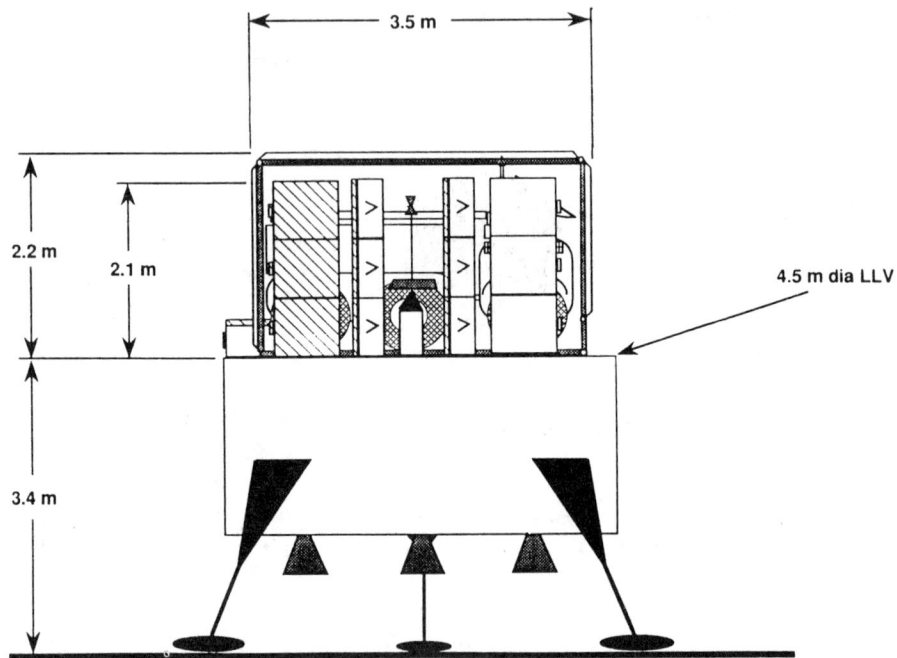

Fig. 7. LLV Payload Integration. The three section ramp is shown above the LLV shelf in its stowed position.

After the LLV has landed, a 7.9 meter ramp deploys to enable the rover to drive onto the lunar surface. Once the rover removes the sensor elements, the central element appendages are deployed. These include the communications mast which in addition to the bicone antenna has rover radio navigation equipment and a camera, the 4 meter diameter wrapped-rib antenna for primary communications to the relay satellite, and 5 m^2 of retractable solar array. Sodium sulfur batteries provide the 120 W peak power during night operation. The wrapped rib antenna and a lower gain backup horn antenna are gimballed to track the relay satellite. The deployed central element is shown in Figure 8.

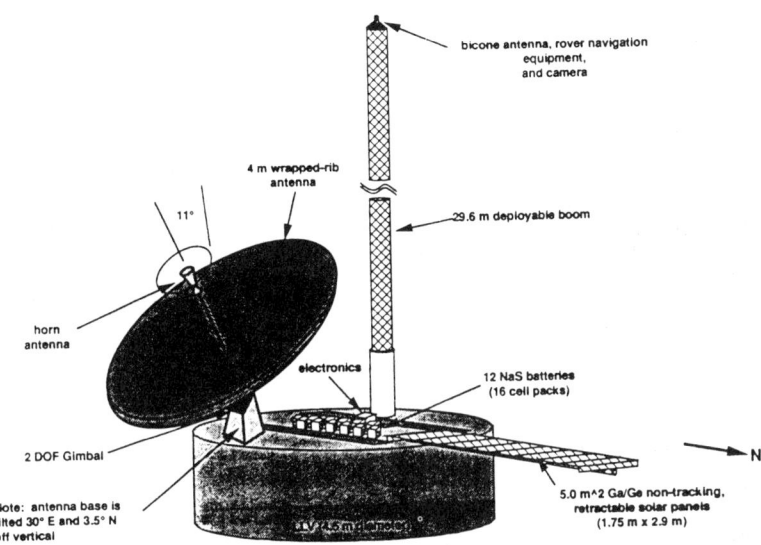

Fig. 8. The Central Element collects data and telemetry from each of the 40 sensor elements and maintains the communications link to the relay satellite.

All telemetry and command links are either S-band or X-band, and are compatible with the NASA Deep Space Network. The rover, central element, and relay satellite have separate telemetry and command links. Sensor element telemetry and command links interface through the central element. There are also Ku-band links for rover video data and the primary 400 Mbps composite mission data.

The telerobotically operated rover removes the sensor elements from the LLV, transports them 15 at a time to their appropriate sites, and orients them so the sensing elements are coalligned (see Figure 9).

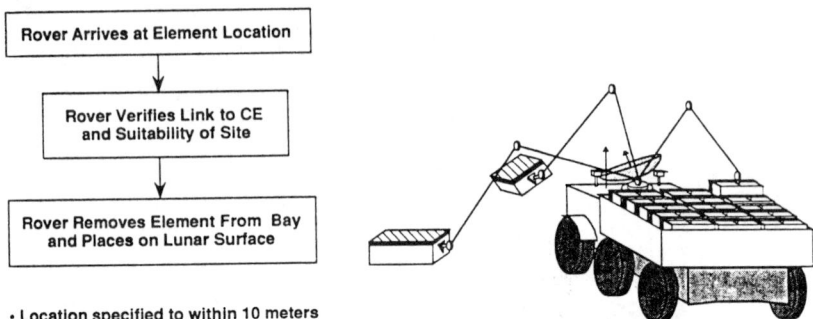

- Location specified to within 10 meters
- Suitability of site
  - Slope less than 10°
  - Attitude specified within 10°
  - Sufficient area for deployed element
- Place each element with antenna on southern side
- Final calibration provides precise element locations

Fig. 9. Sensor Element Set-up. The rover places each element on the lunar surface with its dish antenna on the southern side and its attitude specified within 10 degrees.

The entire process takes about 11 Earth days and can thus be accomplished with continual sunlight. The rover has one passively articulated joint (made up of a series of flexible rods), six wire mesh wheels, a six DOF manipulator arm with a 4 meter reach, two cameras for stereoscopic vision, and autonomous hazard detection. Sodium sulfur batteries provide mobility power; a 10 m^2 retractable solar array periodically recharges the batteries. The rover determines its orientation using gyrocompasses and defines its position relative to the central element by radio navigation. The sensor element setup is shown in Figure 10.

AN UNMANNED LUNAR RADIO OBSERVATORY

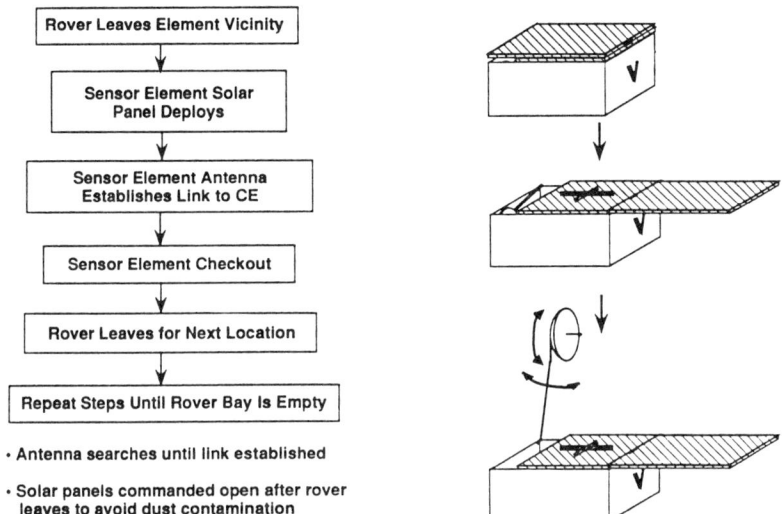

- Antenna searches until link established
- Solar panels commanded open after rover leaves to avoid dust contamination

Fig. 10. Sensor Element Setup. After the rover leaves and the dust settles, the element's panel unfolds and then the dish antenna deploys.

The engineering parameters are summarized in Figure 11.

Mission mass	
LLV/central element	4900 kg
40 sensor elements	1603 kg
Rover	750 kg
Relay satellite + Adapter	<u>948 kg</u>
Total	8201 kg
Operating power	
LLV/Central element	120 W
Sensor element (1)	18 W
Rover	1300 W
Relay satellite	360 W
Data rate	
Mission	400 Mbps
Rover video	1 Mbps
T & C	610 bps
Sensor to central	22 Mbps
Orbit summary	
Low Earth orbit altitude	185 km
Lunar transfer time	6 days
Lunar parking orbit altitude	150 km
Halo orbit radius	3500 km
altitude	64000 km

Fig. 11. Engineering Parameters.

SCHEDULE

The ALLFA mission offers the nation an early SEI precursor mission which both has scientific merit and develops tools for follow-on robotic or manned exploration missions. Development is not paced by technology as all critical elements are currently available. The development milestone schedule, shown in Figure 12, is driven by funding considerations.

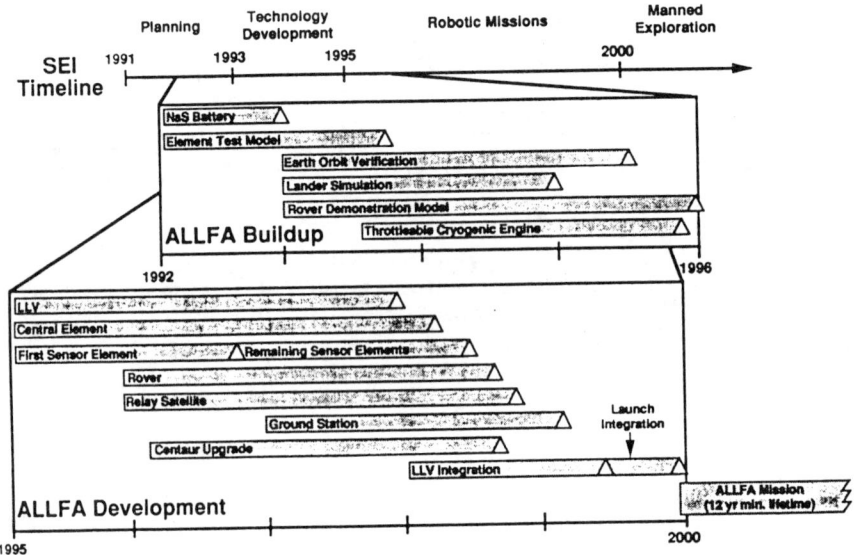

Fig. 12. Schedule.

REFERENCES

Landecker, P. B., Caylor, M. A., Choi, D. U., Drean, R. J., Edelsohn, C. R., Gurley, J. G., Hagen, F. A., Su, G. W., Tillman, M. L. and Wassgren, C. R. 1991, these proceedings.

AFTERTHOUGHTS

VIRGINIA TRIMBLE
Physics Department, University of California, Irvine CA 92717, and, Astronomy Department, University of Maryland, College Park MD 20742

ABSTRACT A summary is given of the state of the art of remote, automatic, and robotic observatories, their scientific contributions, advantages and disadvantages, image in the astronomical community, and future potential, as described by speakers at the symposium.

INTRODUCTION

In one sense, remote observing has a very long history, going back at least to Herschel, who used an integrated hard/software package named Caroline. Participants did not agree on any very precise definition of remote, automated, and robotic observing, and I shall here take them collectively to mean techniques significantly advanced beyond those I learned at Mts. Wilson and Palomar in the mid 60's. Those techniques were even then almost out of date, for, as Code reminded us, Wisconsin operated its observatory at Pinebluff remotely from about 1961 onward. The record for longest, continuous, successful operation of a remote observing facility probably belongs to the International Ultraviolet Explorer satellite, and we can perhaps learn something from NASA's successes as well as its failures.

THE TRADE OFFS

Automatic, remote observing is unquestionably the method of choice in several contexts. First, some projects can only be carried out that way, either because the observatory is intrinsically inaccessible, or because repetitive tasks must be performed with a degree of reliability beyond human capability. Second, if labor costs of a project (including wear and tear) are a significant fraction of the total, then robotic techniques may make economic sense. Finally, if portions of an observing program can be made routine and reliable, then automating them will leave the (always scarce) trained people freer to work on other parts of the problem.

But there is a price one pays. Historical examples are fairly easy to imagine. Suppose Mullard Radio Astronomy Observatory had been using an automated radio source scintillation monitor in 1967. Would it or its overseers have reacted in the same way that Jocelyn Bell did on seeing bits of "scruff" directly on the strip charts? If, in February 1987, a supernova had peaked just below naked eye visibility, and only automated telescopes had been operating in the southern hemisphere, how long would it have been before one of them found 1987A? And if the optical spectrum of 3C 273 had been first recorded in a robotic survey of stellar radial velocities, it would probably have been dropped from the data base as noise.

On the basis of such trade-offs, certain kinds of projects seem particularly well matched to automated techniques. These include:
1. Intense campaigns on a few difficult or prototypical objects for instance, V477 Cyg, Bopp's duplicitous stars, and 3C 273
2. Long-term synoptic studies where stability is particularly important or difficult to achieve, e.g.,
    a. Supernova light curves, because there are no standard stars for spectra dominated by emission lines
    b. Star spots and cycles, active galactic nuclei, and perhaps optical behavior of X-ray binaries
    c. Multiple periodicities of widely varying time scales, e.g. RS CVn stars and cataclysmic variables
3. Dense temporal coverage over long periods of time, as is required in searching for supernovae, optical flashes, MACHOS, lensing by Oort cloud comets, and occultations by extra-solar-system planets
4. Acquisition of very large numbers of objects for statistical purposes, e.g. retyping faint variable stars and following accretion behavior in T Tauri variables
5. Monitoring of artificial satellites, at least in the way that Messier studied nebulae, to keep from confusing them with the intended objects of study

THE STATE OF THE ART

Automated, remote observatories have existed in some form for nearly 30 years, and a few of their operations have become routine, or nearly so. At the other end of the spectrum, several speakers discussed projects about which my mother would have said, "And if the sky falls, we'll all catch larks," meaning that she didn't expect it to happen very soon (even supposing the consequence to be a desirable one). We can lay out existing and contemplated projects along a sort of continuum between the here-and-now and the far future:

Routine	Multicolor aperture photometry
	Monitoring of sky conditions
	Small, jointly-scheduled networks
Now operating	CCD imaging
	Areal photometry
	Infrared interferometry
to	High Speed Photometer on HST
	Monitoring & responding to rapid changes
	Narrow band photometry
	Polaroid polarimetry
Almost operating	APTs for high school & college education
Components exist,	Fiber optic spectroscopy
	Fabry-Perot spectroscopy
some assemblage	Infrared array imaging
	Radial velocity spectrometers
	Dedicated polarimetry
required	Specialized imaging (e.g. comets)
This decade	Large, formal arrays & networks, communal scheduling
to	Mega-star capability (2MASS, MACHO search)
	Submillimagnitude photometry (space only?)
	(Space interferometry etc.)
When the sky falls	Lunar observatories

SCIENTIFIC ACHIEVEMENTS

The first test of a new technique is whether it can recover previously-known results. As several speakers made clear, automated telescopes have passed this test by rediscovering that nights get shorter in the summer, stars set in the West, and Murphy was an optimist.

A number of other astronomical discoveries made wholly or partly by automatic techniques were mentioned in the talks by Dukes, Bopp, Hall, Strassmeier, Filippenko, Perlmutter, Honeycutt, and Schmidt. These are some of the items that were new, at least to me:

1. BQ Cep may be the first triple-mode Cepheid, with periods of 4.26, 2.98, and 2.27 days.
2. Distortions of lobe-filling binary components by the Roche geometry can yield a "pointed end" effect, in which one minimum is significantly deeper than the other, due to an exagerated form of limb darkening.
3. The onset of stellar activity strong enough to produce luminosity fluctuations at the $0.^m01$ level is very sharp and nearly independent of luminosity class, if Rossby number (which is

small for rapid rotation and strong convection) is chosen as as the X axis. The sun falls just below the sharp onset at $\log R_o = -2/3$, and would presumably be much more active if it rotated faster by even a factor two.
4. The amount of stellar differential rotation, as traced by starspot-induced luminosity variations, is strongly correlated with angular frequency. This is apparently not understood, but it may not have been much though about either. The sun is again normal in this correlation.
5. The lifetimes of star spots vary with size. Durations of larger ones can be calculated if you assume they are torn apart by differential rotation, while spots smaller than about $15°$ die faster than calculated, the discrepancy increasing toward the smallest measurable spots. The sun does not fall on the mean correlation, but the problem may be that the sizes are not defined in quite the same way.
6. Pre- and post-main-sequence stars occupy the same areas in a plot of $\Delta T/T$ vs. spot area, and spots that affect total luminosity set in at a later spectral type than do H and K emission, except for W Uma and T Tau stars, for which accretion contributes to the variability; the Algol SW Cyg must also fall in this category, since it continues to vary during total eclipse.
7. Spots can prefer certain longitudes as well as latitudes (in e.g. FK Comae), and activity cycle effects on star size can be large enough to make measurable changes in binary orbit periods, accounting thereby for some otherwise incomprehensibly large changes, both positive and negative.
8. The albedo of Pluto, which at 0.5 is surprisingly high for solid methane exposed to cosmic ray hits, may be a direct result of surface activity that shows up as short term variability.
9. The Berkeley supernova search has found about 20 events and has essentially caught up with Robert Evans (but Zwicky with 127 discoveries to his name will take some beating). Of the events found, 1990E had an unusually rapid rise (in less than five days). And the rate of Type Ibc supernovae being found is about three times that in the Evans search. This could possibly be a luminosity effect, since the Berkeley search goes somewhat deeper in the galaxies it monitors.
10. A large fraction of the faint variable stars in the General Catalogue are wrongly typed. Many that are listed as RRc (overtone RR Lyraes) are really eclipsing binaries. The difference quickly shows in APT measurements, because V and V-R will be strongly correlated in one way for temperature-varying pulsators, but can go any old way for eclipses, depending on which star is in front.

THE APT (ETC.) IMAGE PROBLEM

A number of symposium participants expressed feelings that they operate in relative isolation from most of the astronomical community and in an atmosphere where their accomplishments and goals are not fully appreciated. Like most cases of mild paranoia, these feelings have, I think, a basis in fact. There are misperceptions of atomatic, remote, and robotic observing among non-involved astronomers and some possible cures. There are however, also real, long-range potential ill effects of the methods, and thought needs to be given to overcoming them.

A widespread misconception is that automatic techniques are too specialized, are not applicable to the vast majority of astronomical investigations, and cannot address fundamental problems. The use of APT as a catch-all name may have been partially responsible, and a change could be in order. But the main cure is prompt publication of data from remote (etc.) observatories, together with thorough interpretation of those data, in the standard journals, and not just in special-purpose publications or by email. Recognition that automatic photometric telescopes (broadly interpreted) can now monitor 3C 273 and T Tauri stars should also help to improve this part of the image.

Closely related is the feeling by outsiders that the automated community is a closed semi-secret club in which they are not very welcome. I suspect that this image derives to a considerable extent from the ubiquity of acronyms and other jargon for facilities and data handling systems. Regular users of APTs may find it hard to believe that a good many of their otherwise reasonably intelligent colleagues do not know what to do about FITS, flops, FLAIR, and FOCAP or how to get on a VMSbus, but, as Ed Schmidt said, "the people who don't believe in these things aren't here." One way to show that the doors are open is with clear, self-contained, readable documentation. This is hard work, but not impossible (The ROSAT manuals are very good indeed; even I was able to prepare a proposal following them.)

As remote (etc.) observing becomes more common, inevitably astronomy will attract slightly different kinds of people with different ideas of what constitutes both legitimate work and fun. But here lies the chief danger. It is not necessary for every member of the user community to know how to carry out the modern equivalents of rotating the telescope by hand, focussing the spectrograph, developing plates, and precessing coordinates. But somebody had better know how, and I am, at any rate, not alone in the belief that users with hands-on familiarity with their equipment will get better science out of it.

In summary, then, please publish your data promptly, document your widgets and programs, and try not to let your students forget which end of the screwdriver one is supposed to look through.

LOOKING AHEAD

Short Term Fixes

The problems presented by birds' nests at BAIT and mice at the four-college APT strongly suggest the need for installation of CATs (not an acronym); while a potentially important source of noise at the Australian MACHO search facility might be remedied by a (robotic) fly swatter.

A more important short term need is the addition of autoguiding to most automated telescopes if they are to achieve limiting exposures (the CCD equivalent of the night sky fogging your plates seems to be having the dark current fill up your wells). The rapid, flexible scheduling of observations envisioned by Russ Genet in his first talk would seem to require robotic time assignment committees and peer review panels. These are certainly not impossible. Ninety percent or so of the decisions now made by such groups (the ones I belong to anyhow) could be anticipated by a fairly simple sorting algorithm.

The Next Few Years

The most urgent need clearly will be intelligent archiving of data. In the words of Art Code, "Archiving is easy; retrieval is difficult." I have two rather simple thoughts on the subject. The first is that the more purposes you try to serve, the less well you will serve any one of them (compare card catalogues to the on-line system in your institutional library). Thus we need to think about exactly who is going to want to access various kinds of data and for what purpose. It might make sense to put the first 5% or so of what will eventually be an enormous assemblage (for instance the HST image archives) into several different kinds of systems and let the first generation of consumers choose among them on marketing principles. After all, one generally tries out a few phototubes (or whatever) before buying a thousand and expecting them to run unattended.

Standardization of equipment, languages, and programs is another self-evident desideratum. Several speakers suggested, however, that it needs to occur only at a rather high level of the heirarchy. A potential user of archived data ought, for instance, to be able to access it from any reasonbly powerful workstation and download the necessary image processing software along with the data. I also have a strong feeling that one ought to give it back when one is through, but this may not be true!

Long Range

It was clear from the evening discussions that there is no concensus on what sorts of major facilities should come next. Several people mentioned a national (large) robotic telescope, but others foresaw it rapidly going the way of other large national facilities (so that your project would still get only 3 nights a year, even if they were a bit more efficiently scheduled than current ones and did require your immediate presence).

Another possibility is networking of 8-20" APTs currently owned by individual full- and part-time astronomers. The main investment required is a good deal of effort from one or two people willing to be responsible for scheduling and prioritizing in one of Genet's Simplified Management Structures.

Several participants envisioned a much more versatile network, consisting of a large number of $1-2^m$ telescopes, individually optimized for photometry, spectroscopy, imaging, polarimetry, and so forth, but operated and scheduled as a coherent whole. The cost of this (particularly if 24 hour coverage is desired) puts it in the national or international cosortium categogy, with the usual attendant stresses and strains.

In light of these divergent opinions, the era seems ripe for letting a thousand robotic telescopes bloom, through individuals, institutions, and ad hoc collaborations each doing what they think is most important and achievable -- and publishing it promptly!

This free-market approach unfortunately will not work for a moon-based observatory. A surprisingly wide range of kinds of telescopes are under active consideration for the first generation of lunar observatories. These include an optical transit telescope; a megahertz radio antenna array; optical and infrared interferometers on baselines up to many kilometers; a passively cooled IR observatory in a polar crater; a radio antenna to operate in VLBI mode with earth-based facilities; an array of $1-2^m$ telescopes optimized for different imaging and spectroscopic tasks; and monitors of variable X- and gamma-ray sources. Little was said about ultraviolet facilities, but the airless lunar surface is presumably also suitable for them. One of the more carefully costed projects, the MHz interferometer, can just be squeezed into a single Titan payload and a billion dollar budget. One cannot afford very many such programs per decade; so the traditional, rather burdonsome, beaurocracy of concensus will have to function.

I cannot begin to predict what the next major advances in remote, automated, and robotic observing will be, or what new astronomical phenomena they will reveal. But there is one safe bet: these advances will be achieved primarily as the result of efforts by a few people who take the projects very seriously and care about them more than their friends and relations sometimes think is entirely rational. The rest of us can only be grateful that there continue to be such people in astronomy.

AUTHOR INDEX
(Indexed by first page of article)

Alcock, C.: 171, 193
Axelrod, T. S.: 171, 193
Barthelmy, S. D.: 137
Bennett, D. P.: 193
Bester, M.: 213
Bless, R. C.: 261
Bopp, Bernard W.: 19
Borucki, William J.: 153
Boyd, Louis J.: 289
Burns, Jack O.: 273
Caylor, M. A.: 335, 347
Choi, D. U.: 335, 347
Cline, T. L.: 137
Code, Arthur D.: 3
Cook, K. H.: 171, 193
Danchi, W. C.: 213
Degiacomi, C. G.: 213
Deleo, D. V.: 97
Doty, John P.: 123
Drean, R. J.: 335, 347
Drummond, Mark: 289
Dukes, Robert J., Jr.: 9
Edelsohn, C. R.: 335, 347
Filippenko, Alexei V.: 55, 105, 115
Freeman, K. C.: 193
Genet, David R.: 289
Genet, Russell M.: 153, 241, 289
Greenhill, L. J.: 213
Griest, K.: 193
Griffin, R. F.: 223
Gupta, Ranjan: 249
Gurley, J. G.: 335, 347
Hagen, F. A.: 335, 347
Hall, Douglas S.: 27
Hine, Butler: 289
Honeycutt, R. Kent: 77
Jewitt, David: 183
Kaitchuck, Ronald H.: 89
Kleinmann, S. G.: 203

Landecker, P. B.: 335, 347
Lester, Dan: 325
Lubin, Philip: 253
McGraw, John T.: 305
Muller, Richard A.: 67
Mutel, R. L.: 97
Nelson, M. J.: 261
Newberg, Heidi J. M.: 67
Park, H.-S.: 171, 193
Pennypacker, Carlton R.: 67
Percival, J. W.: 261
Perlmutter, Saul: 67, 193
Peterson, B. A.: 193
Quinn, P. J.: 193
Ramsey, Lawrence W.: 227
Richmond, Michael W.: 105, 115
Ricker, George C.: 123
Rogers, A. W.: 193
Sasseen, Timothy P.: 67
Schmidt, Edward G.: 73
Smith, Craig K.: 67
Strassmeier, Klaus G.: 39
Stubbs, C. W.: 193
Su, G. W.: 335, 347
Talent, David L.: 289
Teegarden, B. J.: 137
Tillman, M. L.: 335, 347
Townes, C. H.: 213
Treffers, Richard R.: 105, 115
Trimble, Virginia: 359
Trueblood, Mark: 289
Turner, George W.: 77
Vanderspek, Roland K.: 123
van der Veen, Janet: 253
Van-Vegchel, John J.: 235
von Rosenvinge, T. T.: 137
Wassgren, C. R.: 335, 347
White, R. L.: 261